Textile Science and Clothing Technology

Series Editor

Subramanian Senthilkannan Muthu, SgT Group & API, Hong Kong, Kowloon, Hong Kong

This series aims to broadly cover all the aspects related to textiles science and technology and clothing science and technology. Below are the areas fall under the aims and scope of this series, but not limited to: Production and properties of various natural and synthetic fibres; Production and properties of different yarns, fabrics and apparels; Manufacturing aspects of textiles and clothing; Modelling and Simulation aspects related to textiles and clothing; Production and properties of Nonwovens; Evaluation/testing of various properties of textiles and clothing products; Supply chain management of textiles and clothing; Aspects related to Clothing Science such as comfort; Functional aspects and evaluation of textiles; Textile biomaterials and bioengineering; Nano, micro, smart, sport and intelligent textiles; Various aspects of industrial and technical applications of textiles and clothing; Apparel manufacturing and engineering; New developments and applications pertaining to textiles and clothing materials and their manufacturing methods; Textile design aspects; Sustainable fashion and textiles; Green Textiles and Eco-Fashion; Sustainability aspects of textiles and clothing; Environmental assessments of textiles and clothing supply chain; Green Composites; Sustainable Luxury and Sustainable Consumption; Waste Management in Textiles; Sustainability Standards and Green labels; Social and Economic Sustainability of Textiles and Clothing.

Syeda Rubab Batool · Sheraz Ahmad ·
Yasir Nawab · Muzzamal Hussain
Editors

Circularity in Textiles

Springer

Editors
Syeda Rubab Batool
School of Engineering and Technology
National Textile University
Faisalabad, Pakistan

Sheraz Ahmad
School of Engineering and Technology
National Textile University
Faisalabad, Pakistan

Yasir Nawab
School of Engineering and Technology
National Textile University
Faisalabad, Pakistan

Muzzamal Hussain
School of Engineering and Technology
National Textile University
Faisalabad, Pakistan

ISSN 2197-9863 ISSN 2197-9871 (electronic)
Textile Science and Clothing Technology
ISBN 978-3-031-49478-9 ISBN 978-3-031-49479-6 (eBook)
https://doi.org/10.1007/978-3-031-49479-6

This Springer imprint is published by the registered company Springer Nature Switzerland AG
The registered company address is: Gewerbestrasse 11, 6330 Cham, Switzerland

Contents

Introduction

Syeda Rubab Batool, Sadia Razzaq, and Yasir Nawab

Abstract This chapter provides an overview of the concept of circularity, and its application in the textile industry and highlights the necessity for adopting circular practices. Further discussion continues with the benefits associated with circularity, such as reduced environmental impact, enhanced resource utilization efficiency, and others. Various initiatives and innovations aimed at promoting circularity in the textile industry are examined. Additionally, this chapter briefly addresses the challenges and opportunities that arise in the pursuit of circularity, considering factors like technological limitations and market dynamics. The key elements such as eco-innovation, circular fashion, eco-design of textiles and sustainable raw materials, etc., are required to establish circularity within the textile industry are briefly discussed. Lastly, this chapter concludes by emphasizing the importance of transitioning towards circularity to achieve sustainable and resilient textile production and consumption systems.

1 Introduction

The textile industry stands as one of the leading manufacturing sectors globally, contributing to the production of approximately 100 billion clothing items annually [1]. This industry has shown a great technological advancements and rapid growth over time, playing a significant role in the economies of nations. The other side of the story is the criticism that this sector receives due to its extensive consumption of resources resulting in adverse socio-environmental consequences. Generally, this industry follows a conventional linear model of production and consumption, commonly known as the "take-make-dispose" approach [2]. In this model, raw materials are extracted, converted into textiles, and eventually discarded as waste after a

S. R. Batool · S. Razzaq
School of Engineering and Technology, National Textile University, Faisalabad 37610, Pakistan

Y. Nawab (✉)
National Center for Composite Materials, National Textile University, Faisalabad 37610, Pakistan
e-mail: ynawab@ntu.edu.pk

S. R. Batool et al. (eds.), *Circularity in Textiles*, Textile Science and Clothing Technology, https://doi.org/10.1007/978-3-031-49479-6_1

short period of use. This linear model has caused severe environmental consequences such as depletion of resources, emission of greenhouse gases, climate change, global warming, air pollution, water pollution, and soil damage [3].

In the past few decades, an unparalleled increase in clothing manufacturing and consumption has been driven by the rapid evolution of fast fashion and unnecessary consumer lifestyles. [3]. The elevated production levels have simultaneously caused a notable increase in waste generation. Every year, over 92×10^6 tonnes of clothing waste are generated, with a mere 14% undergoing recycling while the rest is deposited in landfills. It is estimated that by 2030, this figure is anticipated to increase to 134 million tonnes [4]. Unfortunately, the textile industry contributes to nearly 10% of global carbon emissions, recognized as a leading driver of global warming [5]. Its excessive water usage ranks second on a global scale. For instance, the cultivation of cotton requires intensive water usage along with 16% pesticides and 4% fertilizers leading to water scarcity and chemical pollution. Textile dyeing, printing, and finishing operations entail the emission of hazardous chemicals into the atmosphere and water, which adds to further pollution [6]. Also, textiles take a long time to disintegrate and produce greenhouse gases as they do so, disposing of textile waste in landfills worsens the environment. Moreover, the societal effects of the textile industries cannot be disregarded. Exploitative labor practices, more frequent in developing nations, are a result of the goal of rapid fashion and low-cost production. Long hours, poor pay, and hazardous working conditions are all common for garment workers, who also have limited access to labor laws and safeguards [3] These statistics highlight the urgent need for sustainable practices and a shift towards circularity in the textile industry. Circularity in textiles aims to reduce environmental impact and increase resource efficiency across the sector. The ideas of the circular economy, which have gained popularity recently, offer a framework for implementing this change.

In this chapter, we explore the intertwined notions of circularity and the circular economy. Subsequently, we explain the need for embracing circularity and explore initiatives and innovations within the textile industry. We also address the challenges and opportunities encountered throughout the processes of collection, sorting, reuse, recycling, and the economic facets of circularity within the textile industry. Ultimately, we conclude by outlining the fundamental elements covered within this book.

2 Circularity and Circular Economy

Circular economy (CE) has gained significant attention both from scholars and practitioners. It is a widely investigated concept but is still subject to interpretation and debate. CE and circularity are very often used together and are closely related concepts, but they are not the same thing. They are complementary concepts. CE is introduced as a concept focused on reducing environmental impacts, resource use, and waste generation [7]. It contains various strategies and activities, such as eco-design, material and energy efficiency, waste hierarchy principles, industrial symbiosis, and innovative business models [8]. There are 114 definitions of CE found in the literature [9], whereas some of these definitions connect it to recycling and some link it to sustainable development. It is a broader concept that refers to an economic system designed to be restorative and regenerative by design. It seeks to eliminate waste and pollution and to circulate products, materials, and nutrients within the economy [10]. It shifts away from the notion of "end-of-life" to emphasize restoration, moves towards the use of renewable energy sources, removes harmful chemicals that hinder reuse, and tries to eradicate waste through the creation of advanced materials, products, systems, and business models [11]. The main objective of CE is to tackle global challenges (waste, pollution, climate change, etc.) and accomplish sustainable development, help the environment, and economy, and make things better for present and future generations.

On the other hand, circularity refers to the adoption of sustainable methods that promote a closed-loop system by lowering waste and raising resource value [2]. It is about keeping materials and resources in use for as long as possible. A CE cannot be achieved without circularity, and circularity can only be achieved within a CE. The transition to a CE is a complex challenge, but it is one that we must address if we want to protect the environment and create a sustainable future. Circularity is an important part of this transition, and it is something that we can all do to make a difference. According to Jacopo Zotti, circularity indicates the existence of internal energy and material flows within the economy. This arrangement enables the control of matter and energy within the economic system, thereby postponing their release back into the environment. This concept of circularity emphasizes the core principle of CE and is distinct from the traditional concept of CE. It is coherent with the materials-energy balance model, thermodynamically compatible, simple, parsimonious, and independent from specific implementation instruments [12].

3 Linear Versus Circular Model

The production and consumption of textile products follow a traditional linear model (take, make, use, discard). In this model, fibers are first spun into yarn and then yarn is knitted or woven into fabric. Once the fabric is ready, it is used to make the ultimate textile products for end-use applications (Fig. 1a). Geissdoerfer and his colleagues, in their study, reported that the linear economic model is currently followed by our industrial system, in which the end-product is disposed of, which ultimately harms the environment [13]. According to Lenzing (2017), 63% of textile fibers are extracted from petrochemicals, which are responsible for significant carbon dioxide (CO_2) emissions. Whereas the remaining 37% of textile fibers are predominantly composed of cotton, which requires substantial amounts of water and involves the use of harmful pesticides during cultivation, resulting in water depletion and hazardous contamination from pesticides [14]. Consequently, we are depleting natural resources extensively while generating waste, resulting in significant environmental and social harm.

The circular model is considered a possible way to tackle the wasteful use of resources (Fig. 1b). Unlike the linear model, it aims to enhance sustainability and eliminate waste by designing products that are durable, reusable, and reparable [15]. The concept of a circular economy is based on three fundamental principles outlined by the Ellen MacArthur Foundation [11]:

- The first principle emphasizes the preservation and enhancement of natural capital by managing finite reserves and balancing renewable resource flows. Instead of extracting resources from nature, the circular system aims to utilize renewable resources more effectively through technological innovation, thereby contributing to nature's regeneration.

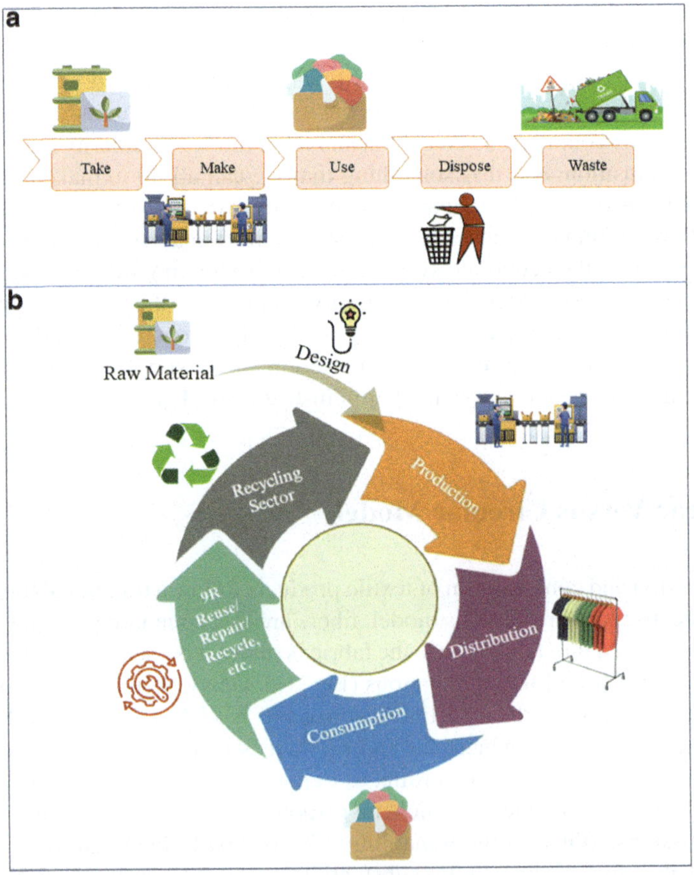

Fig. 1 Schematic representation of **a** Linear model and **b** Circular Model

- The second principle is based on optimizing resource yields by ensuring that products and materials are always in their highest utility state. This involves creating products that are designed for easy remanufacturing, repair, updating, or recycling when they reach the end of their functional life. By doing so, components and materials continuously circulate within the system, minimizing waste generation, extending product lifespans, and promoting reuse.
- The third principle focuses on enhancing system efficiency by identifying and eliminating negative externalities during the design phase. This includes reducing harm in areas such as education, health, leisure, and food, as well as managing resource use to eliminate various forms of pollution, including acoustic, air, water, and toxic substances.

To develop a circular economy, these three pillars are essential. It additionally involves the incorporation of concepts like sustainable design tactics, zero-waste design, elongation of product lifespan, resource reclamation, repair, and remanufacturing services. It can be stated that the CE framework is guided by the principles of the 3R approach (reduce, reuse, recycle), which should be implemented across the entire spectrum of production, consumption, and resource return [16]. However, the 3R framework is prominent and has been associated with waste management policies. In 2017, Kirchherr, Reike, and Hekkert proposed a 9R strategy that encompasses "Refuse," "Rethink," "Reduce," "Reuse," "Repair," "Refurbish," "Remanufacture," "Repurpose," "Recycle," and "Recover [17]. It's essential to recognize that these "R" frameworks represent a progression from a linear economy to a circular one, with "refuse," "rethink," and "reduce" strategies being identified as particularly effective in enhancing product use and manufacturing efficiency. However, the varying levels of circularity highlight that "refuse" (preventing the use of raw materials) represents the initial step towards circularity, while "recover energy" or "recover" represents the final step in extracting value from resources [18]. According to the Ellen MacArthur Foundation [11], the main features of a circular economy include:

i. Ecosystem Regeneration
ii. Extending the lifespan of products through reuse as second-hand items
iii. Optimization of production processes and increased efficiency by improving the performance of manufactured products
iv. Reusing product components for remanufacturing or creating new products
v. The application of new technologies that facilitate the above aspects, allowing products to adapt to new transformation processes

The core of CE lies in the concept of extracting value from significant resources by employing a more focused closed system of reuse and refurbishment. This approach has the potential to enhance both economic and environmental outcomes compared to traditional methods like recycling and energy retrieval. In this context, Van Wassenhove and Guide (2009) defined closed-loop supply chains as follows: "Closed-loop supply chains involve the strategic planning, management, and execution of a system that aims to optimize value generation throughout a product's entire life-cycle, involving the ongoing recovery of value from various types and quantities of returns as time progresses" [19].

4 Need for Circularity

The textile industry is considered the biggest industry that employs millions of people globally, it mainly fulfills the fundamental needs of human beings [20]. The global textile market size is expected to reach 1412.5 billion US dollars by the year 2028, as reported in the latest Grand View Research Inc. Report. The market size is also anticipated to increase at a CAGR of 4.4% from 2021–2028. The textile industry has a considerably positive influence on the economy. However, in the long term, it faces many ecological and social problems [21]. It is one of the most polluting industries in the world, releasing toxic substances into the air, water, and soil [22]. These toxic substances are harmful to workers' health and other inhabitants who live near the industry. Textile waste is produced in each stage of production, particularly fast fashion culture promotes mass consumption and disposal of clothes. This creates problems for the environment and other sectors of society. Therefore, there is a need to control fast fashion, towards a non-wasteful and green economy [23]. CE aims to address these issues by focusing on recycling and upcycling, adding value to end products, and minimizing waste. The concept of the CE goes beyond mere product recycling and extends to upcycling, a process that enhances the value of the product. This approach distinguishes itself from traditional recycling, which primarily involves reutilizing resources without actively seeking to enhance the quality of the resulting new product. Further, by reusing and recycling textiles, we can decrease the strain on raw materials and energy consumption. Embracing circular practices not only reduces the industry's ecological footprint but also drives economic efficiency through reduced waste management costs and novel revenue streams. Extending the lifespan of products through repair and refurbishment cuts down on frequent replacements and enhances sustainability. Consumer engagement with circular strategies, like second-hand purchasing and rental, further fosters responsible consumption. Collaboration across stakeholders drives innovation in materials, processes, and business models. Ultimately, the adoption of circularity aligns with the growing demand for sustainable products and regulations, making it an essential pathway for a greener textile industry.

5 Benefits of Circularity

In textile industry, the concept of circularity is gaining increasing recognition and offers a range of benefits as depicted in Fig. 2. The enhancement of resource efficiency is one of the foremost benefits. By extending the lifespan of clothing through repair, refurbishment, and recycling, the industry minimizes resource depletion and lowers the demand for virgin materials, consequently reducing environmental strain. Moreover, reuse practices of the resources lead to a decrease in the extraction of raw materials, subsequently resulting in a reduced environmental footprint for the countries supplying these resources. In addition, it actively contributes to the reduction or elimination of pollutant emissions that result from corporate economic activities. Through practices like recycling and upcycling, circularity conserves natural resources, minimizes waste, and mitigates the negative impacts of textile production, such as excessive water usage and pollution. Further, circularity aligns with the goal of achieving net-zero emissions by reducing energy consumption during production and extending the lifespan of textiles, lessening the carbon footprint associated with manufacturing new textiles and thus contributing to carbon neutrality. Circularity also plays a pivotal role in lowering greenhouse gas (GHG) emissions by reducing emissions from textile production processes, decreasing transportation-related emissions, and diverting textiles from landfills, which emit GHGs when textile waste decomposes [24, 25].

Fig. 2 Schematic illustration of the benefit of circularity

CE principles in textiles also open up new economic opportunities. The circular model fosters innovation, entrepreneurship, and business growth. Strategies like recycling, upcycling, and circular supply chains stimulate job creation and the emergence of new markets, thereby enhancing economic resilience and diversification within the industry. Circular practices prioritize longevity, favoring durable and high-quality products that withstand trends. This approach contrasts with fast fashion's ephemeral nature, ensuring that textiles and clothing remain relevant and functional over extended periods, contributing to reduced consumption rates. In addition to these economic and environmental benefits, circular economy principles also have a positive impact on consumer engagement. Ethically conscious consumers resonate with the environmentally responsible practices associated with circularity. Brands that adopt circularity position themselves as responsible contributors to sustainable practices, thereby enhancing their brand reputation and fostering customer loyalty among a growing consumer base focused on ethical consumption [19]. Lastly, the shift towards circularity necessitates innovation and adaptability within the industry. These changes inspire creative thinking and foster adaptability, ensuring the textile sector's long-term viability and competitiveness. In embracing circularity, the textile industry not only benefits from a more sustainable and resource-efficient approach but also opens up a wealth of opportunities for economic growth and positive consumer engagement, all while driving innovation and adaptability for a brighter and more sustainable future [23].

6 Initiatives and Innovations for Circularity in the Textile Industry

Textile Industries have started working on creating more sustainable and circular products. They offer services where you can purchase worn items and recycled materials. For example, Patagonia, an American clothing brand, has a history of embracing secondhand items, encouraging customers to purchase worn wear. They are planning to launch the Tee-Cycle™ T-Shirt embodies a zero-waste approach by using recycled cotton from old T-shirts, aiming for landfill-free production. Similarly, Reformation_ another brand, its latest impact report sets circularity goals, aiming to become a circular fashion brand by 2030, building on its "Climate Positive by 2025" pledge from 2020. H&M has launched initiatives such as their "Conscious Collection" and in-store garment collection for recycling, along with their "Close the Loop" campaign to promote garment recycling. Additionally, Adidas, Eileen Fisher, Levi's, Nike, etc., are just a few examples of textile industries and brands that are actively engaged in circular textile initiatives. Many other companies across the industry are also making efforts to shift towards a more sustainable and circular approach to textile production and consumption. Brands, retailers, suppliers, and other members of the textile industry are coming together to create and implement circular processes through organizations like the Sustainable Apparel Coalition and the Ellen MacArthur Foundation.

Governments are also taking initiatives to promote circular textiles. Some governments are presenting legislation requiring firms to collect and recycle textiles, or they're offering financial incentives to companies that use circular business models. For instance, the European Union has a textiles regulation that requires brands to collect and recycle textiles when they are no longer usable. Similarly, the UK has a tax on waste textiles, which encourages companies to recycle materials rather than dump them in landfills. Here are a few more examples of government initiatives to foster circularity including France's Anti-Waste Law, Netherlands Circular Textile Program, Sweden's Textile Strategy, CE Promotion Law in China, Japan's Cool Biz Campaign, United States EPA Initiatives, and so on.

7 Current Challenges Faced in Circularity

Recently, a fashion group reported that the addition of 57 million tons of waste materials generated annually could potentially result in an approximate 60% surge in commercial waste from 2015 to 2030. Subsequently, the total amount of waste from this fashion domain will expand to 148 million tons by the year 2030 [26]. This sector is not fully mature enough to attain the main objective of circularity [27]. The major challenge within the circular system is to establish a complex and globalized supply chain [28]. Insufficient Awareness and knowledge regarding sustainability, as well as a limited understanding of the ecological consequences of decisions, along with inadequate policies, significantly inhibit the transition from a linear to a circular perspective [19]. Additionally, another barrier is the lack of awareness regarding circularity, leading people to adhere solely to the traditional linear model [29]. They are unaware of how to return used items to producers and establish a circular sustainability relationship [30]. There are several challenges associated with achieving circularity in the textile industry. This chapter addresses some of these challenges, specifically focusing on collection, sorting, reuse, recycling, and economic issues (Table 1).

7.1 Collection

This term mainly consists of stores, charities, donation boxes, retail centers, etc. The number of these collection centers may differ globally. Typically, a relatively small portion of waste (approximately 15%) is usually recovered, while the majority is disposed of in landfills. Therefore, it is very crucial to spread awareness and educate customers about post-used textile items. An efficient collection system is necessary to support the recycling of textile waste. Textiles cannot be added to the existing recycling facilities due to contamination concerns and the frequent lack

Table 1 Key Challenges in the textile Industry and their prospects in terms of circularity

	Challenges	Prospects/opportunities	References
Collection	No effective framework	Requires important evolutionary modification	[47]
	No consistent textile collection principles	Requires consistent collection standards for better quality of textile	[31]
	High transport expenses	Provide awareness regarding best procedures	[48]
Sorting	Costly manual sorting	Manufacturing of automatic sorting machines	[36]
	A manual system cannot recognize the composition of fibers	Improvements in sorting tools, digital recognition, & Improvements in artificial intelligence (AI) algorithms	[49]
	Deficiency of devoted textile sorting services	Formation of devoted local sorting services	[47]
Reuse & repair	Underappreciated by consumers, & the industry	Enhance knowledge of garment protection, reuse, and repair advantages	[50]
	Lacks public awareness, equipment, & time for garment repair	Workshops/training resources are required	[37]
	Fast fashion clothing feature is inferior	Customer awareness of the real impact of fast fashion	[51]
Recycling	Need for high temperatures and time	Improvement in chemical recycling methods	[38]
	Needs consistent, high-capacity feedstock	Development of clear rules	[40, 52]
	Does not work well for blends	Improvements in the separation of blended textile products	[53]
	Emission of dangerous chemicals	Evaluation of alternative chemicals	[39]
	Total ecological impact	Quantification of the ecological impact of this	[41, 43]
Economic issues	Deficiency of decision-making capabilities	Improvement of education, awareness seminars, & lectures	[45]
	Shortage of corporation strategies, training, as well as sufficient planning approaches	Formation of regulations & laws	[19]
	Governmental, supply chain, & market problems	Microplastic emission estimation system, & market data assessment practices	[46]

of municipal recycling capabilities to separate textiles properly. They should be collected separately but this approach makes the collection method more expensive [31]. Shifting towards a circular framework requires financial support, effective brand organization, charitable programs, and appropriate regulations.

7.2 Sorting

Another major challenge faced in circularity is the lack of effective sorting methodologies. This issue can be addressed by focusing on resolving accuracy errors and enhancing processing speed to develop innovative sorting techniques. However, this approach requires a high cost, more workforce, and new methodologies. To address these issues, an effective information system is needed for making good decisions and creating effective plans [32]. The textile items that are not intended for resale will undergo processing and certification to ensure their suitability for future utilization in commercial sectors. This process involves categorizing and identifying these items based on their quality and condition [33]. A variety of sorting techniques are employed in the US, and a significant portion of these operations are carried out in Central America, the UAE, Pakistan, and India due to the more favorable pricing structures [34]. Currently, manual sorting and grading are the predominant methods. However, Automatic sorting technologies are valuable for reducing manual labor costs and improving the accuracy of fiber classification [35]. Near-infrared (NIR) Spectroscopy is frequently used in automatic sorting systems across various segments of the recycling sector, such as the PET recycling process [36]. Another approach to enhancing the speed and efficiency of textile sorting is the implementation of QR codes (Quick Response) on fabric items.

7.3 Reuse and Repair Centers

Utilizing previously used textile items holds the most significant value compared to alternative routes (such as repurposing or recycling) and has the least environmental impact [34]. A staggering 85% of textiles from consumers are disposed of in regular communal garbage, thereby eliminating the possibility of achieving circularity. Consumers lack knowledge about the long-term value and potential for the reuse of certain items. This often leads buyers to opt for purchasing new items instead of repairing broken ones due to a lack of awareness and time. The restoration industry plays an important role in achieving circularity. However, repairing branded textiles on a large scale can be costly due to factors such as transportation expenses, lengthy restoration procedures, and a decline in apparel quality. To address this problem, some brands have introduced clothing repair services, which have been positively embraced by consumers [37].

7.4 Recycling

Recycling transforms waste into new or reusable products. It's essential for modern waste management and circularity. There are two types of recycling which include mechanical and chemical recycling. Mechanical recycling is simpler, cheaper, and more environmentally friendly. In mechanical recycling textile waste items are cleaned and then cut into small fragments. When fibers undergo mechanical recycling, their length naturally decreases, which can lower their strength and flexibility and ultimately decrease their value. This method is most effective for textiles made of a single fiber type, and the resulting recycled fiber can be incorporated into new fabrics at a rate of about 30% [38]. On the other hand, chemical recycling converts high molecular-weight polymers into low molecular weight substances. The obtained substances can be used as reactants for preparations of other chemicals and polymers. Different chemical recycling techniques, such as gasification, pyrolysis, pretreatment, and enzymatic hydrolysis, have been described in the literature. The resulting products from these processes have been utilized during developmental stages as well as start-up. New fibers can be created through chemical recycling, which can serve a variety of purposes. [39]. The chemical recycling process typically demands a feedstock purity ranging from around 80% to 95%, depending upon the specific chemical methodology used [40].

The recycling of textiles is often encountered as a hurdle due to its high cost compared to new materials [29]. The use of manual sorting systems due to insufficient equipment for material identification and sorting leads to costly labor participation, resulting in economic disadvantages and limited market opportunities for recycled materials [41]. Recycling textiles poses challenges due to the variety of components, colors, and finishers. Lack of technology, scaling, and economic feasibility are also hindrances to achieving circularity [42]. Using chemicals and machines for recycling can be expensive and not very effective. In modern times, customers expect comfortable and flexible clothing, which can mean adding small amounts of elastic. Recycling this material is challenging due to its difficulty in removal and recovery. Various recycling methods exist, including upcycling, downcycling, open-loop, and closed-loop recycling. Upcycling transforms textile waste into higher-value items, while downcycling converts textile waste into lower-value items. Open-loop recycling produces different items from waste materials, while closed-loop recycling creates identical items.

7.5 Economic Issues

The existing system for circular textiles is uneconomical due to high transportation, labor, and fast fashion expenses. Large-scale reuse and repair are prevented due to these problems. Customers often refuse to pay the premium cost associated with achieving circular outcomes [43]. Extended Producer Responsibility (EPR) has the potential to support circular business models through mechanisms like eco-modulated fees and consumer-driven projects such as sustainable fashion and social media campaigns. These efforts can stimulate the economic aspects of circularity within the textile sector. Because of the globalization of the textile sector, the production of clothing has shifted from developed to developing nations, especially in Southeast Asia. This change gives opportunities to promote circular textile practices and revive abandoned industrial areas, but it also presents challenges to supply chain efficiency and transportation [44]. The same opportunities for increased circularity have also been detected in other origins as well [45, 46]. Kazancoglu classified the primary barriers to achieving circularity as administrative capability, high labor cost, design tasks, materials, principles, knowledge of circularity, association, price analysis, as well as framework [45].

8 Key Elements of Circularity in the Textile Industry

In this chapter, we provide a brief overview of key elements of circularity in the textile industry, which will be elaborated on in later chapters of this book.

- Eco-Innovation: It plays a key role in promoting circularity in the textile industry. It is defined as the implementation of novel products, methods, technologies, marketing strategies, and organizational frameworks, which leads to ecological improvements [54]. It involves implementing sustainable and creative methods to rethink how textiles are designed, produced, and used.
- Eco-design of Textiles: The initial stage of developing a product is eco-design, which helps to generate ideas and a framework for achieving circularity. It allows for the integration of environmental and economic solutions throughout a product's entire lifecycle. It is the proactive approach to reducing the environmental impact of textile products from the initial concept to their final creation [55].
- Sustainable Raw Materials: These materials are very important for protecting our environment and fighting climate change. Sustainable raw materials are based on natural fibers, and synthetic fibers. Natural fibers include flax, hemp, jute, ramie, kenaf, abaca, banana, bamboo, whereas synthetic fibers include lyocell, Poly lactic acid, Polyhydroxyalkanoates, Polycaprolactone etc. The use of sustainable raw materials is essential for achieving long-term environmental sustainability and meeting the needs of a growing global population without depleting finite resources.
- Sustainable Production Practices: it is an important key element of circularity in textile industry. Indeed, sustainable production practices are a crucial component of achieving circularity in the textile industry. To create a truly circular system, where materials are reused and waste is minimized, it's essential to start with sustainable production. By adopting eco-friendly processes, reducing waste, and ensuring ethical labor practices, the textile industry can lay the foundation for a

more responsible and environmentally conscious approach. Sustainable production not only reduces the negative impact of textile manufacturing but also aligns with the principles of circularity, where resources are conserved, reused, and recycled to create a more sustainable and resilient industry.

- Life Cycle Assessment (LCA) of Textile Products: It evaluates the utilization of resources and environmental impacts throughout the life cycle of a product, for example, extraction of raw material, processing, useful life, and waste management (disposal and recycling). It is a significant tool, useful for policymakers, and consumers [56]. It facilitates manufacturing industries in recognizing opportunities for ecological reduction, promoting effective practices and sustainable raw materials to reduce carbon footprint [57].
- Recycling of fiber is a fundamental element of achieving circularity in the textile industry. This process involves collecting and reprocessing textile materials, such as old clothing and fabric scraps, to create new fibers or products. By recycling fibers, the industry reduces its reliance on virgin materials and minimizes waste. It not only conserves resources but also lowers the environmental impact associated with textile manufacturing. Recycling of fiber plays a pivotal role in promoting sustainability, resource efficiency, and the long-term viability of the textile sector within a circular economy framework.
- Digitalization of the textile sector in circularity: Real-time abilities and digitalization lead the textile industry toward circularity Ecological impact regarding the life cycle of products collected by the Internet of Things (IoT) can help different approaches to promote circularity and lower carbon footprints. Digitalization can also help the textile industry to minimize waste material and optimize manufacturing approaches, resulting in more sustainable practices.
- The circular business model: It is defined as a framework that works within a closed material loop or in a closed resource loop. The transition to a circular economy can be facilitated by implementing circular business models. This model aims to achieve circular objectives such as prolonged use, reuse, and recycling. Additionally, this model can also help with the 9R strategies' execution, which aims to increase the circularity of the economy as they depend on smarter product usage, producers, and efficient use of materials.
- Awareness regarding generation, assessment as well mitigation of microplastics (MPs): Microplastics are available everywhere in the environment and pose a great challenge due to their potential environmental and health impacts. it is defined as very small particles having a size less than 5 mm, and many other plastic components having a size less than 1 μm are known as nano-plastics [58]. These particles are generated because of the degradation of macro-plastics such as bags, plastic bottles, and packaging, as well as the shedding of microfibers when textile items are cleaned. Circular practices in textiles involve not only sustainable production and recycling but also the responsible management of materials throughout their lifecycle. Preventing the release of MPs from textiles is vital for maintaining the integrity of the circular economy. It ensures that textile materials, once in use, do not contribute to plastic pollution in the environment. Thus, awareness

and action regarding MPs are integral to creating a truly sustainable and circular textile industry.

9 Conclusions

In this chapter, we discussed the difference between circularity and circular economy and emphasized the urgent need for circularity in the textile industry to address its significant environmental and social impact. We highlighted the benefits of adopting a circular approach, such as reduced resource consumption, less waste, and potential economic growth. Additionally, we explored the innovative initiatives and practices that are driving circularity in the textile sector and discussed key challenges encountered in circularity and last, we briefly discussed the key elements of circularity in textiles. Each key element contributes to the primary goal of creating a more sustainable, eco-friendly, and circular textile industry, ultimately paving the way for a brighter and more responsible future in the world of textiles.

References

1. Bour A, Budde Christensen T, Hunka AD et al (2023) Implications of circular textile policies for the future regulation of hazardous substances in textiles in the European Union. Science of The Total Environment 896:165153. https://doi.org/10.1016/j.scitotenv.2023.165153
2. Schumacher KA, Forster AL (2022) Textiles in a circular economy: an assessment of the current landscape, challenges, and opportunities in the United States. Front Sustain 3:146
3. Chen X, Memon HA, Wang Y et al (2021) Circular economy and sustainability of the clothing and textile industry. Mater Circular Econ 3:12. https://doi.org/10.1007/s42824-021-00026-2
4. John EP, Mishra U (2023) A sustainable three-layer circular economic model with controllable waste, emission, and wastewater from the textile and fashion industry. J Clean Prod 388:135642
5. da Silva PC, de Oliveira Neto GC, Correia JMF, Tucci HNP (2021) Evaluation of economic, environmental and operational performance of the adoption of cleaner production: survey in large textile industries. J Clean Prod 278:123855
6. Okafor CC, Madu CN, Ajaero CC et al (2021) Sustainable management of textile and clothing. Clean Technol Recycl 1:70–87
7. Bastianoni S, Goffetti G, Neri E, et al (2023) LCA based circularity indices of systems at different scales: a holistic approach. Sci Total Environ 897:165245. https://doi.org/10.1016/j.scitotenv.2023.165245
8. (2022) European Commission
9. Kirchherr J, Reike D, Hekkert M (2017) Conceptualizing the circular economy: an analysis of 114 definitions. Resour Conserv Recycl 127:221–232. https://doi.org/10.1016/j.resconrec.2017.09.005
10. Murray A, Skene K, Haynes K (2017) The circular economy: an interdisciplinary exploration of the concept and application in a global context. J Bus Ethics 140:369–380
11. MacArthur E (2013) Towards the circular economy. J Ind Ecol 2:23–44
12. Zotti J, Bigano A (2019) Write circular economy, read economy's circularity. How to avoid going in circles. Economia Politica 36:629–652. https://doi.org/10.1007/s40888-019-00145-9
13. Geissdoerfer M, Savaget P, Bocken NMP, Hultink EJ (2017) The circular economy–a new sustainability paradigm? J Clean Prod 143:757–768

14. Lenzing AG (2018) The global fiber market in 2016. Lenzing com From http://www.lenzingcom/en/investors/equity-story/global-fiber-market html
15. Ghisellini P, Cialani C, Ulgiati S (2016) A review on circular economy: the expected transition to a balanced interplay of environmental and economic systems. J Clean Prod 114:11–32
16. Koszewska M (2018) Circular economy—challenges for the textile and clothing industry. Autex Res Jo 18:337–347
17. Van Buren N, Demmers M, Van der Heijden R, Witlox F (2016) Towards a circular economy: the role of Dutch logistics industries and governments. Sustainability 8:647
18. Okorie O, Salonitis K, Charnley F et al (2018) Digitisation and the circular economy: a review of current research and future trends. Energies (Basel) 11:3009
19. Jia F, Yin S, Chen L, Chen X (2020) The circular economy in the textile and apparel industry: a systematic literature review. J Clean Prod 259:120728. https://doi.org/10.1016/J.JCLEPRO.2020.120728
20. Desore A, Narula SA (2018) An overview on corporate response towards sustainability issues in textile industry. Environ Dev Sustain 20:1439–1459. https://doi.org/10.1007/s10668-017-9949-1
21. Roy M, Sen P, Pal P (2020) An integrated green management model to improve environmental performance of textile industry towards sustainability. J Clean Prod 271:122656. https://doi.org/10.1016/J.JCLEPRO.2020.122656
22. Leonas KK (2017) The use of recycled fibers in fashion and home products. Textiles and clothing sustainability: recycled and Upcycled textiles and fashion, 55–77
23. Gardetti MA (2019) 1—Introduction and the concept of circular economy. In: Muthu SSBT-CE in T and A (ed) The textile institute book series. Woodhead Publishing, pp 1–11
24. Leal Filho W, Ellams D, Han S et al (2019) A review of the socio-economic advantages of textile recycling. J Clean Prod 218:10–20
25. Khalifa AA, Ibrahim A-J, Amhamed AI, El-Naas MH (2022) Accelerating the transition to a circular economy for net-zero emissions by 2050: a systematic review. Sustainability 14:11656
26. Eder-Hansen J et al (2017) Pulse the fashion industry; Consulting, Global Fashion Agenda & The Boston Group
27. Claxton S, Kent A (2020) The management of sustainable fashion design strategies: an analysis of the designer's role. J Clean Prod 268. https://doi.org/10.1016/j.jclepro.2020.122112
28. Ki CW, Chong SM, Ha-Brookshire JE (2020) How fashion can achieve sustainable development through a circular economy and stakeholder engagement: a systematic literature review. Corp Soc Responsib Environ Manag 27:2401–2424. https://doi.org/10.1002/CSR.1970
29. Kirchherr J, Piscicelli L, Bour R et al (2018) Barriers to the circular economy: evidence from the European Union (EU). Ecol Econ 150:264–272. https://doi.org/10.1016/J.ECOLECON.2018.04.028
30. Ellen MacArthur Foundation (2012) Towards the circular economy
31. Brasch J (2021) Legal and regulatory barriers and opportunities for a textiles circular economy. NIST Workshop: Facilitating a Circular Economy for Textiles.
32. Su B, Heshmati A, Geng Y, Yu X (2013) A review of the circular economy in China: moving from rhetoric to implementation. J Clean Prod 42:215–227. https://doi.org/10.1016/J.JCLEPRO.2012.11.020
33. Adler M (2021) Textile circularity in the US: current landscape, bottlenecks, and critical gaps. In: NIST workshop: Facilitating a Circular Economy for Textiles.
34. King J (2021) Textile reuse and recycling: A SMART perspective. In: NIST workshop: facilitating a circular economy for textiles
35. Fibersort (2021) Valvan baling systems introduces fibersort. In: (Valvan Baling Systems)
36. Barker K (2021) Optical sorting advances are driving the MRF of the future. In: Recycling product news
37. Feitelberg R (2018) The repair economy gains momentum. In: Women's wear daily
38. Johnson S, Echeverria D, Venditti R, et al (2020) Supply chain of waste cotton recycling and reuse: a review. AATCC J Res 7:19–31. https://doi.org/10.14504/AJR.7.S1.3
39. Aquafil Global (2022) ECONYL R Pure. In: (Aquafil Global)

40. Bender D (2021) Challenges to mechanical and chemical recycling of textilles. In: NIST workshop: facilitating a circular economy for textiles
41. Sandvik IM, Stubbs W (2019) Circular fashion supply chain through textile-to-textile recycling. J Fash Mark Manag 23:366–381. https://doi.org/10.1108/JFMM-04-2018-0058/FULL/XML
42. Pedersen ERG, Earley R, Andersen KR (2019) From singular to plural: exploring organisational complexities and circular business model design. J Fash Mark Manag 23:308–326. https://doi.org/10.1108/JFMM-04-2018-0062
43. D'Adamo I, Lupi G (2021) Sustainability and resilience after COVID-19: A circular premium in the fashion industry. Sustainability (Switzerland) 13:1–5. https://doi.org/10.3390/SU13041861
44. Bressanelli G et al. (2022) Circular economy in the textile industry: evidence from the prato district. In: (a) Role of circular economy in resource sustainability (Cham: Springer)
45. Kazancoglu I, Kazancoglu Y, Yarimoglu E, Kahraman A (2020) A conceptual framework for barriers of circular supply chains for sustainability in the textile industry. Sustain Dev 28:1477–1492. https://doi.org/10.1002/SD.2100
46. Huang YF, Azevedo SG, Lin TJ et al (2021) Exploring the decisive barriers to achieve circular economy: Strategies for the textile innovation in Taiwan. Sustain Prod Consum 27:1406–1423. https://doi.org/10.1016/J.SPC.2021.03.007
47. AL SK and F, An (2022) Textiles in a circular economy: landscape, assessment of the current Challenges, and opportunities in the United States. Front Sustain 3:1038323. https://doi.org/10.3389/frsus.2022.1038323
48. Atkar A, Pabba M, Sekhar SC, Sridhar S (2021) Current limitations and challenges in the global textile sector. Fundamen natural fibres and textiles, 741–764. https://doi.org/10.1016/B978-0-12-821483-1.00004-8
49. FIBERSORT (2021) Valvan baling systems NV—PDF catalogs | technical documentation | brochure
50. Schumacher K, Forster AL (2022) Facilitating a circular economy for textiles workshop report. NIST Spec Publ 1500:207
51. Chen H-Y (2021) The effect of fast fashion brand awareness on purchase intention: a study of fashion clothing. Int J Organ Innov 14
52. Wang PY (2022) Interviewee, professor, school of materials science and engineering. In: Georgia institute of technology
53. Damayanti D, Wulandari LA, Bagaskoro A, et al (2021) Possibility routes for textile recycling technology. Polymers 2021 13:3834.https://doi.org/10.3390/POLYM13213834
54. OECD (2009) Eco-innovation in industry—enabling green growth. In: Paris: OECD publishing
55. Wang W, Bessède JL (2015) Life cycle assessment of equipment for electricity transmission and distribution networks. Eco-friendly innovation in electricity transmission and distribution networks, 123–133. https://doi.org/10.1016/B978-1-78242-010-1.00006-9
56. Henry BK, Russell SJ, Ledgard SF, et al (2015) LCA of wool textiles and clothing. Handbook of life cycle assessment (LCA) of textiles and clothing, 217–254. https://doi.org/10.1016/B978-0-08-100169-1.00010-1
57. Khan WM, Siddiqui S (2017) Estimation of greenhouse gas emissions by household energy consumption: a case study of Lahore, Pakistan. Pakistan J Meteorol, 14
58. Blackburn K, Green D (2022) The potential effects of microplastics on human health: what is known and what is unknown. Ambio 51:518–530. https://doi.org/10.1007/S13280-021-01589-9

Eco-innovation: A Case Study of the Kenyan Textile and Apparel Industry

Josphat Igadwa Mwasiagi, Nonsikelelo Sheron Mpofu, Elizabeth Kariuki, and Kefa Chepkwony

Abstract Eco-innovation is the act of creating, developing, and applying new or improved processes and services that have a positive impact on the environment, while at the same time optimizing positive social and economic impacts. The Textile and Apparel (T&A) sector is one of the key industries, that provide essential products to mankind, but could also lead to adverse effects on the environment. This research work concentrated on the state of the Kenyan T&A sector. Apart from desktop research, data and information were collected from key stakeholders in the T&A sector in Kenya. Results obtained in this research work indicated that there is a need for more work on the Kenyan, policy framework. Other key areas that need to be addressed to ensure the Kenyan T&A sector can operate in a sustainable manner include production of natural fibers (cotton and sisal), energy management, wastewater management, financial policies, and management of post-industrial and post-consumer waste. Based on the results obtained in this research work, it is recommended that a more in-depth analysis that will include, environmental, economic, and social hotspots and product Environment Footprint (PEF) should be undertaken for the Kenyan T&A sector.

1 Introduction

The organization of the manufacturing process involves the interaction of several factors which include material (raw materials), man (human resource), and technology (methods, machine, and measurement) [1]. The earth is generally considered to be the main source of raw materials and also a support for all other activities, so it can be included as an important factor. Other factors like energy and climate change have also taken center stage and need to be considered. Traditionally, most

J. I. Mwasiagi (✉) · N. S. Mpofu · E. Kariuki
School of Engineering, Moi University, Eldoret, Kenya
e-mail: igadwa@gmail.com

K. Chepkwony
School of Business, Moi University, Eldoret, Kenya

S. R. Batool et al. (eds.), *Circularity in Textiles*, Textile Science and Clothing Technology, https://doi.org/10.1007/978-3-031-49479-6_2

researchers concentrate on one or two factors to improve the manufacturing process [2, 3]. There is a paradigm shift where a more holistic view of the manufacturing process has pushed the agenda of sustainability, forefront of humanity discussion, to slow down rapid degradation of the resources on earth. Businesses, governmental departments, researchers, civil societies, and international organizations have continually worked on strategies to introduce sustainable methods in manufacturing. While several approaches have been suggested, Eco-innovation has recently been proposed as a viable strategy that could bring the desired change if properly implemented.

Eco-innovation (Eco-i) can be considered as the act of creating, developing, and applying new or improved processes and services that have a positive impact on the environment. Eco-i has recently gained increasing attention from governments, industries, businesses, and civil society organizations to address the environmental challenges facing our planet Earth. This chapter provides an overview of the concept of eco-innovation, its drivers, and its impacts, drawing on a range of academic and non-academic sources. UNEP [4] describes eco-innovation as a new business approach that promotes sustainability throughout the entire life cycle of a product, while also boosting a company's performance and competitiveness. Eco-i can support enterprises to access new and expanding markets, increase productivity, attract new investment into the business, increase profitability across the value chain, and help companies stay ahead of regulations environmental standards, and fair trade. Eco-i can help address key issues of development and environment, where development is done in a sustainable manner without adversely affecting the environment. Therefore, the most popular approach of considering Eco-i is to include environmental, social, and economic impacts in a practical and optimized manner (see Fig. 1). In the process of implementing Eco-i; circularity, which includes the maximization of the usage of material in the Textile and Apparel (T&A) sector (see Fig. 2) is a critical component, which can be used to drastically reduce the amount of virgin fibers needed to provide T&A goods to humanity. UNEP emphasizes the importance of policy support for eco-innovation. Eco-innovation can help achieve multiple environmental objectives, such as reducing greenhouse gas emissions, improving resource efficiency, and reducing pollution [5]. However, eco-innovation by itself is not sufficient to address broader systemic challenges required for effective and efficient sustainable business. There are three main drivers of eco-innovation namely: environmental regulations, customer demand, and company culture [6].

Enterprises that prioritize eco-innovation are more likely to invest in research and development and to collaborate with other stakeholders. Another important aspect of eco-innovation is its impact on business performance. They tend to perform better financially, attract, and retain customers, and attract and retain employees [7]. It can also contribute towards cost reduction mechanisms by firms and thus increase competitiveness. However, eco-innovation can also face several challenges such as lack of access to finance, market failures, and regulatory uncertainty [8]. These barriers can prevent companies from investing in eco-innovation and policy support is needed to address them.

In a nutshell, eco-innovation is an important concept, with the potential to help address some of the most pressing environmental challenges facing our planet.

Fig. 1 Factors to be considered in Eco-innovation models of production

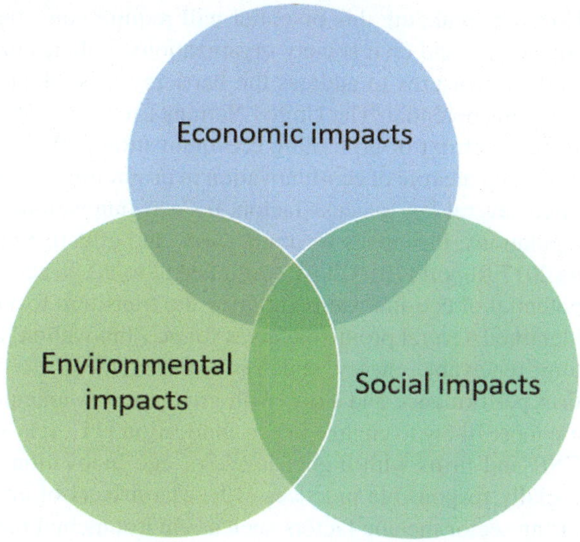

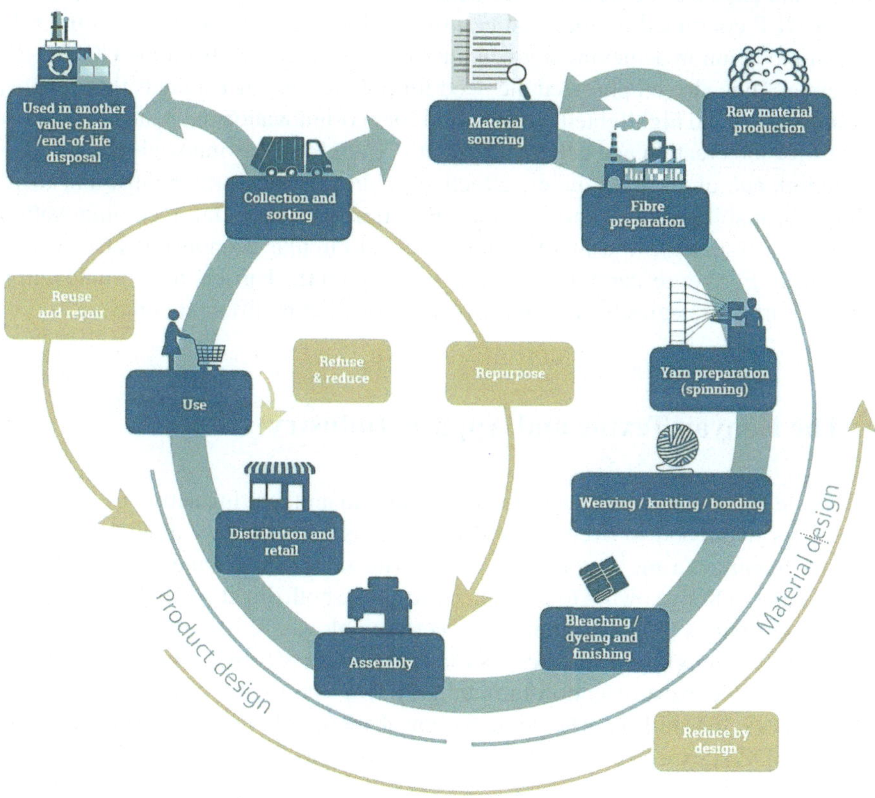

Fig. 2 Circularity in the textile and apparel sector (*Source* UNEP, 2020)

However, realizing this potential will require concerted efforts from governments, businesses, and civil society organizations. Policies that support eco-innovation, as well as measures to address the barriers to its adoption, will be essential to realizing this potential. The United Nations Environment Programme (UNEP) has been at the forefront of promoting eco-innovation globally. In a 2014 report [9], UNEP discussed the role of eco-innovation in promoting sustainable development and identified several key success factors for eco-innovation, including strong policies and regulations, incentives for innovation, and effective public–private partnerships. In its 2017 Report [10], UNEP continued to focus on eco-innovation as it highlights the potential of eco-innovation to drive the transition to a circular economy. The report identified several promising areas for eco-innovation, including sustainable agriculture, renewable energy, and waste management. Eco-innovation positively affects firm performance and firms with greater environmental awareness and capabilities are more likely to engage in eco-innovation [11]. It is also positively associated with CSR and firms with higher levels of eco-innovation are more likely to engage in socially responsible practices [12]. The drivers of eco-innovation in the manufacturing sector include factors such as environmental regulations, technological capabilities, and customer demand that influence the adoption of eco-innovation [13]. In 2021, UNEP continued its focus on eco-innovation [14], which discussed the role of eco-innovation in achieving a green and inclusive recovery from the COVID-19 pandemic. They also emphasized the need for policies and investments that support eco-innovation and highlighted the potential for eco-innovation to create new opportunities for job creation and economic growth. Therefore, eco-innovation highlights the importance of strong policies, effective public–private partnerships, and technological capabilities in promoting eco-innovation. The potential of eco-innovation to drive the transition to a more sustainable and circular economy is also widely recognized. Further research is needed to understand the barriers to eco-innovation and to identify strategies for promoting eco-innovation in different contexts.

2 The Kenyan Textile and Apparel Industry

The textile and Apparel sector is one of the major manufacturing activities in Kenya. It provides livelihood to over 600,000 households and employs about 30% of the national manufacturing sector labor force. The sector contributes about 14% of employment in Kenya owing to its capacity to create both direct and indirect employment opportunities [15]. Generally, the sector employs over 2.5 million people, of which 84% are employed in Micro SMEs [16] while 8% are employed indirectly and over 1.6% consist of cotton farmers [17, 18]. The total turnover of the sector is about US$ 564 million. The contribution of the textile manufacturing sector to Kenya's GDP is expected to continue growing from 10% in 2019 to 15% by 2022 [19].

Kenya's textile and apparel sector has the potential to anchor its transformation into a middle-income economy. It is a good prospect for wealth and employment creation. The sector offers opportunities for value chain development through eco-innovation as well as capacity development of stakeholders, spurring manufacturing and supporting the balance of trade and income as well as income diversification. Kenya's national trade policy objectives include transforming the country into a competitive export-led and efficient domestic economy thus moving towards a more open trade regime and further integration into the world economy, strengthening, and increasing overseas market access for Kenyan products including textiles, agro-based products and processed goods [20]. The garment sector of Kenya is principally driven by exports to the US under the AGOA initiative. Through the African Growth and Opportunity Act (AGOA), Kenya's apparel exports to the USA increased to USD 495 million (KES 59.6 billion) in 2021, employing over 57,500 workers in EPZ alone [21]. There are approximately 170 large-scale garment manufacturing units operating in Kenya today, dominated by EPZ [22]. The garment manufacturing units constitute 29% of all EPZ enterprises, 78% of total EPZ local employment, 56% of EPZ exports, 52% of total EPZ sales, and 30% of all EPZ private investment in Kenya.

Several Government of Kenya (GOK) documents indicate that the textile industry is one of the industrial sectors that can be used to spur economic growth [23, 24]. The Kenya Industrial Transformation Programme (KITP) of 2015 which is anchored to the Kenya Vision 2030, aims to launch flagship projects in textiles to support Kenyan SMEs and create an enabling environment to accelerate industrial growth. Under this program, the Government aims to develop an integrated textile industrial park, also referred to as a textile city, in Naivasha to attract investors. The government has also put in place a strategy to promote development under the Big Four Agenda [25] where textile and leather sectors are key manufacturing pillars. Kenya Industrial Estates (KIE) is also a government agency created to offer incubation, accelerator, and business advisory services, financing, and development of industrial parks. KIE has identified textile and apparel value chains as one of the key sectors of focus. As of 2019, KIE had hosted several incubators and financed over 66 textile and apparel SMEs. The government through Kenya Export Promotion and Branding Agency has developed a 'Made in Kenya Brand Mark' which aims to promote locally manufactured products in the local and global markets. Kenya Industry and Entrepreneurship Project (KIEP) has also been initiated through the Ministry of Industrialization, Trade, and Enterprise Development to strengthen the innovation and entrepreneurship ecosystem, increase productivity and innovation at the firm level, and provide project implementation and monitoring and evaluation support for different industry sectors. Fibre Crops Directorate was also established in 2014 to regulate, develop, and promote fiber value chains such as cotton and sisal in the country.

SMEs are critical to the Kenyan economy; they constitute about 80% of Kenyan businesses and employ around 78% of the labor force (about 14.9 million). Approximately 7.4 million SMEs in Kenya collectively contribute about a third of the country's GDP. The manufacture of textile apparel constitutes about 41.6% of licensed SMEs or about 74,000 firms [26]. Almost 90% of firms in the textile sector are categorized as 'micro'-sized SMEs and they are involved in all segments of the

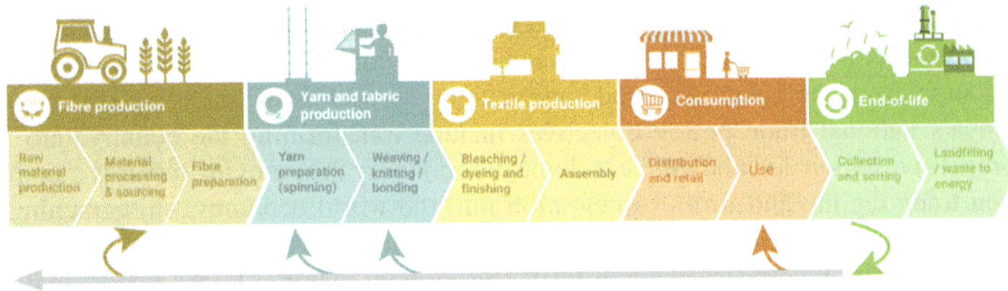

Fig. 3 Linear textile and apparel value chain (UNEP, 2020)

textile value chain. About 21 SMEs operate in the EPZ, employing an average of 1800 staff per company. Some of these SMEs are part of multi-national value chains and exports to the US, regional, and EU markets [27]. Despite several institutional frameworks being in place to promote the textile and apparel industry in Kenya, low investment in textile and apparel manufacturing value chains remains a key impediment to the growth of the sector. The main textile industrial processes that take place in Kenya include fiber production, ginning, spinning, thread production, weaving and knitting, textile processing (including bleaching, dyeing, printing, and finishing), and apparel design and production. The Kenyan textile industry can be represented using Fig. 3 (designed by UNEP for a typical T&A sector) which is therefore relatively diverse, and can be divided into four broad categories whose respective key products are:

- cotton growing—raw cotton
- ginning—cotton lint and seeds
- yarn and thread production—spun yarn and threads
- fabric and apparel manufacture—woven, knitted, and non-woven fabrics (bedsheets and shirting/dress material, outerwear textiles including cotton and synthetic knitted shirts, t-shirts, sweaters, woven pants and shorts, woven shirts, and dresses) (See Fig. 4).

Economic activities in Kenya were notably subdued in the first quarter of 2019, relative to the performance recorded in the same quarter of 2018. During this period, the economy expanded by 5.6% compared to 6.5% in the corresponding quarter of 2018 according to the Kenya National Bureau of Statistics. As stated earlier on, the T&A industry is a core industry, which includes; researchers, ginners, farmers, spinners, input suppliers, textile manufacturers, and extension service providers. It is estimated that approximately 40,000 farmers are involved in cotton farming, while the overall sector provides livelihood to approximately 200,000 households.

About 50,000 smallholder farmers are engaged in the growth of cotton in arid and marginal regions, under rain-fed conditions on small land holdings of about one hectare. The current production of cotton lint in Kenya is approximately 7,000 tons versus a potential production of 200,000 tons of lint or 750,000 tons of seed cotton. Production has been volatile for the last few years and has not been sufficient

Fig. 4 Typical garment-making operations in Kenya

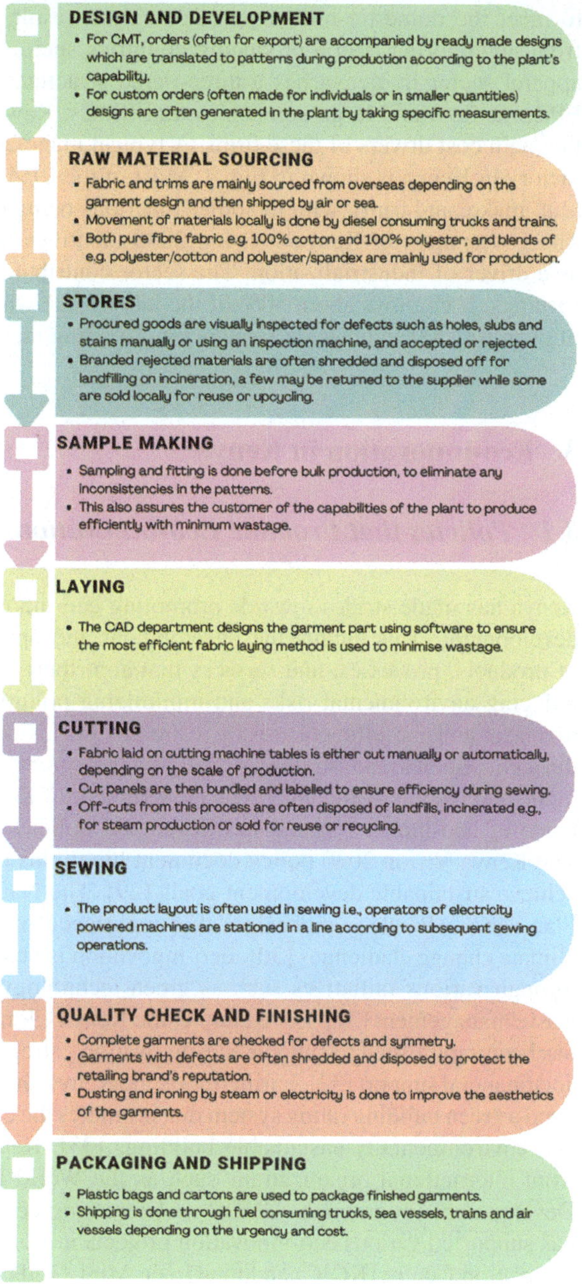

to meet the domestic mill requirement. As a result, Kenyan firms import cotton from neighboring cotton-producing countries such as Uganda and Tanzania. The apparel sector in Kenya has a three-tiered structure: in the EPZ, and outside the EPZ, medium and large companies and SMEs. Raw materials and overheads are the main cost drivers of these firms. A typical manufacturing process for a garment factory in Kenya is shown in Fig. 4. Most of the EPZ factories use the CMT model (cut, make, and trim), with the other functions being done abroad (outside Kenya). The government of Kenya's Vision 2030 identifies the garment and textile sector as a driver of industrialization. The sector contributes 7% of the country's export earnings. It employs about 30% of the labor force while supporting the livelihoods of over 200,000 small-scale farmers by providing markets for cotton.

3 Eco-innovation in Kenya

3.1 Policies that Promote Eco-innovation

Kenya has made strides towards promoting eco-innovation to promote sustainable development. This entails the development and application of a new generation of products, processes, and services that contribute to sustainable development by reducing environmental risks and minimizing pollution while improving competitiveness and the efficient use of resources [28]. Eco-innovation can be achieved through policies and strategies that promote green growth, sustainable development, and resource efficiency. The Government of Kenya has recognized the importance of eco-innovation and has developed policies and strategies to promote it. The Kenya Vision 2030 policy document highlights the need for eco-innovation to achieve sustainable development goals [29]. The National Climate Change Action Plan (NCCAP) also emphasizes the importance of eco-innovation in addressing climate change challenges [30]. Eco-innovation is being implemented in the country through various initiatives such as green technology, sustainable agriculture, and waste management [31]. For example, the Kenya Climate Innovation Centre (KCIC) has been established to support eco-innovation entrepreneurs by providing technical and financial support [32]. Furthermore, the Kenya Green Building Society has developed a green building rating system that promotes the construction of energy-efficient and environmentally sustainable buildings [33]. Kenya has also received support from international organizations such as the World Bank and the United Nations Development Programme (UNDP) in promoting eco-innovation. The World Bank has supported various eco-innovation projects in Kenya, such as the Kenya Climate Innovation Centre (KCIC) and the Green Mini-Grids Program [34]. The UNDP has also supported eco-innovation initiatives in Kenya, such as the Low Emission and Climate Resilient Development (LECRD) project [35].

In addition, the East African Community (EAC) has developed policies and strategies to promote eco-innovation in member countries, including Kenya. The EAC

Climate Change Policy Framework highlights the importance of eco-innovation, in addressing climate change challenges in the region [36]. The EAC has developed a regional strategy for eco-innovation that aims to promote eco-innovation by strengthening policy frameworks, building capacity, and promoting collaboration among stakeholders. At the county level, some counties in Kenya have developed policies and strategies to promote eco-innovation. For example, the Makueni County Climate Change Action Plan highlights the need for eco-innovation in addressing climate change challenges in the county [37]. Kiambu County government has developed a sustainable waste management plan that involves recycling, composting, and waste-to-energy initiatives [38]. Similarly, the Mombasa County government has developed a sustainable transport policy that aims to reduce greenhouse gas emissions by promoting non-motorized transport modes such as cycling and walking [39]. Kisumu County Integrated Development Plan highlights the need to promote eco-innovation in the county [40]. The Nairobi County Integrated Development Plan also recognizes the importance of eco-innovation in addressing environmental challenges in the county [41]. Overall, Kenya has made significant progress in promoting eco-innovation through policy development and implementation, as well as support from international organizations. However, there is still a need for more investment in eco-innovation to achieve sustainable development goals and address climate change challenges.

3.2 *Opportunities and Challenges for Promoting Eco-Innovation in T&A Sector*

This section focuses on broad global trends about various aspects of eco-innovation and circularity and supplements it with stakeholder inputs on the potential opportunities in the Kenyan T&A sector. Stakeholders from industry, government, academia, design fraternity, financial sector, and government were consulted in 2021, and the following issues were highlighted; design and circularity, materials, chemicals, waste management, energy, water usage and pollution, innovative business models and financial policies. These issues are discussed in the following subsections.

3.2.1 *Design and Circularity*

The factories in the Export processing zones use cut, make, and trim (CMT). The designs, fabric sourcing, and marketing are often done abroad (outside Kenya). The Kenyan T&A industry needs to;

- Build capacities of local designers to work with alternate materials, quicker turnaround on design options, and increased awareness about the potential for circular design.

- Work on clear labeling and content laws to enable better management of the garment at the end of life.

3.2.2 Materials/Fibre

Globally, there is increasing pressure on the apparel and textile sector to address its adverse environmental and social footprint through its choice of materials. While the Kenyan T&A sector imports fibers, yarn, fabric, and accessories, this section will concentrate on fibers. Materials like cotton are extremely water-intensive and require pesticides and insecticides. Polyesters on the other hand though easier to produce and manage, trace their origin to fossil fuels and lead to the menace of microplastics which is causing marine pollution at an alarming rate. Further, there is scope to reduce the consumption of virgin resources and increase the percentage of recycled content.

Against this backdrop, brands and large retailers are making commitments to shift to nearly 100% sustainable cotton [42, 43]. Several companies convert plastic waste into new garments. In fact, a leading garment cluster in India has recorded the manufacture of T-shirts made from recycled plastic for sportswear [44]. Increasingly, alternative natural fibers are being explored and offered by innovators. These are at a niche stage and more research and development are needed to improve suitability for large-scale fabric production. Designers in Kenya are experimenting with alternate fibers as well, with an emphasis on cotton, sisal, and silk. Some of the key recommendations from the stakeholders include.

i. ***Cotton***

   - Increase indigenous production of cotton to reduce dependence on imports and support a growing textile industry. Cotton production stands between 25,000–30,000 bales, well below the requirement of approximately 200,000 bales.
   - Consider the selling price of cotton which is currently very low and discourages more cotton cultivation. However, industry players expressed concern that determining a Minimum Support Price will affect the cost of production and make the products uncompetitive both in the domestic and global markets.
   - Focus on sustainable cotton which will entail capacity building for farmers, support adherence to accepted certification, and promote high yielding variety of seeds for higher productivity. Industry actors suggested that farmers could be supported with a subsidy program to move to more sustainable practices.
   - Support farmers with access to low-cost farm inputs and high yielding variety of seeds including seeds of BT cotton, that address the challenges of pest infestation.
   - Adopt a value chain approach and ensure that there is a sufficient supply of high-quality locally grown cotton.
   - Ensure traceability and transparency in cotton value chains, especially with a view to contain pest infestation. This could be done through proper mapping

of seeds used by the farmers. There is a policy on traceability but needs to be better implemented.

ii. ***Silk***

The sericulture sector in Kenya has great potential but has lagged due to poor policy coordination, lack of R&D, and dependence on external donor support. Some notable issues from stakeholders include;

- A high percentage of waste is generated in silk reeling. Therefore, there are opportunities to minimize and utilize waste. Some SMEs work with silk waste and have an opportunity to scale up such businesses.
- SMEs operating in this space also offer immense scope to empower women and youth as they provide employment and skilling opportunities to these vulnerable groups.

Box 1: An example of partnering with local communities for economic empowerment and minimising waste

Alisam Products development and design is an enterprise legally registered in Kenya whose main business is to rear silkworm, produce silk yarn and silk materials. Alisam established over 10,000 mulberry trees on a 4-acre land to produce leaves as feeds for worms and also for climate mitigation purposes. 7 women are fully employed to take care of the plantation. The project works in partnership with community groups in ASAL areas of Nyanza, whereby the community members are provided with free mulberry seedlings, trained on silkworm rearing, and supply Alisam with cocoons. Over 800 farmers are benefiting from the program. In the process of yarn production, the business generates wastes in fibre form. The wastes are processed into pillows, for local market.

3.2.3 Chemicals

The Apparel and textile sector uses many chemicals through various processes across the value chain starting from fiber production to wet processing to post-consumption care. Some of these chemicals are toxic and lead to water pollution and related health hazards. Research and innovations are identifying greener and cleaner chemicals for the sector. Brands and retailers are regulating the use of chemicals through prescribed Restricted Substance Lists. Initiatives such as Zero Discharge of Hazardous Chemicals (ZDC) [45] aim to move towards zero discharge of harmful chemicals in the leather, textiles, and footwear sector. Developments of new techniques such as waterless dyeing help reduce both water intake for processes and contain the associated water pollution as well.

The aspect of chemicals was not discussed in detail in this round of stakeholder consultations. However, industry actors did indicate that they had moved to adopt some cleaner chemicals and designers had also indicated working with natural and organic dyes. Experience from other countries has highlighted that awareness and incentives will play a large role in facilitating a shift to greener chemicals. It is also important to adopt a life cycle approach and examine the use of chemicals at each stage including in the consumption stage where processes such as dry cleaning could contribute to chemical pollution.

3.2.4 Waste Management

One of the biggest impacts of the T&A sector is the generation of huge quantities of waste. The waste can be classified into two, post-industrial and post-consumer waste. While data for post-industrial was not available, the issue of importation of high amounts of second-hand clothes was highlighted, as a possible contributor to the amount of post-consumer waste in Kenya. Figure 5 gives the amount of waste imported into Kenya. The fact that the bulk of this waste makes its way to the landfill with dismally low rates of reuse or recycling compounds the issue. Tackling the solid waste generated by the sector would not only address environmental concerns but will also enable resource efficiency through recycling and reuse. Other aspects of waste are the hazardous chemicals and the wastewater that is generated through wet processing.

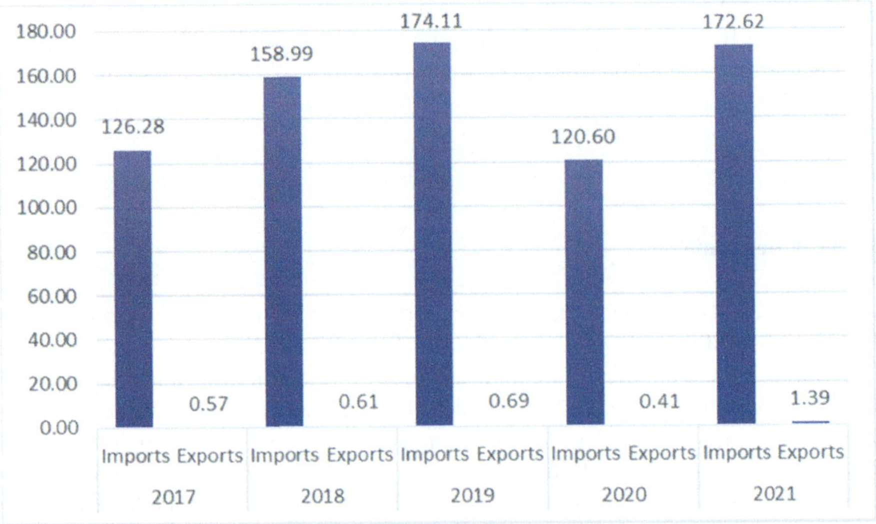

Fig. 5 Trade of Second-hand clothes in Kenya

Box 2: The Trade of second-hand Clothes in the Kenyan T&A Sector

The Kenyan population spends 2.5% of private consumption on clothes and footwear which adds up to USD 40 per person per year for clothing and footwear. The purchase of second-hand clothes and footwear accounts for 40% of the expenditure. However, there is little information about the end of life for imported second-hand clothes and therefore there is a possibility that the clothes are dumped in landfills or incinerated. There was a ban on the importation of second-hand clothes by the East African community, which Kenya backtracked on. In 2020, the was another ban on the importation of second-hand clothes during the COVID-19 pandemic. The ban was however lifted a few months later. The importation of second-hand clothes by Kenya has caught the attention of other plays which include Secondary Material and Recycled Textiles Association [SMART] of the USA, the UK's Textile Recycling Association (TRA), the European Recycling Industries' Confederation [EuRIC] and the Bureau of International Recycling [BIR], who reported that Kenya plays a critical role in the recycling of T&A products at a global level. In Kenya the trade of second-hand goods is governed by several policies, which are not specific, but generic in nature. Kenya Bureau of Standards has however given protocols to be followed when importing second-hand clothes. The Kenya Revenue Authority and other government agencies provide guidelines on taxes and other levies to be paid.

The issue of wastewater management is covered in more detail under the sub-heading of water usage and pollution. Industries comply with regulatory requirements when it comes to the discharge of hazardous materials but did indicate the need for more organizations/vendors that handle hazardous waste. Some key messages from stakeholders included:

- At present there is low recycling/reuse of textile and clothing waste in Kenya.
- Export Processing Zones regulations do not allow for second-hand clothing to be brought into EPZs thus restricting the potential for new circular business models.
- No incentives for waste reduction or utilization. Fully integrated units can fully utilize their waste. Such practices should be recognised and incentivized.
- Limited coordination amongst industries to utilize waste between industries. Waste resources from own operations as well as from other businesses can offer a rich resource base. There is a need to build understanding and awareness amongst businesses to understand and capitalize on such opportunities.
- No Extended Producer Regulations for garments and clothing thus causing most of the apparel waste to be sent to landfills. There is scope for new businesses that offer innovative solutions for the reuse/repurpose of textile and apparel waste.
- Need to develop collection mechanisms for used and old clothes.

- Capacity and skills of designers should be enhanced to work with waste materials and create upcycled and recycled products. This also presents opportunities for exports.
- Need for consumer awareness and branding to increase acceptance of recycled clothes.
- There is a need for research and development to support fiber to fiber recycling.
- Technology upgradation will help reduce wastages at firm levels. Need to create awareness about such technologies and support with affordable finance. Upgradation of hand looms should be undertaken. There is scope to reduce yarn wastage as well, but this needs to be supported through training and skilling.

3.2.5 Energy

Most of the countries around the world are committing to measurable and tangible targets for transitioning to cleaner and more efficient energy systems. In fact, the global narrative is quickly shifting towards more ambitious net zero emission targets, and various governments are laying down commitments to be achieved by 2050 [46]. Kenya's action plan on climate change aims to reduce GHG emissions by 30% by 2030. In addition to government commitments, international brands and large retailers have made public sustainability commitments and are laying down clear environmental standards for the supply chain actors to adopt. There is a clear business case for organizations to shift to cleaner systems as it improves the bottom line and business competitiveness. The textile sector in Kenya can significantly contribute to these clean energy commitments and ensure its resilience as high-power costs continue to be a limiting factor for the growth of the sector in Kenya. Stakeholder consultations highlighted the following:

- Modernisation of mills and shift to Energy Efficiency (EE) and Renewable Energy (RE) systems needs to be increased. This will need to be supplemented with awareness programs as surveys highlighted that there is limited knowledge about technology upgradation for sustainable production.
- Non EPZ firms do not get a 100% rate and financing clean energy systems becomes expensive. Earlier 100% rebate on CAPEX was available to the industry, but that has been reduced to 50% outside EPZ and that too is available in a phased manner. Industry stakeholders have expressed the need for regulatory incentives and rewards to augment the initiatives that the private sector is spearheading on its own. One policy win has been to waive the VAT on imported solar panels thus encouraging businesses to make investments in clean energy systems.
- Power tariffs are high and remain a large part of the costs of production despite partial offset from clean energy production. Within EPZ—power costs with financing solutions @ 5–6 cents per unit- govt charges 17 cents. Night shifts remain a challenge as solar systems too expensive with storage.
- Turnkey solutions are needed for the industry and Kenya Association of Manufacturers is working on the same. However, at times the setup of contractual agreements and payment structures delays the process, especially in EPZ areas.

- SMEs lack the knowledge and financing to transition to cleaner energy systems. There is low awareness about incentives that may be available to them. Further, SMEs may also lack resources to own and install machinery individually and shared machinery models can be promoted.

3.2.6 Water Usage and Pollution

The water footprint of the apparel and textile sector is under scrutiny. The high consumption of water across fiber production and wet processing, wastewater generated, and water used in post-consumption care are all contributing to water scarcity. Further, it is also essential to understand the source of water used and the associated pollution caused by the excessive use of pesticides, unmeasured use of water, and non-compliance with discharge of toxic wastewater. The rhetoric is shifting towards less water-intensive crops or sustainable agriculture practices, stressing on water use efficiency during processing cycles, investing in innovations relating to waterless techniques or greener chemicals, use of more water-efficient technology, increasing the usage of recycled wastewater (whether it be from own industrial operations or exploring treatment and use of other wastewater such as from the domestic sector). There are strict regulations in manufacturing sectors about treatment and discharge of wastewater and set up of effluent treatment plants (ETPs), common effluent treatment plants (CETPs) and Zero Liquid Discharge Systems (ZLDs) is the norm across manufacturing sites. However, it is interesting to note that over the past few years, there is a marked shift on understanding and rationalizing water consumption across the value chain and not just focusing on managing wastewater. Key messages from stakeholder consultations in Kenya highlighted the following:

- Large processing units are complying with the industrial water management wastewater treatment requirements as laid down in the regulations. However, there is an urgent need to revisit the regulations and charges as water treatment is an expensive proposition and requires high capital investment. Businesses are not only bearing the cost of treating and releasing water into CETPs, but certain regulations also charge them for water being released from CETPs thus increasing the costs of production.
- Most of the SMEs did not have a significant water footprint due to the nature of their operations and the processes they handle. There is a case, however, to increase awareness about the water footprint and explore affordable solutions for the same.

3.2.7 Innovative Business Models

Eco-innovation and practices of circular economy advocate for a life cycle approach and close collaboration between stakeholders across the value chain. Availability of finance, affordable technology, continuous skilling, and conducive policies and regulatory regimes are critical to encourage and support the adoption of sustainability practices and promote eco-innovation. There is potential for livelihood opportunities through new businesses/start-ups offering sustainable solutions as well as reskilling of labor to support transition to sustainable business models. The key messages from stakeholders included.

- Need to increase awareness about Circular Economy, Sustainability, Eco-innovation principles, and new opportunities offered by these.
  - There is a need to establish a common definition and understanding of circular economy principles for the Kenyan context.
  - Many stakeholders indicated that academic curricula at various levels should integrate these concepts.
  - Industry stakeholders suggested that the awareness amongst SMEs on these issues is very low, as a result, they are unable to capitalize on new opportunities to improve their competitiveness and access new markets.
  - SMEs can benefit from dedicated centers on circularity and eco-innovation to access information. Knowledge-sharing platforms and case studies can boost the adoption of such practices.
  - Build awareness about new business models such as focusing on rental, repair, recycling, sustainable packaging, waste utilization, upcycling, etc.
- Skilling gaps must be identified, and appropriate programs must be developed in consultation with industry actors. Many industry stakeholders stressed upon the lack of an adequate, skilled workforce for growing operations in the textile sector and cautioned that this could impede the growth of the sector.
  - It was suggested that a detailed skilling needs assessment should be done in partnership with industries and relevant skilling programs be offered. Labor costs in Kenya are some of the highest in the region and it is imperative that labor efficiency and productivity are considerably improved through relevant skilling programs.
  - SMEs should be supported in capacity building for their employees.
  - Tie-ups with academia and industry can help foster innovations that underpin the adoption of CE practices and eco-innovations.
  - Training programs will have to be designed and customized for different nodes in the value chains. While there is a large base of weavers and artisans, they need to be trained in modern techniques and potential circular models. Large firms not only face a shortage of labor but also require access to managerial talent.
- Access to Research and Development: Designers and SMEs indicated that they had limited access to new research for material or process innovation due to non-disclosure agreements in place, thus limiting their ability to implement more sustainable practices.

- Financing for sustainability, circularity, and eco-innovation. In general, there is limited awareness in the financial sector about these concepts and consequently, not many financial products are customized or considered for promoting circularity or eco-innovation.
  - SMEs face many challenges in accessing financing for day-to-day operations and it is nearly impossible to avail financing for sustainable practices.
  - Innovations for sustainability are expensive and start-ups and SMEs need to be adequately supported to make these transitions.
  - An SME fund is being set up by the Ministry of Industrialisation, Trade, and Enterprise Development that could be tapped into to provide financing for sustainable SMEs.
  - Financial institutions should also focus on ease of financing and make finance more accessible and affordable.
  - From the financiers' viewpoint, appropriate risk management instruments will have to be designed and offered to enable financing for new business models and concepts.
- Market access for sustainable products—Forums could be created for buyers and sellers to meet. Proper branding and labelling will help. Green labelling should be introduced for green products.
- Awards and Recognition mechanisms—It was suggested that SMEs can be recognised and motivated through awards for cleaner production and more circular practices.
- Policy Implementation—While stakeholders indicated that there is a need for incentives and conducive regulations to support circularity, they also acknowledged that there is also a requirement for improved implementation of existing policies.
  - Multi-stakeholder collaboration and consultations should be organized to assess the effectiveness of existing schemes and policies and to identify suitable adjustments.
  - Improve interdepartmental coordination. This is especially a pain point for SMEs who lack the time and resources to navigate through cumbersome procedures.
  - Industry actors also highlighted the need for enhanced coordination and cooperation between government departments at the national, regional, and county levels. Oftentimes the decision-making, and information remain concentrated at the national level thus making it challenging for businesses to access incentives and information under offered schemes.
  - Policies should sharpen focus on women entrepreneurs, youth, and marginalized communities.
  - The "Buy Kenya, Build Kenya" should be better implemented to boost local production and to enable local manufacturers to compete with cheap imports.

3.3 Financial Policies

The major financial hurdles for T&A companies in the Eastern region of Africa include [47, 48]; high cost of finance, lack of collateral, low valuation of machineries when used as collateral, and longer business cycles (180 to 240 as opposed to 90 days (preferred financial institutions)), limited awareness of trade finance instruments, presence of illegal imports (which are cheaper and can be imported within a short time), and the limited understanding (by financial) of how the T & A industry operates. Additionally, the T&A sector in Kenya lacks import/export duty incentives and investment financing [48], common in other developing countries such as Bangladesh and Vietnam. It is worth noting that companies that manufacture products within the supporting industries framework are entitled to import/export duty incentives under the current law on import/export duties. Another issue to note is that over 90% of the profits in the T&A in the region go to international banks [49]. Despite the hurdles, there are several opportunities in the region, which include possibilities of using reverse factoring and improvement in the availability of finance. The stakeholders opined that the government should consider setting up special financial infrastructure, which should include cheaper finances for the T&A sector.

4 Recommendations

Based on the results obtained in this research work, we recommend that a more in-depth analysis that will include, environmental, economic, and social hotspots and product environment Footprint (PEF) should be undertaken for the Kenyan T&A sector. Models of textile waste recycling and upscaling should also be developed and implemented to avert environmental degradation caused by T&A post-consumer and post-industrial waste in Kenya.

5 Conclusions

The chapter works on eco-innovation, with special reference to the Kenyan Textile and Apparel sector. The definition of eco-innovation, regarding the Kenyan Textile and Apparel sector was discussed. Interviews with stakeholders (government and industries) indicated that the Kenyan T&A sector value chain needs to investigate; policies, design of T&A products, materials/fibers, chemical management, waste management, energy, water, business models, and financial management. Based on the results of this research work, it can be concluded that; the Kenyan T&A sector needs to work on its policies so that they can promote eco-innovation that promotes green growth, sustainable development, and resource efficiency. The need for Kenya to work towards the production of natural fibers (cotton and sisal) to support the local industry was earmarked as a key strategy that will promote eco-innovation. Innovation chemical, waste, financial and energy management, were recognized as key areas that need concerted efforts from policymakers, manufacturers, and researchers to ensure optimized utilization of resources while avoiding adverse environmental effects. The business models in the Kenyan T&A sector need a paradigm shift towards sustainability and circularity.

Acknowledgements This publication would not have been possible without the training on Eco-innovation received from the "Innovative Business Practices and Economic Models in the Textile Value Chain (InTex)" project of the United Nations Environment Programme (UNEP) with the support of the European Union. We are grateful to the International Trade Center—SITA project, who co-funded the eco-innovation work and facilitated the data collection from stakeholders in 2021. Our sincere thanks go to Ms. Devan Hari, of Centre for Responsible Business, New Delhi, India, Mr. Soumyajit Kar, and Ms. Anda Valla Efendija of ITC Kenya. We are immensely grateful to all stakeholders who have engaged actively in this study and have generously contributed their time, insights, and expertise.

References

1. Groover MP (2020) Fundamentals of modern manufacturing: material, processes and systems, 7th edn. Wiley, New Jersey
2. Ortiz-Bailon M, Vera-Espino R, Quiroz-Flores JC, Alvarez J (2021) Improvement for production management and control using lean manufacturing tools in the manufacturing of posts and accessories. In: Human interaction, emerging technologies and future. Springer Nature, Cham, pp 560–565
3. Guldenpfennig M, Hald KS, Hansen A (2021) Productivity improvement and multiple management controls: evidence from a manufacturing firm. Int J Oper Prod Manag 41:991–1017. https://doi.org/10.1108/IJOPM-09-2020-0667
4. UNEP (2022) Eco-innovation
5. UNEP (2020) Global environment outlook—GEO-6: Healthy planet, healthy people
6. Horbach J, Rammer C (2013) The role of energy efficiency in the context of eco-innovation: concepts and empirical evidence. Energy Econ 36:99–108
7. UNDP (2019) SDG innovation: generating and scaling solutions for the sustainable development goals
8. Leitner SM (2018) Eco-Innovation: drivers, barriers and effects—a European perspective
9. UNEP (2014) Eco-innovation for resource efficiency and sustainable lifestyles
10. UNEP (2017) The business case for the circular economy: opportunities in the built environment
11. Jang WY, Choi YR, Kim HY (2015) The effect of eco-innovation on firm performance: a dynamic capabilities perspective. Sustainability 7:6797–6816
12. Goncalves AL, Marques RC, Gomes R (2021) Eco-innovation and corporate social responsibility: empirical evidence from the Portuguese context. J Clean Prod 294:126151
13. Nnaji CC, Igbuku C (2019) Drivers of eco-innovation in the Nigerian manufacturing sector. J Clean Prod 240:117907
14. UNEP (2021) Global status report for buildings and construction
15. CEIC (2022) Kenya employed persons
16. ITC (2017) ITC trademap HS code
17. ISAAA (2018) Kenya Government banks on Bt cotton to revive textile industry. In: Crop Biotech Updat

18. ISAAA (2020) Long wait over as Kenya finally commercializes Bt cotton. In: Crop Biotech Updat
19. KAM (2018) Manufacturing in Kenya under the "big 4 agenda": a sector deep-dive report
20. (2017) Kenya national trade policy
21. UN (2022) UN Comtrade database
22. Varun V (2011) Competitive supply side analysis of cotton, textile and apparel sectors in East Africa-Kenya, Sudan, Tanzania and Uganda
23. GOK (2015) Kenya industrial transformation programme
24. GOK (2020) Implementation status of the big four agenda, 2018/2019
25. GOK (2018) Third medium term plan 2018–2022; Transforming lives: advancing socio-economic development through the big four
26. Krishnan A, Willem te Velde D, Were A (2019) Integrating Kenya's small firms into leather, textiles and garments value chains. Background paper on creating jobs under Kenya's Big Four agenda.
27. KNBS (2015) Kenya economic survey
28. UNEP (2014) Annual report
29. GOK (2017) kenya green economy strategy and implementation plan 2016–2030
30. GOK (2013) Kenya national climate change action plan 2013–2017
31. Kibua TN, Opiyo FE, Kavale SM (2020) The state of green innovation in Kenya. In: Rao PS, Asif M, Boughanmi NE (eds) Green innovations and energy management. Springer, pp 295–312
32. World Bank (2017) Project appraisal document on a proposed credit in the amount of SDR 133.0 million (US$ 180.0 million equivalent) to the Republic of Kenya for the Kenya Climate Innovation Centre Project
33. UNDP (2016) Climate business innovation network: Kenya green building society
34. World Bank (2021) Kenya—energy and extractives global practice: green mini grids project (English). Washington, D.C
35. UNDP (2018) Low emission and climate resilient development programme in Kenya, Nairobi, Kenya
36. East African Community (2011) EAC climate change policy Frameork. Arusha, Tanzania
37. Makueni County Government (2020) Makueni climate change action plan. Makueni, Kenya
38. County Government of Kiambu (2017) Kiambu County integrated solid waste management plan
39. County Government of Mombasa (2018) Sustainable transport policy
40. Kisumu County Government (2018) Kisumu County integrated development plan 2018–2022. Kisumu, Kenya
41. Nairobi City County Government (2018) Nairobi City county integrated development plan
42. Textile Exchange (2017) Over 36 major brands pledge to achieve sustainable cotton by 2025
43. Textile Exchange (2022) 2025 Sustainable cotton challenge, fourth annual report
44. Karelia G (2020) What Links Tamil Nadu to Australian Open? 25K Outfits Made from 1.8 Lakh PET Bottles
45. Sustainable Apparel Coalition (2022) Zero discharge of hazardous chemicals programme (ZDHC)
46. Climate Council (2023) What does zero emissions mean?
47. UKAID (2021) Opportunities to enhance trade finance in the textile and garment sector across East Africa and the Horn of Africa
48. Chemengich M, Vaid V, Olweny H, Karuiki FG (2013) Policy research on the Kenyan textile industry, findings and recommendations. Nairobi, Kenya
49. Supporting Economic Transformation (SET) (2018) ODI and ministry of industry, trade and cooperatives: financing large firms in the garment sector in Kenya. In: Work. Proceeding

Eco-design of Textiles

Ali Raza Shafqat and Alberto Saccavini

Abstract Textile and Fashion industry has a significant impact on the environment due to resource depletion, pollution, water contamination, waste generation, and greenhouse gas emission. Seemingly, the best possible strategy to reduce the impact of the textile industry on the environment is to introduce a circular end of life of Textile products. Textile products mostly consist of multilaterals; therefore, their recycling process is very complicated. Designing for circularity is the utmost possibility to make a textile's end-of-life circular. Textile products have gigantic supply chain mostly consisting of various suppliers and stakeholders spread over wide geographic locations due to which compliance and recyclability are complex but traceability and smart labeling can help ensure compliance to principles of eco-design and circularity of textile products. This chapter sheds light on the eco-design of textiles, textile supply chain, role of eco-design in circular economy, key elements of eco-designing, circular business models, future trends, and innovations in eco-designing of Textiles. Eco-design is the only way forward to validate and achieve a circular economy, securing the environment, biodiversification, avoiding the use of hazardous chemicals, transparency, and traceability in the textile sector meanwhile providing verifiable data.

1 Introduction of Eco-design

Eco-design is the first step of the product development process that can help us initiate a product idea and framework that can enable us to achieve circularity. Eco-design helps us integrate solutions to environmental implications and economics of

A. R. Shafqat
School of Engineering and Technology, National Textile University, Faisalabad 37610, Pakistan

Department of Textile Engineering and Design, BUITEMS, Quetta, Pakistan

A. Saccavini (✉)
Sustainability in Fashion Consultant, Milan, Italy
e-mail: ethicalista@gmail.com

S. R. Batool et al. (eds.), *Circularity in Textiles*, Textile Science and Clothing Technology, https://doi.org/10.1007/978-3-031-49479-6_3

the product at the very beginning, keeping in mind the complete lifecycle of the product. European Environmental Agency defines eco-design as.

"The integration of environmental aspects into the product development process, by balancing ecological and economic requirements. Eco-design considers environmental aspects at all stages of the product development process, striving for products which make the lowest possible environmental impact throughout the product life cycle" [1].

Keeping principles of Eco-designing in view before manufacturing the products of the future can be made circular. The key points to consider for eco-designing are preserving resources already in use, preferring sustainable materials, reclaiming materials from waste, designing businesses for a circular economy, designing for longevity and recyclability with end-of-life in mind and integrating traceability and transparency within products, developing collaboration between all stakeholders to reinforce circular economy [2].

1.1 Importance of Eco-design in the Textile Industry

Eco-design is the first step and most important towards achieving circularity in textiles and is the most crucial step towards reducing the environmental impact and sustainability of textiles, the key idea is to reduce the environmental impact of textiles during the design inherently, this idea is the strongest pillar towards achieving circularity and sustainability during the complete lifecycle of textile products.

The textile industry by design involves massive consumption of raw material, water, and energy while the ultimate fate of clothing products is landfill or incineration approximately after three years in developed countries [3]. Discarded textile products are then replaced by new products that again require the same input of raw material, water, and energy thus again marking a significant footprint on the environment. This cycle of discarding old products and manufacturing new ones is not simply arithmetic but exponential as the footprint increases at each turn of this cycle, further the fast fashion trends, increasing demands of society, better living standards, cheap manufacturing, the mass production capacity of industries along with an increase in global population are multiplicative factors that exceptionally increase the impacts of textile industry on the environment. Changing the linear route of the textile make-use-dispose model to a circular model is inevitable if by any means a complete circular economy in the textile sector is achieved through eco-design only.

According to the Cotton Works report the impact of textiles on the environment is huge, during use and after discarding the textile materials pollutes the environment if not recycled or composted. Table 1 mentions the amount of microfiber remaining in various water conditions against time.

Textile products consume 80-billion-meter cube of water, genera oceans andic tons of equivalent CO_2 of GHGs that makes 10% of the total produced globally, contribute half a million tons of microplastics that make 35% of global microplastics released in oceans, and produce 95 metric tons of textile waste every year [5]. It is

Table 1 Microfiber percentage remaining in waste, fresh, and saline water conditions after 32 days [4]

Microfiber percentage remaining in the water		
Water condition	The cotton microfiber percentage remaining (%)	The polyester microfiber percentage remaining (%)
Wastewater	8	94
Freshwater	21	100
Saltwater	52	96

estimated that known fossil fuel reserves which are a major source of raw materials for synthetic fiber manufacturing are unfortunately projected to last for the next 35 to 40 years only [6]. Textile products mainly consist of multiple materials, i.e., sometimes they consist of blends of various fibers, sometimes coatings, dyes, paints, auxiliaries, and closures. This makes recycling textiles very challenging. Due to multi-materiality along with other reasons.

Eco-design of textiles can address and promote sustainable and circular practices in the textile industry in can help in the conservation of natural resources, reduction of GHGs emissions, reduction of chemicals and hazardous substances released in the environment, reduction of water pollution, reduction of waste produced, transparency in the textile supply chain, meet consumer expectations of eco-friendly products, meet regulatory standards set by governments, and provide differentiation edge to producers.

2 Definition of Key Concepts

2.1 *Eco-design*

Eco-design refers to integrating and balancing ecological and economic requirements into all stages of product development to assure that product will have the lowest possible impact on the environment throughout its complete lifecycle [7].

2.2 *Supply Chain*

A supply chain of textile products consists of all transporters, manufacturers, suppliers, retailers, warehouses, and customers. The circle of the supply chain encompasses all departments within each organization, all direct and indirect operations involved to ultimately fulfill the customer demands as represented in Fig. 1. The overall goal of supply chain management is to maximize the overall worth of the product to the customer while minimizing all the costs related to the supply chain minimum possible. The conventional supply chain balances the economics and value to the customer. To impart circularity and sustainability supply chain must account for all environmental impacts of the product to the minimum possible level [8].

For example, Table 2. depicts the sketch material consumed and waste generated in the Moroccan textile industry from the year 2019, the figures are presented in tones. Just a single country, Morocco, consumes an enormous amount of material and produces a huge waste to fulfill the industrial demands, the clothing that is manufactured gets discarded within the next three years [3].

Fig. 1 Major stages of the textile supply chain [8]

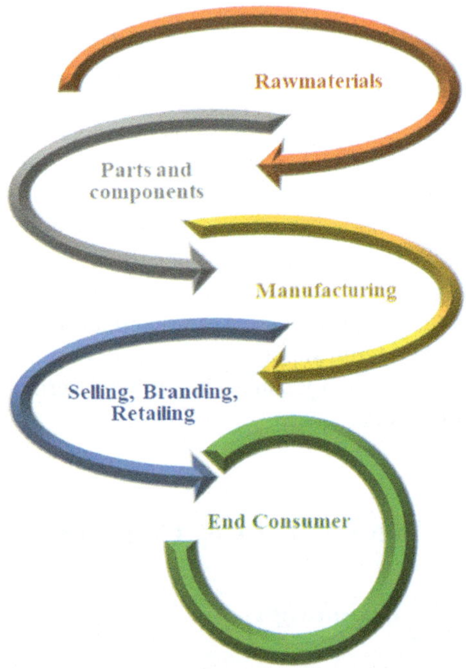

Table 2 Material consumed, and waste generated in the Moroccan textile industry from the year 2019 [9]

Process	Raw material consumed (tones)	Waste generated (tones)
Spinning	52,294	5,012
Fabric Manufacturing	177,777	9,832
Garments production	362,041	51,683
Post-industrial (deadstock, overproduction)	288,971	16,673

2.3 Sourcing

The sourcing process involves locating multiple suppliers to meet the organization's purchasing requirements with quality goods and services in a predetermined timeframe for the smooth running of manufacturing, operations, and delivery of on-time quality services [10].

2.4 The Conventional Model of Economy

The conventional model of economy is a linear model, in which products are produced and consumed linearly. This model broadly consists of take, make, use, and discard in a linear fashion. Eco-design and the Fashion supply chain.

Supply chain definition (source, make, use, dispose of) role consequential for each in design. The fashion supply chain consists of suppliers, manufacturers, consumers, and disposal facilities if any.

2.5 Take, Make, Use, and Dispose off in the Fashion Supply Chain

Make

To make is to manufacture, manufacturing involves the processing of materials either chemically or physically to modify properties, change size, shape, physical or chemical form, characteristics, and appearance, to form a part or to fabricate multiple parts to form complex product by using energy, tools, machines, fixtures, and human resource. Hence manufacturing involves all stages of product manufacturing i.e., between sourcing to end consumer [11].

Use

The purpose of design is to create a product to be used by end-users [12]. The use of the product mainly revolves around the end consumer, the ultimate use and application of the product including maintenance and servicing are all included in the use phase. The use phase in the product life cycle includes all stages after the supply chain and before disposal.

Dispose

After the use of the product by the end consumer, the products are discarded in the trash. This disposal of products happens after the products have completed their utility or are worn off during their lifecycle. In the case of clothing, the products are often disposed of due to changes in consumers' fit and size, become obsolete, out of

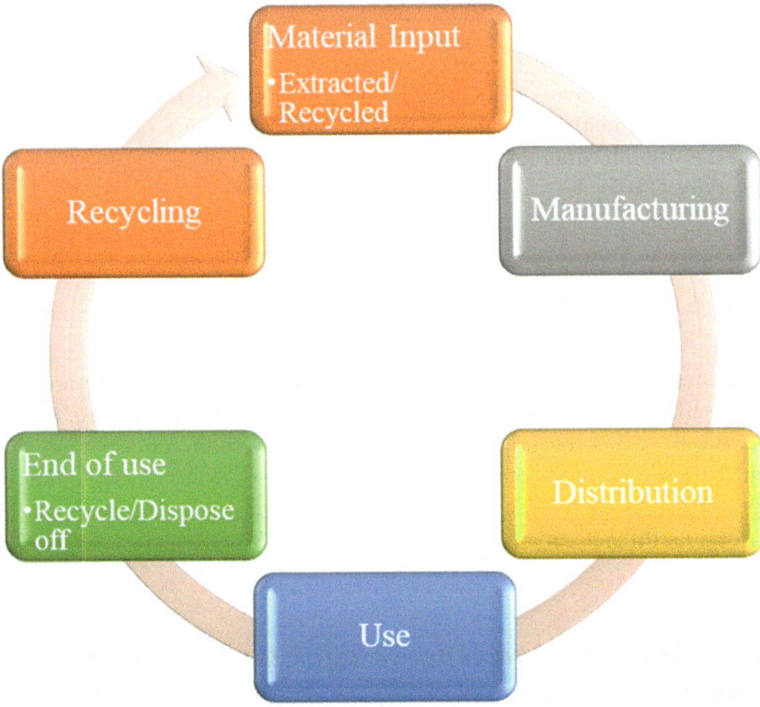

Fig. 2 The circular model of the Textile industry, a visual demonstration [13]

trend, or are just out of fashion or just because of more attractive and new designs launched by brands in their latest collections.

2.6 Circular Model of Economy

Unlike the linear model of the economy, the circular model of the economy keeps the products, materials, and services in a close loop as long as possible. The circular model of the economy is visually presented in Fig. 5 whereas the visual mapping for the role of different actors in the circular textile's economy is given in Fig. 2.

2.7 Roles and Interactions of the Different Actors

The circular model of the economy takes into consideration manufacturing, and environmental impact meanwhile traceability helps in maintaining valuable data that helps identify all roles and their interactions and verify the data. This complete

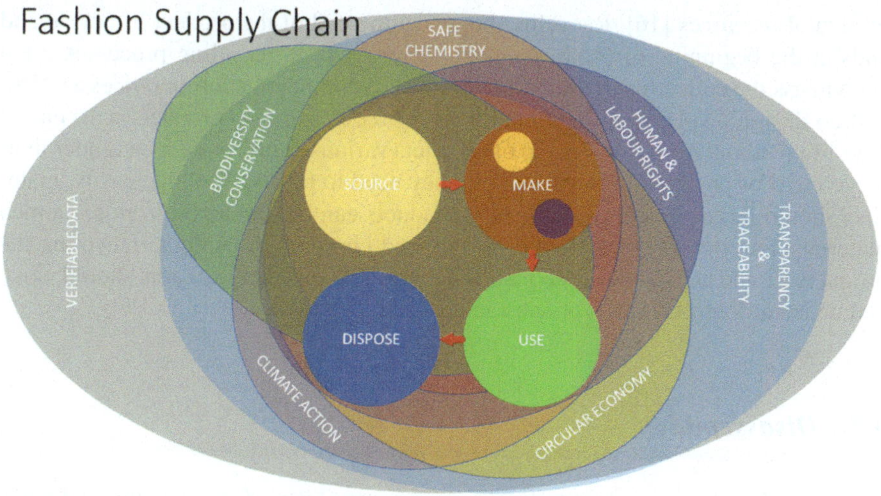

Fig. 3 Roles and interactions of the different actors in the fashion supply chain

picture helps enhance the transparency in the system. The roles and interactions of various actors can be seen in Fig. 3.

3 Design for Circularity

3.1 Biodegradability and Composability

Biodegradability is the intrinsic property of organic materials to undergo biological degradation utilizing living organisms, these microbial organisms convert biodegradable materials into biomass, basic elements, carbon dioxide, water, and methane [14].

The controlled biological degradation of organic materials to form a nutrient-rich compost is known as composability. The process of composability converts organic materials into compost by aerobic biodegradation and disintegration in the presence of oxygen to form a soil-like substance that of rich in organic substances and can be a natural alternative to fertilizers [15].

3.2 Recyclability

Recyclability of a product means that it can be recycled or reused at the end of the product lifecycle, recycling helps minimize pollution, and waste and reduces the use of natural resources [16]. Recycling begins at the end of the product life cycle and ends at the beginning of product remanufacturing [17], recycling processes often encompass the collection of waste, processing, and remanufacturing processes [18].

Eco-designing plays an important role in the recyclability of products at the end of their life, without proper designing of products various complexities can hinder their recycling. For products to be recyclable they need to be disassembled easily, materials should be recyclable, high-quality products can be developed from reclaimed materials, and the products should be safe and circular. To facilitate recycling the products should consist of mono materials, valuable recycling streams should exist, and no hazardous chemicals or substances should be used.

3.3 Disassembly

Disintegrating the parts of a product, to repair, refurbish, or for recycling is known as disassembly. As most of the products consumed by the end user are complex products by design they cannot be reprocessed until disassembled [19]. As textiles are mostly made up of multiple materials, disassembly is required before sorting and recycling materials. FREITAG is a firm working to incorporate circularity into textiles by providing innovative solutions for the circularity trap, disassembly, and recycling of textiles. The accessories they use are easily detachable without intensive processes and labor [20].

The Wear2go's flagship project is smart stitch in which they offer stitching thread made up of biopolymers that melt at comparatively low temperatures making disassembly easy. Their disassembly system is also the world's first thermal disassembly system [21]. An example of a smart disassembly system is given below in Fig. 4.

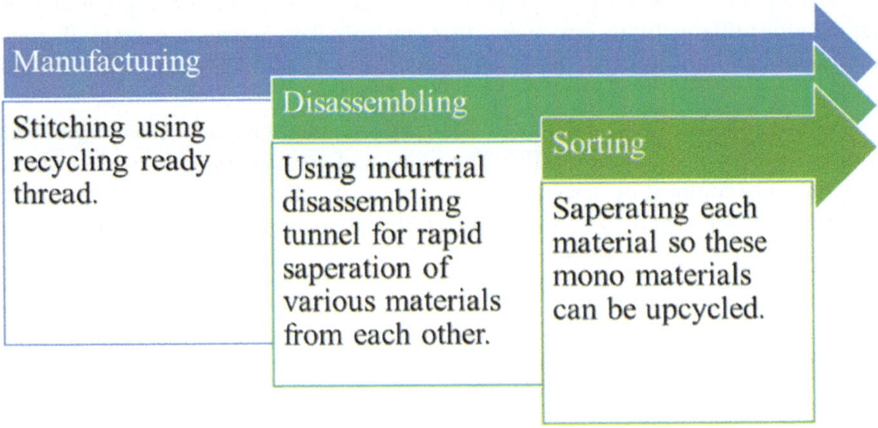

Fig. 4 An example of a smart disassembly system

Disassembly can be enhanced by various methods [22], some of these are discussed below:

- Using nut bolts, screws, and pinup mechanisms instead of using permanent fastening of auxiliaries to the garments
- Avoiding permanent sealants and glues etc.
- Easy dissolvable or degradable binding materials
- Use similar fastening mechanisms throughout apparel.

3.4 Longevity

Designing for longevity means expanding and maximizing the life of products by using materials, processes, etc., or by emotional bonding of the product with the consumer.

Designing for longevity is a strategy for manufacturing long-lasting textiles. This can be achieved by using good quality materials, experienced human resources, precise size, and following garment maintenance instructions. Furthermore, the emotional attachment of end consumers to garments also plays a vital role that helps create value in garments that expand the longevity of clothes, after exploring these two factors business models should be designed that can help maintain the garments and explore opportunities for active use of garments [23].

3.4.1 The Emotional Attachment Between the Consumer and the Garment

- Long-lasting textiles can be created by using durable materials, using the best possible craftsmen, improving fitting, and communicating the product care information explicitly [23].
- Emotional bonding of a product with consumers is created by using classic designs, storytelling, building consumer loyalty, and enabling easy servicing and personalization [23].
- Physical longevity can be enhanced by using durable materials, using modular clothing designs i.e., designing that allows growth allowance for body shape fluctuations like size and fitting, customizable clothing that allows change in shape, style, or color, etc., and, clothing in pieces that allow customizability like changing colors, sleeves and auxiliaries of clothing [24].
  - The longevity of textiles can be achieved in various ways, the first and foremost important method is to enhance the overall quality so the life span can be increased, this can be achieved by using high-quality fibers and other materials with high strength. Longevity can also be increased by reusing textiles, and clothing items that are suitable for use again and again as secondhand garments sold after being discarded or donated by the firsthand user. The after-sales

services like repair, mending, changing style, and changing fitting can also help achieve the longevity of apparel. Better fitting of garments and tailor ability to fit also increases the overall life of the garment, the ability to alter fitting, adaptability, and tailor ability of a garment can be incorporated by using auxiliary parts like buttons, fitting straps, zippers, belts, etc.

- Including timelessness in design can also help boost the longevity of apparel, trendy products have short lifetimes while classic products have much longer life. Timelessness can be incorporated into garments by including customization options like changes in style, color, texture, etc.
- Using a product as a service can also help in the longevity of a garment. These services include sharing garments, renting, or swapping with consumers. Various business models can be built based on these services too.

3.4.2 Design Waste Out

Eco-designing insists designers indulge in a re-thinking process to redesign products that can make products eco-friendly. Such approaches include using new and innovative materials to manufacture textiles, either fibers or auxiliaries. These eco-friendly materials include materials derived from algae, cactus, food waste, mycelium, etc., and using processes like MMC and chemical recycling to bring such materials back into the loop. Further, unsustainable materials can be replaced by non-oil synthetics, etc. One such example of redesigning is a product made by Spintex Engineering i.e. engineered silk that is made of liquid protein and solid silk fiber amalgam that when pulled under stress from silk fiber under room temperature. Under stress, the chains align themselves through self-assembly while water is expelled leaving solid fiber behind. This forms premium-grade silk fiber.

Another example is a technique that converts pre-consumer cotton waste that originated during the cutting process of garment manufacturing into regenerated fiber that is circular and can be recycled again and again. Redesigning the waste into a circular product can be seen visually in Fig. 5.

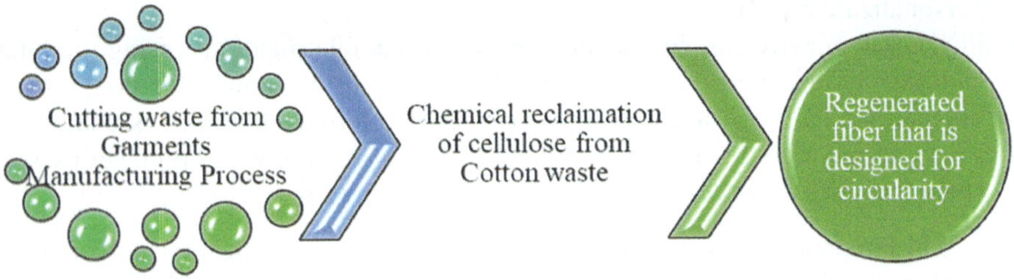

Fig. 5 Technology that converts pre-consumer cotton waste to regenerated fiber [25]

3.4.3 Safe and Circular

Not all textile materials are safe as these may be chemicals, dyes, and pigments so they must not be trapped in circularity. Circular designing emphasizes the use of safe and circular materials, for safety considerations circular designing process considers the user's safety first while designing to keep materials in use again and again [26]. ZDHC is an organization working with more than 150 collaborators including fashion brands, suppliers, and chemical suppliers. They are widely renowned for their roadmap to zero programs, the purpose of this program is to eliminate harmful chemicals in the fashion supply chain, and for this, they are working to develop more sustainable manufacturing processes to protect the workforce, end consumers, and environment. This program is focused on three areas i.e., input of safer chemicals, better processes, and better outputs [27].

3.4.4 Energy, Water, and Resources Use Reduction (Closed Loop Systems, Efficiency...)

The textile industry consumes 20% of the global freshwater making it polluted water after the dyeing and finishing processes of textile manufacturing. There is an eminent need for time to reduce water consumption during the processing of textiles (Parliament—https://www.europarl.europa.eu/, n.d.) [28]. Various technologies already exist in various forms to conserve water during textile wet processing, one such technology is named Ozone treatment, it is a chemical-free process and is highly effective for dyeing and finishing textiles. This is a simple process and helps conserve water, energy, and time thereby reducing water generation [29].

- Other designing approaches include altering processes like conventional wet processing with waterless dyeing that works in a close-loop fashion, producing zero process waste [30].

3.5 Role of Innovative Circular Materials

Circular materials play an important role in the manufacturing of Circular textile products. These materials can help in upcycling of waste materials at the end of the textile's life thereby addressing solutions and promoting the manufacturing of close-loop circular textiles. The UNEP is working to provide strategic leadership and encourage collaboration across all sectors for a transition to a sustainable and circular value chain. The circular supply chain as defined by UNEP can be seen in schematics demonstrated in Fig. 6.

Fig. 6 The circular textile supply chain as defined by UNEP [31]

3.6 Circular Supply Chain Management

3.6.1 Packaging

An average human being produces approximately 2kgs of waste every day, mostly consisting of packaging materials. In the US nearly thirty percent of all solid waste generated consists of packaging materials [32]. Packaging materials are the most rapidly discarded materials with the shortest possible life span and mostly delivered with the product as a necessity while it is in most cases is auxiliary, circular packaging use materials that are designed specifically to reduce environmental impact relative to the conventional packaging materials like plastics, fabrics, and other materials. Circular packaging is made by recycling waste from other products like clothing, food containers, toys, and medical disposables. Using circular packaging materials enhances sustainability, saves packaging costs, offers the competitive advantage of being environmentally friendly, and helps improve brand image and customer relations [33].

Using circular packaging may help reduce carbon footprints by up to 73%, reduce consumption of fossil fuels by 68%, and biobased raw materials by 30% as compared to using plastic packages [34]. The following are ways to develop circular packaging [33]:

- Replace plastics and other non-compostable packaging materials.
- Redesign packaging for recyclability later
- Reduce overall material required for manufacturing packages.
- Develop systems to recover the packaging materials to recycle.
- Include transparency and traceability in packaging materials.

Zara is a brand working to reduce waste by redesigning packaging. They manufacture packaging materials like boxes and hangers from waste.

3.6.2 Traceability

The textile supply chain is extensive, long, and widely distributed where raw material originates in one geographical region, is manufactured at another location, branded, and packaged at some other location, and sold and consumed elsewhere.

The core concept of traceability revolves around the ability to track the complete lifecycle of a product right through all stages of its life including raw material extraction, production, consumption, discarding, and recycling phases. The importance of traceability in the textile industry can be well understood by the following arguments mentioned below [35]:

- Traceability helps identify the complete environmental impacts the product may have on the environment at each stage of its lifecycle.
- Traceability is a necessity of sustainability in the textile industry:
- The sustainability claims are just word of mouth without proper trackability, without a proper system in place including information from all stakeholders the claims of sustainability are vague.
- The textile industry is one of the most polluting sectors of the modern world, despite all the efforts, technology, and reports related to environmental concerns textile industry is thriving, so customers must be made aware of the environmental impacts caused by consuming textiles. Awareness of textile product consumers provides the choice to choose eco-friendly products and may help in reducing the use of highly polluting materials.
- Better working conditions and the well-being of laborers can be achieved with data of all stakeholders trackable. The textile sector is well known for the illegal utilization of child labor and forced labor; with proper systems of traceability, this issue can be tackled.
- With traceability in place, the risk of making wrong decisions may be reduced.

Smart labeling

- Traceability can be achieved by either label marking or chemical or physical marking of textiles [36].
- One of the most emerging and promising technologies for traceability is the use of Blockchain. Using blockchain can help provide real-time data to all stakeholders during each phase of the product lifecycle. Blockchain can be integrated into textiles by certifying each batch of material with twin fiber coins, that may be linked with a digital token i.e., fingerprint, ensuring a unique identity. Blockchain data cannot be manipulated or tampered with, and its history lasts forever. Even though the cycle continues circularly the data remains preserved and available readily in real-time.

- Such examples of traceability are Textile Genesis and Aura Blockchain Consortium which use blockchain to enhance traceable supply chains in the fashion industry.
- Fiber trace as another example of traceability in the fashion industry uses capsule denim that depicts transparency through all lifecycle stages of the denim industry [37].

Smart labels

The use of smart labels will help boost traceability and therefore help in providing information related to the circular economy of products, data for reuse and recyclability, assembly instructions, materials, and their circularity index, and Environmental friendliness score of products among many other types of embedded data (see Fig. 7).

Fig. 7 Smart labeling data set examples [38]

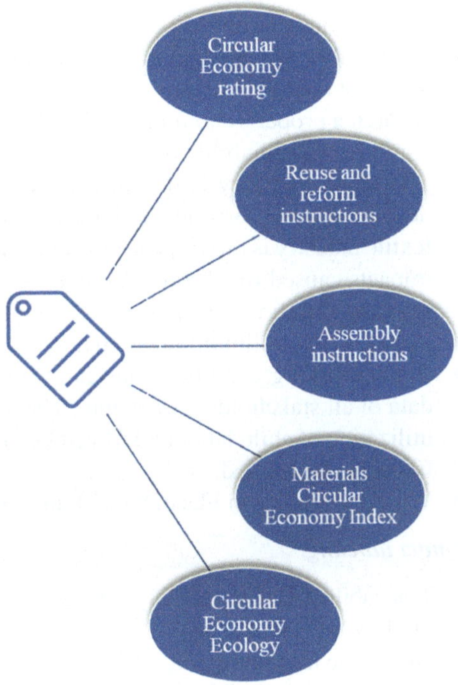

4 Circular Business Models (Take Back, Lease, Rent, Reverse Logistics…)

Circular business models offer innovative solutions to keep the textile materials and value created by textile processes in a closed loop. The conventional textile business model includes the extraction of materials, production of textile products, and use and end at discard as waste.

Circular business models include all monetary efforts designed into businesses with the end goal of avoiding the extraction of new materials and later landfill and incineration as shown in Fig. 8.

The circular business models include reuse, repair, redistribute, refurbish, and remanufacture of textile goods. Table 3 provides the overall sketch of circular textile businesses.

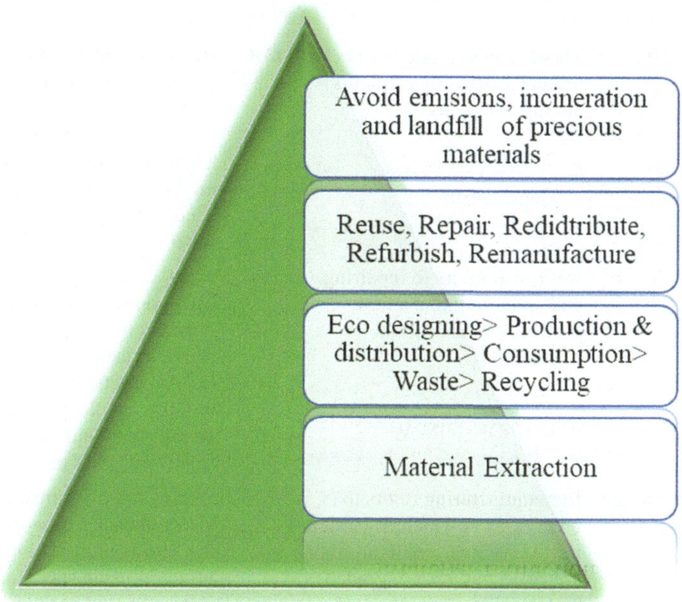

Fig. 8 Circular business models to avoid extraction, landfill, and incineration [39]

Table 3 Circular business models in the Textile industry

S/n	Business Type	Definition	Type	Example
1	Reuse	Reuse mainly revolves around the sole principle of longevity, the durable materials increase the life of textile products thereby enabling them to be reused. Whereas the business models are designed to acquire used garments and resell them at lower prices	B-C	The secondhand textile product market is the best example
2	Repair	The business models are designed to acquire used garments, repair them, and after proper amendments resell them at relatively cheaper prices	B-C	The secondhand textile product market is where small businesses acquire, repair, and resell used textiles
3	Redistribute	This business model works by mutual sharing of garments among consumers	C–C	Businesses that work on recommence principles are included in this business model, companies like ThreadUp, TheRealReal, swap, and Patagonia enable consumers to sell their used clothing and purchase from other consumers [40]
4	Refurbish	Refurbish refers to repairing and amending like a new product by the original manufacturer	B-C	Many companies offer discounted offers on refurbished items. In this business model companies provide cashback or promos for new products by patching in gently used old garments. Thereby refurbishing and selling them at relatively lower prices
5	Remanufacture	Remanufacturing refers to businesses in which discarded post-consumer textile products are collected, sorted, graded, and then recycled into new products	B-B	Due to strict compliance by governments, many organizations are now manufacturing new products using specific percentages of reclaimed fibers from post-consumer waste. This business model acquires garment waste that is reclaimed and reused in specific proportion with virgin material

5 Scope of Eco-design in Textiles

5.1 *Challenges and Opportunities in the Field of Eco-design of Textiles*

- Eco-designing of textiles is a complex phase, and the following are the challenges and associated opportunities that come with these challenges.
- Textile materials mostly consist of multi-materials, often in the forms of blends, whereas most of the time accessories are used like buttons, zippers, etc. that restrict designers from employing multi-materials thereby creating hindrances in the recyclability of materials.
- Selection of sustainable and eco-friendly material is also a great challenge when it comes to eco-designing, for example, polyester fiber on the one hand is fossil fuel-based on the other hand single kilogram of cotton fiber requires thousands of liters of water before harvesting. Furthermore, the selection of sustainable and renewable materials imposes limitations in terms of functionality and performance.
  - The use of hazardous chemicals is common in the textile industry, although solutions to alternative routes exist most of these solutions require high operational or capital costs.
  - The textile industry is a resource-intensive industry, that currently consumes huge amounts of energy and water.
  - The textile sector is a large producer of solid waste, the supply chain is complex, vast, and widely distributed over various geographic locations. Efficient waste management systems are required to collect and reclaim material from waste.
  - Textile products are trendy and are mostly discarded within five years, fast fashion and cheap prices altogether reduce the lifecycle of textile products too short, and most of the time textile products are used as disposables. Innovative solutions that help maintain emotional bonding, fitting, adaptability, and alterability can help counter the rapid discarding of textile products.
  - Transparency and traceability are other huge issues when it comes to the eco-design of textiles. Due to the vast supply chain that included various suppliers spread over various countries, accountability becomes very difficult, and regular audits are difficult to manage, even with strict compliance the stand-alone database within each organization cannot explain the complete social and environmental impacts especially when the supplier of material has another supplier whose supplier is someone else and the list continues. Many suppliers also are less interested in audits, transparency, and traceability. Only aware consumers of textile products having access to traceability data can make better- and well-informed decisions when choosing eco-friendly textile products.

6 Future Trends and Innovations

Innovations in sustainable fibers and other materials required for textile manufacturing will thrive in the next decade, these include exploring new Bio-based materials, biosynthesis, and expanding the use of already available Bio-based materials. i.e., Fibers with higher sustainability performances like state-of-the-art mechanically recycled fibers and materials, chemically recycled materials, more sustainable natural fibers, closed-loop manmade cellulosic fibers, etc.

Biodegradable and compostable textile materials are highly sought after, and exploring opportunities in the development of such materials, systems, and supply chains around biodegradability and composability will be an interesting area of focus in the near future.

Traceability and transparency are major challenges in implementing a circular economy in the textile sector, and the development of novel solutions will be very helpful for upcoming startups in this niche.

Development of new business models of circular economy are required that already have huge potential in today's world, new startups based on these models will be supported by government legislation and consumer awareness.

Technologies for water and energy conservation are highly required, the businesses that explore opportunities can thrive keeping in view the water scarcity issues and switching to renewable energy mode. This would specifically help in meeting the current and future water needs and reducing the carbon footprints of the Textile industry.

7 Conclusion

Eco-designing of textiles is impossible without collaboration among all stakeholders of the textile sector, collaboration among designers, researchers, manufacturers, and end consumers is the only way forward for implementing a circular economy in the textile industry.

References

1. eco-design—European Environment Agency. https://www.eea.europa.eu/help/glossary/eea-glossary/eco-design. Accessed 19 May 2023
2. Circular Economy for Dummies—Uncover how to reimagine your supply chain, manufacturing, designs, and more to avoid waste and the environment. https://circulareconomyfordummies.com/. Accessed 4 Jul 2023
3. Niinimäki K, Peters G, Dahlbo H et al (2020) The environmental price of fast fashion. Nat Rev Earth Environ 1:4 1:189–200. https://doi.org/10.1038/s43017-020-0039-9
4. Li L, Frey M, Browning KJ (2010) Biodegradability study on cotton and polyester fabrics. J Eng Fiber Fabr 5:42–53. https://doi.org/10.1177/155892501000500406

5. The impact of textile production and waste on the environment (infographic) | News | European Parliament. https://www.europarl.europa.eu/news/en/headlines/society/20201208STO93327/the-impact-of-textile-production-and-waste-on-the-environment-infographic. Accessed 3 Jun 2023
6. Shafiee S, Topal E (2009) When will fossil fuel reserves be diminished? Energy Policy 37:181–189. https://doi.org/10.1016/J.ENPOL.2008.08.016
7. eco-design—European Environment Agency. https://www.eea.europa.eu/help/glossary/eea-glossary/eco-design. Accessed 5 Jul 2023
8. Delgado C, Castelo BM (2013) Supply chain management. Encyclopedia of corporate social responsibility, 2349–2357.https://doi.org/10.1007/978-3-642-28036-8_139
9. Textile waste mapping in Morocco and Tunisia
10. (2000) SOURCING. Encyclopedia of production and manufacturing management, 716–717. https://doi.org/10.1007/1-4020-0612-8_902
11. Segreto T, Teti R (2014) Manufacturing. CIRP encyclopedia of production engineering, 828–830.https://doi.org/10.1007/978-3-642-20617-7_6561
12. Friedrich T (2008) Use. Design dictionary, 429–431. https://doi.org/10.1007/978-3-7643-8140-0_294
13. Setting environmental goals for supply chains-BSR contribution to OECD workshop on climate change
14. Goswami P, O'Haire T (2016) Developments in the use of green (biodegradable), recycled and biopolymer materials in technical nonwovens. Adv Techn Nonwovens, 97–114.https://doi.org/10.1016/B978-0-08-100575-0.00003-6
15. Ecozema. https://ecozema.com/en/why/biodegradability-compostability/. Accessed 5 Jul 2023
16. Kubba S (2012) Green building materials and products. Handbook of green building design and construction, 227–311.https://doi.org/10.1016/B978-0-12-385128-4.00006-8
17. Recycling and the circular economy: what's the difference? https://ellenmacarthurfoundation.org/articles/recycling-and-the-circular-economy-whats-the-difference. Accessed 5 Jul 2023
18. The U.S. Recycling System | US EPA. https://www.epa.gov/circulareconomy/us-recycling-system. Accessed 5 Jul 2023
19. Vanegas P, Peeters JR, Cattrysse D et al (2018) Ease of disassembly of products to support circular economy strategies. Resour Conserv Recycl 135:323–334. https://doi.org/10.1016/J.RESCONREC.2017.06.022
20. The circular truck tarpaulin | FREITAG. https://www.freitag.ch/en/circular-tarp. Accessed 8 Jul 2023
21. The benefits—WEAR2GO. https://wear2.com/en/corporate-workwear/the-benefits/. Accessed 5 Jul 2023
22. Circular Economy for Dummies—Uncover how to reimagine your supply chain, manufacturing, designs, and more to avoid waste and the environment. https://circulareconomyfordummies.com/. Accessed 5 Jul 2023
23. Design for Longevity—Redress Design Award. https://www.redressdesignaward.com/academy/resources/guide/design-for-longevity. Accessed 5 Jul 2023
24. Design for Longevity—Redress Design Award. https://www.redressdesignaward.com/academy/resources/guide/design-for-longevity. Accessed 9 Jul 2023
25. Refibra—SNT TEXTILE. https://snttextile.com/refibra/. Accessed 5 Jul 2023
26. Safe & Circular. https://www.circulardesignguide.com/safe-circular. Accessed 8 Jul 2023
27. Roadmap to Zero. https://www.roadmaptozero.com/?locale=en. Accessed 5 Jul 2023
28. https://www.europarl.europa.eu - The impact of textile production and waste on the environment (infographics) | News | European Parliament
29. Ozone in the Textile Industry | Absolute Ozone. https://absoluteozone.com/ozone-applications/industrial-ozone-applications/ozone-in-the-textile-industry/. Accessed 5 Jul 2023
30. Jeanologia presents first Jeans finishing plant to guarantee zero contamination—The Textile Magazine. https://www.indiantextilemagazine.in/jeanologia-presents-first-jeans-finishing-plant-to-guarantee-zero-contamination/. Accessed 5 Jul 2023

31. Sustainable and circular textiles | UNEP—UN Environment Programme. https://www.unep.org/explore-topics/resource-efficiency/what-we-do/sustainable-and-circular-textiles. Accessed 5 Jul 2023
32. Circular Packaging Explained | Greener Ideal. https://greenerideal.com/guides/circular-packaging-explained/. Accessed 5 Jul 2023
33. Product Packaging for a Circular Economy: Everything You Need to Know—Meyers. https://meyers.com/meyers-blog/circular-economy-packaging/. Accessed 5 Jul 2023
34. What is Circular Packaging ? | Sustainable Packaging Makes a Difference. https://circulesolutions.com/circular-packaging/. Accessed 5 Jul 2023
35. Kumar V, Agrawal TK, Wang L, Chen Y (2017) Contribution of traceability towards attaining sustainability in the textile sector. Textiles Clothing Sustain, 3https://doi.org/10.1186/S40689-017-0027-8
36. Traceability in clothing industry: what it is and why it is important. https://rifo-lab.com/en/blogs/blog-di-rifo/tracciabilita-industria-abbigliamento. Accessed 5 Jul 2023
37. Sustainability, circularity and traceability in textiles & apparel industry—Fibre2Fashion. https://www.fibre2fashion.com/industry-article/9396/sustainability-circularity-and-traceability-in-textiles-apparel-industry. Accessed 5 Jul 2023
38. Chae Y, Hinestroza J (2020) Building circular economy for smart textiles, smart clothing, and future wearables. Mater Circular Econ 2:1 2:1–4. https://doi.org/10.1007/S42824-020-00002-2
39. Challenges in shifting from a linear towards a circular plastics system—European Environment Agency. https://www.eea.europa.eu/media/infographics/challenges-in-shifting-from-a/view. Accessed 5 Jul 2023
40. Reuse examples to reduce textile waste—Rheaply. https://rheaply.com/blog/textile-waste/. Accessed 5 Jul 2023

Sustainable Raw Materials

Farooq Azam, Faheem Ahmad, Sheraz Ahmad, and Amino ddin Haji

Abstract The demand for sustainable and circular textiles has led to increased interest in exploring alternative raw materials that minimize environmental impact and promote circularity in the textile industry. This book chapter provides an overview of sustainable raw materials for circularity in textiles, focusing on natural fibers such as flax, hemp, jute, ramie, kenaf, abaca, banana, bamboo, synthetic fibers including alginate, lyocell, Poly lactic acid, Polyhydroxyalkanoates, Polycaprolactone, and silk and sustainable dyes. The chapter highlights the unique properties, cultivation methods, and processing techniques of each fiber, emphasizing its potential for sustainable textile production. Furthermore, it discusses the environmental benefits and challenges associated with these raw materials, including their biodegradability, recyclability, and carbon footprint. The chapter also explores future directions in the field, such as the exploration of new natural fibers, advancements in fiber processing technologies, sustainable synthetic fibers, circular design and recycling, and life cycle assessment. By providing a comprehensive overview of sustainable raw materials, this chapter aims to contribute to the promotion and adoption of circular and environmentally conscious practices in the textile industry.

1 Introduction to Sustainable Raw Materials for Textiles

1.1 Principles of Sustainability

Sustainability refers to the capability of a system, community, or society to maintain itself over time, devoid of compromising the demands of future generations. Sustainability principles have been developed to guide human activities toward a more sustainable future, recognizing the interdependence of economic, social, and

F. Azam · F. Ahmad · S. Ahmad (✉)
School of Engineering and Technology, National Textile University, Faisalabad 37610, Pakistan
e-mail: sheraz@ntu.edu.pk

A. Haji
Department of Textile Engineering, Yazd University, 8915818411 Yazd, Iran

S. R. Batool et al. (eds.), *Circularity in Textiles*, Textile Science and Clothing Technology, https://doi.org/10.1007/978-3-031-49479-6_4

environmental factors. One of the fundamental principles of sustainability is the need to guarantee that human activities do not surpass the planet's holding capacity. This means that we need to adopt strategies that preserve natural resources and maintain the health and productivity of ecosystems. This includes sustainable agriculture, conserving energy and water, reduction in emissions of greenhouse gases, protecting biodiversity, and forestry practices [1].

To promote sustainability, it is important to ensure that the production of organic textiles covers all aspects of the supply chain, from the cultivation of raw materials to the distribution of final products in the form of fibers, yarns, fabrics, and garments, as well as home textile products. The garment business can achieve sustainability by adhering to the standards set forth by the Global Organic Textile Standard [2]. This standard establishes basic requirements for environmentally and socially responsible manufacturing practices and also evaluates the quality of organic fibers based on rigorous social and environmental standards. It includes measures such as organic certification, labeling, and licensing to provide consumers with credible assurance that the textiles have been responsibly produced, from the harvesting of raw materials to the final manufacturing stage.

1.1.1 Raw Materials

Textiles can be manufactured using raw materials sourced from either natural or synthetic fibers. Natural fibers sourced from plant fibers can be obtained from various parts of the plants such as stem (hemp and flax fibers), seed (cotton fibers), leaf (sisal fibers), and fibers from coconut husk. Meanwhile, animal fibers are obtained from animal wool, secretions such as silk, and hair. Furthermore, abaca is an alternative resource of rope, which can be utilized as a substitute for glass fibers for energy conservation in automobiles. Coir, derived from the husk of coconuts, is a multifunctional material employed in various fields such as ropemaking, mattress production, brush manufacturing, geotextiles, and automotive seating. Additionally, angora wool, which is produced from angora rabbits, is a high-quality material known for its silky white appearance, softness, and fine texture. Lastly, alpaca wool is a luxurious fabric that yields 5000 tons per year, and it is highly sought after in the fashion industry [3].

The utilization of natural fibers in the garment industry is still dominated by world-renowned cotton, which is composed of pure cellulose. Flax, on the other hand, is recognized as one of the toughest fibers globally, possessing great strength. Cashmere, known for its insulation properties and soft texture, is an ideal choice for many consumers. Additionally, hemp is a top-notch alternative to cotton production, offering exceptional quality. Silk, often referred to as the 'queen of fabrics,' has been utilized since ancient times, exhibiting a luxurious and royal demeanor. Mohair, a lustrous fiber, exhibits excellent dye absorption capabilities, as well as possessing softness and sheen. Jute, a fiber known for its farmer-friendly qualities, serves as a vital source of income for countless small-scale farmers due to its durable fibers commonly used in the production of sackcloth. Ramie, like flax, is also a strong natural fiber, possessing silky luster, absorbency, and density like silk. Sisal, known for its ability to substitute coarse glass fibers in textile composites, automotive components, and furniture, offers exceptional strength and longevity. Wool, the world's foremost textile fiber, has limited production but unique characteristics. However, over the last fifty years, synthetic or man-made fibers have gradually replaced natural fibers in the textile industry, household furnishings, agriculture, and other industries. Polyester, Acrylic, polypropylene, and nylon are examples of synthetic fibers manufactured in large quantities from petrochemicals. These fibers are produced with consistent lengths, strengths, and colors, allowing for easy customization to suit specific needs. Their cost-effectiveness and ease of processing have contributed to their success [3].

The improvement in the 'green' economy is based on the adoption of energy-efficient methods, the utilization of renewable resources for polymer products, and the incorporation of recycling in industrial processes to minimize carbon emissions. Carbon fibers are developed using polyacrylonitrile precursor with the remaining percentage utilized for rayon or petroleum pitch. The molecular structure of these materials adheres to strict specifications, with variations of precursor molecules. The polymerization process involves the conversion of ethylene, the main constituent, into polyester during the manufacturing of this synthetic material. Polyester, commonly utilized in paper products, is a large molecule characterized by an ester bond structure. However, in the context of fibers, it specifically refers to polyethylene terephthalate (PET). It is widely known for its water and heat resistance and high strength. Spandex is created from a long polyurethane chain and is ideal for stretchy clothing, like sportswear [4].

The strength of the textile industry lies in its wide-ranging production base, encompassing a variety of natural and synthetic fibers, containing cotton, jute, wool, silk, polyester, viscose, nylon, and acrylic. However, the industry's exponential growth in response to rising requirements for textile goods has led to a significant increase in textile mills and their associated wastewater, resulting in a major pollution problem worldwide. The use of chemicals in textile manufacturing processes has been associated with adverse environmental and health impacts, with dyes being a particularly significant source of pollution in textile wastewater. The presence of hydrosulphide in the effluent drastically reduces oxygen levels and light penetration in water bodies, which negatively impacts the water ecosystem. Textile wastewater is a major contributor to environmental degradation and human health problems. Harmful chemicals can enter our bodies through inhalation or skin absorption, leading to allergic reactions and other health issues, even in unborn children. Textile effluent contains heavy metals that can accumulate in the organs of the body, leading to long-term health issues. Thus, untreated, or inadequately treated textile effluent can adversely affect aquatic and terrestrial ecosystems, causing long-lasting health effects [5].

The textile industry encounters various obstacles because of growing environmental consciousness and the potential consequences of environmental regulations affecting the entire supply chain. The textile manufacturing process requires substantial amounts of resources, including water and fuel, and various chemicals are used, generating considerable waste. The textile industry faces challenges from technological and machinery obsolescence, which can pose a threat to its operations. Poor infrastructure, inadequate labor skills, poor productivity, structural anomalies, economic norms, high power consumption, and social commitments have contributed to the low productivity of the garments industry [6].

To overcome these challenges, fiber creators have suggested several solutions. The authors suggested that fostering knowledge transfer from academic institutions to industry could have a positive impact on productivity improvement and environmental sustainability. Academic research institutions need to focus on maximizing the utilization of knowledge and ideas generated by students, scientists, and researchers fostering an environment that encourages collaboration with textile and clothing companies. This interaction can lead to innovative experimentation and transform scientific facts into applications on a larger scale, driving sustainability and innovation. Furthermore, the implementation of sustainable initiatives such as water and energy conservation, as well as the increasing significance of technical textiles, play a vital role. Collaboration throughout the supply chain is essential to accomplish these objectives [5, 7].

1.1.2 Impact on Environment

The use of pesticides in farmland can have adverse effects on wildlife, contaminating other products and ultimately entering the food chain. Similarly, the chemicals used to bleach and dye textiles, such as sulfuric acid, sodium chlorite, sodium hypochlorite, sodium hydrosulfite, and optical brightening agents, can harm the environment and pose health risks to people, potentially resulting in slow poisoning, health disorders, and even death. Discarding old clothes in landfills results in the release of methane, a harmful greenhouse gas that contributes significantly to global warming and takes up precious space in these sites. Textile machinery also produces noise, sound, and air pollution, which can be detrimental to the environment and human health. In addition, the application of dyes and chemicals in the production of garments and footwear can result in soil contamination and have detrimental effects on both surface and groundwater. The textile and clothing industry contributes to ecological imbalance through the excessive use of natural resources and lack of recycling facilities. The industry is associated with substandard working conditions, while animal exploitation in intensive farming practices adds to environmental harm [5].

2 Sustainable Natural Fibers

Sustainable living refers to the practice of utilizing natural resources from the environment without depleting them beyond their natural replenishment rate, while also avoiding overloading the environment's capacity to purify and rejuvenate itself through natural processes. Sustainable resources do not get exhausted, for example, wind energy can be harnessed continuously while oil resources are gradually diminishing. Regarding fibers, a fiber can be considered sustainable if it is manufactured using completely renewable chemicals and non-fossil-fuel-based energy during its production processes. The use of renewable sources for polymeric materials is a viable solution for promoting sustainable development and ecologically attractive technology [8]. Vink et al. [9] outlined several characteristics that an ideal sustainable material should possess, including the ability to:

- substitute an existing product and perform just as well, if not better, while also being competitively priced.
- have a minimal environmental impact throughout all stages of production, including upstream and downstream processes.
- be produced using sustainable resources and exclusively incorporate ingredients that are safe for both human health and the environment.
- not have any adverse effects on food supplies or water resources.

Vink's criteria showcase a deep understanding of the importance of addressing environmental concerns, with a focus on poly (lactic acid)'s potential positive impact on the manufacturing process [10]. The concept related to a fully green life cycle is critical for a genuinely eco-friendly material (such as fiber), which entails innovative approaches to utilizing biopolymers, safeguarding raw materials based on fossil fuels, minimizing fossil fuel usage during fiber processing, reducing waste volume, enabling disposability in the environment, promoting biological degradability, reducing carbon dioxide emissions, and eliminating any environmentally detrimental chemistry throughout the product's life cycle [11].

2.1 Flax

Flax fiber, classified as a bast fiber, is obtained from the stem of the linseed/flax plant (Linum usitatissimum L.). The matured harvest of the flax plant is depicted in Fig. 1a. The flax plant is well-known for its two main components: flax fiber, extracted from the stem, and linseed oil derived from the plant's seeds, which find applications in various industries. Flax's attributes of strength and nobility were already recognized as early as 6000 BC. The Phoenicians, renowned as skilled traders and navigators, brought flax from Egypt and distributed it to Ireland and Britain, thereby introducing flax fiber to Europe. Babylon began cultivating flax around 3000 BC, and burial chambers from around 650 BC depict the cultivation and use of flax fibers in clothing [10]. The flax plant has a tap root system that reaches a depth of approximately 62–102 cm. The size of the root system is greatly influenced by its growing conditions. In favorable soil and moisture situations, the root system is not as robust. In Fig. 1b, the scanning electron microscope (SEM) image of flax fiber reveals cross-marks (deformed zones), commonly identified as kink bands on the fiber's surface. The plant's stem consists of three primary components: the root neck, a non-branched stem (in the case of fiber flax) or a branched stem (in oil flax), and an inflorescence

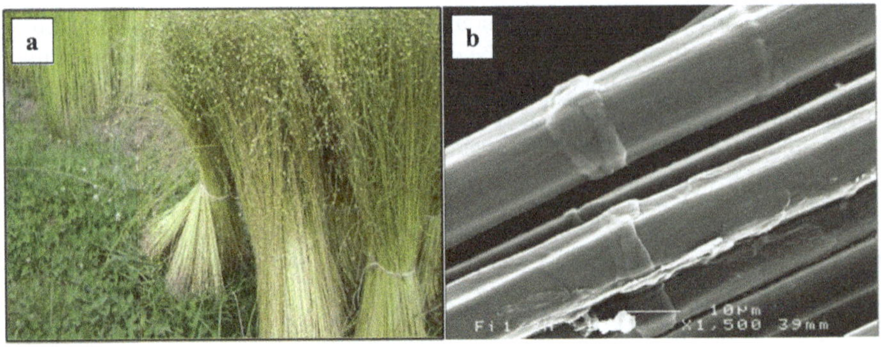

Fig. 1 **a** Field of ripe fiber flax [13], **b** SEM image of flax fibers [14]

known as a panicle. The stem's cross-section view of the flax plant is given in Fig. 2. The flax plant has small, narrow, lanceolate-shaped leaves that grow spirally along the stem, without petioles. Generally, the number of leaves decreases as you move up the stem. The plant's leaves display three prominent parallel veins and possess a delicate wax coating [12].

The stem of flax consists of eight main elements when viewed in cross-section as shown in Fig. 2. The stem of a plant is comprised of various layers and structures that are essential for its overall shape and functionality. Beginning with the outermost layer, the epidermis consists of a single layer of epidermal cells that are protected by

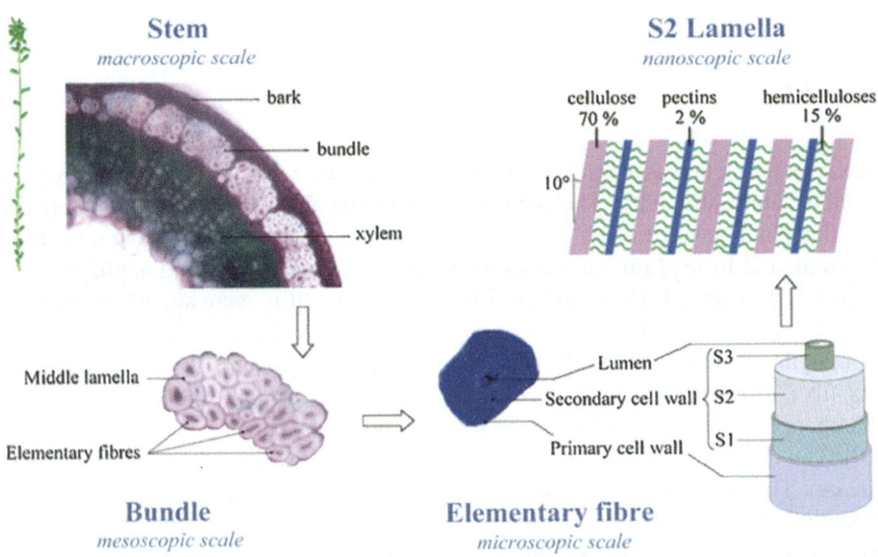

Fig. 2 Cross-sections and schematic representations of flax at different scales, from the stem to the cellulose fibrils [15]

a cuticle. This cuticle acts as a shield against external elements. Directly beneath the epidermis, the primary cortex is formed by 2–7 layers of parenchyma cells. These cells provide support and contribute to the stem's flexibility one distinctive feature of the stem is the presence of a ring of bast fibers, known for their remarkable tensile strength and frequent use in textiles and other material production. As we move inward, we encounter the cortex, consisting of a secondary cortex layer comprised of small cells that make up the vascular bundles. These bundles are instrumental in transporting water, nutrients, and other essential substances throughout the plant.

The cambium, positioned between the cortex and the xylem, is a layer of parenchyma cells responsible for generating new cells. This continual cell production enables the stem to increase its diameter over time. Below the cambium lies the xylem, a thick layer of cells characterized by sturdy walls. The xylem performs a central role in the transportation of water and minerals within the plant. In the center of the stem, we find the core, which represents the inner part of the xylem. The core contributes significantly to the stem's strength and stability. Finally, there is the core channel, a space that extends along the entire length of the stem. Although it lacks specific structures, the core channel serves as a central hollow area within the stem. The layer of the cortex contains clusters of fibers called fiber bundles. Fiber flax is distinguished by its abundance of densely packed bundles, whereas oil flax has fewer bundles that are loosely arranged.

2.1.1 Economic Importance of Flax

Flax plant is predominantly cultivated in damp, temperate climates and has a significant presence in European agriculture. China and Egypt also cultivate fiber flax, although to a lesser extent. The textile industry highly values flax fiber as an eco-friendly and valuable raw material. Linen fabrics produced from flax fiber possess exceptional aesthetic, hygienic, and health qualities, which keep them in fashion. These fabrics allow for breathability, absorb perspiration, and efficiently release heat, making them ideal for use during hot weather. Moreover, flax fiber does not accumulate an electrostatic charge, providing psychophysical comfort and promoting good health. Conventionally, linen was largely exploited to manufacture tablecloths, upholstery, and bed linen fabrics. Now, it has become increasingly popular as a clothing raw material and has applications in geotextiles, floor coverings, and automobiles. Processing flax also yields shive, which constitutes 50% of the harvested biomass and serves as an excellent raw material for the furniture and construction industries.

Flax offers a wide range of valuable raw materials that have applications in the food sectors and pharmaceuticals. Adding flaxseed to bakery items enhances their flavor, nutritional value, and extends their shelf life. Linseed oil, renowned for its polyunsaturated fatty acids, is a quick-drying oil commonly used as a key ingredient in the production of paints and varnishes. Furthermore, it is a valuable dietary and curative product that aids in peristaltic action, inhibits skin issues, and demonstrates anti-sclerotic properties. Oil flax is typically cultivated in warm and dry climates, with the primary crop being seed. While fiber serves as a secondary material, it is occasionally utilized for the high-quality production of pulp and various other applications. Currently, flax faces stiff competition from cotton and synthetic fibers. To overcome this challenge, breeders are working towards improving flax's characteristics by creating better cultivars. The traditional approach of developing new cultivars typically requires around 16 years, but advancements in molecular biology and biotechnology have accelerated this process.

2.1.2 Production Method

Cultivation of Flax

The ideal weather conditions for the growth of flax plants are high temperature and humidity. During the vegetation period, the sum of rainfall should fall between 110 and 130 mm. 1-g dry matter of the flax consumes almost 400–600 ml of water. The most suitable soils for flax cultivation are loess soils with favorable fertility, as well as medium-heavy clay, sandy, and sandy-loam soils with a strong structure that can retain and release water during dry conditions.

For flax grown for fiber, the optimal time for harvesting is when the stems are yellow, roughly one-third of their length, are devoid of leaves, and the bolls on the plant start to exhibit a yellow coloration. When cultivating flax for its seeds, it is recommended to harvest the crop at the yellow maturity stage, which is characterized by fully yellow stems, two-thirds of the leaves have fallen off, and the bolls are yellow with fully developed and turning brown seeds. Pulling machines are utilized for harvesting flax in the first stage, which is typically self-propelled and extracts the straw, leaving it in the field for retting. To ensure even retting, the straw is flipped over when it reaches the halfway point. This process is facilitated by specialized turning machines that not only aid in the retting process but also assist in seed harvesting. Occasionally, the act of pulling the straw may also result in seed harvesting [16].

Degumming of Flax

Degumming refers to the procedure of separating the fiber bundles from the surrounding tissue by breaking down the natural adhesive pectin. This process allows for the extraction of the fiber bundles with greater ease. There are various methods for degumming, but the most used methods are water and dew retting.

In the past, industrial water retting of flax was conducted in special basins called retting mills. The straw was soaked in warm or cold water and the retting process was facilitated by bacteria in anaerobic conditions. However, due to the high prices associated with this method, including the energy required to heat water to 30 °C and a large amount of wastewater generated (20 tonnes of water per tonne of straw and an additional 10 tonnes for washing and rinsing), this method poses a significant environmental threat and is no longer used in many countries.

Nowadays, dew retting is the most common method used to ret flax straw in most countries. This process engages the fungi's action, mainly Cladosporium herbarum, and is aerobic. Bacteria also play a role in rainy conditions. Regardless of the organisms involved, the objective of the retting process is to decompose the pectin that binds the individual fibers together, thereby allowing them to aggregate into cohesive strands alongside woody fractions. The timing of the retting method is critical to the fiber's quality, as stopping it too soon will result in rough, stiff fiber that is hard to detach from the hurds while continuing the process for too long will weaken the fiber and cause it to break into shorter fragments.

Chemical degumming involves exposing the straw to chemicals such as ethylene, sodium carbonate, sulfuric acid, etc. In addition to traditional methods, various physical techniques have been devised to effectively separate the fibers from the surrounding tissue during the degumming process. These include steam explosion, ultrasound oscillation, steam hydrolysis, osmosis degumming, etc. Osmosis degumming is particularly noteworthy as the only method that can be used more widely. This process involves treating fibrous plants with continuously flowing water to extract the fibers without affecting their natural properties. The best results are obtained at a temperature of 25–35 °C, with a controlled water flow for 72–168 h.

Straw Processing

Once the retting process is complete, the dried stems are gathered, and only the straw with a moisture content of less than 19% is suitable for subsequent processing stages. The primary goal of processing the straw is to separate the fiber from the shive by weakening its bond through a series of operations, such as breaking, shaking, and scutching. This process is repeated until the fiber is effectively extracted and separated from the shave. Based on the intended fiber type, the straw can undergo processing on a scutching line to yield long fibers and short waste fibers (tow), or on a tow-producing unit to produce homogenous fibers.

To achieve the objective of producing tow, there are three main steps in the production line: breakers, scutchers, and shakers. The breakers are comprised of two cylinders with ridges that crush the straw, loosening the stem and allowing for the removal of the woody part (shive). Scutchers, which can be like scutching drums or have specially shaped teeth, knock off the shive from the crushed stems. Shakers utilize needles and sieves to separate the loosened hurds from the remaining fiber after scutching.

2.1.3 Applications

Flax fiber is a versatile material that finds use in both clothing and non-clothing applications. For clothing, it is a popular and healthy choice, often used in exclusive woven and knitted garments. Non-clothing applications include curtains, tablecloths, and bed linen, some of which may require fire retardant treatment. Luckily, cellulosic fibers like flax are relatively safe compared to other polymers in this regard. The production and processing of flax also yield significant by-products like shive, seeds, and waste fibers. Flax seed is rich in valuable nutrients like fatty acids, lignans, and mucilage, making it a source of valuable agro-fine chemicals.

Shive, accounting for approximately 50% of the processed straw, finds widespread applications, with particle board manufacturing being one of its primary uses, offering versatility in terms of different density requirements. Flax fibers are also used not only in textiles but also in other areas like composites and floor covering. Due to their exceptional strength and superior tenacity when compared to glass fiber, natural fibers are highly suitable for applications in the automotive industry. Furthermore, composites reinforced with natural fibers are biocomposites, making them less of a burden on the environment.

2.2 *Hemp*

Hemp, a type of cannabis plant is commonly known for its psychoactive properties. However, it is also used for other purposes. For example, its seeds can be used as fishing bait and its fibers can be used in textile applications. Refined hemp fibers are highly suitable for manufacturing fabric, particularly for summer clothing. This fabric offers exceptional breathability and moisture absorption qualities, along with antibacterial and UV protection properties [17]. However, the rigid and rough nature of hemp fibers makes them challenging to handle, resulting in clothing made from untreated fibers appearing wrinkled and feeling prickly.

During fabric manufacturing of hemp, the short fibers tend to protrude from the yarn's surface, resulting in a hairy texture of the fabric. Although hemp is not as well-recognized in the textile industry as flax, ramie, and pineapple fibers, it is still used in this industry. Aside from clothing textiles, Chen et al. [18] conducted a comparative analysis of composites utilizing hemp fibers and found them to be equivalent to composites fabricated from other natural fibers. In another study, Pietak et al. [19] subjected hemp fibers to various treatments, including alkaline, steam, combined steam-alkaline, and enzymatic processes, followed by acetylation to reduce the fibers' polarity. Sodium hydroxide (NaOH) was chosen as the alkaline treatment agent. Following the chemical treatment, the fibers exhibited enhanced wettability, contributing to the increased strength of the resultant composite material.

The anatomical structure of the hemp nut includes a seed shell, ovary, and embryo containing apical bud, endosperm, radicle, and cotyledons. The primary chemical constituents found in hemp seeds include carbohydrates, fats, and proteins. The depth and branching of the root system depend on soil type and water level. The stem of the hemp plant is given in Fig. 3a. The stem's diameter varies with sowing density, and it consists of collenchymas, bast (primary and secondary), cortex, and wood. The middle portion of the stem contains the greatest concentration of fiber; the cross-section view of hemp fiber bundle is shown in Fig. 3b. The quality and amount of fiber are influenced by genetic potential, soil type, sowing density, and time of harvesting.

Fig. 3 **a** Stem of hemp plant [16], **b** cross-section morphology of hemp fiber bundle [17]

2.2.1 Economic Importance of Hemp

The global cultivation of hemp, a cosmopolitan plant, dates back to 5000 years in China and then spread throughout the world. Hemp is mainly grown for fiber and seed and was used for technical items before cotton and chemical fibers became available. Subsequently, the cultivation of hemp was prohibited in Western Europe and the United States of America due to its psychoactive properties. Nevertheless, there is renewed interest in hemp due to the pursuit of alternative food crops in Europe and the growing demand for renewable resources. New areas of application for hemp include composites, building material, extraction of cellulose, clothes, and textiles, Agro textiles, oil-based products, etc.

2.2.2 Production Method

Cultivation

Hemp has high soil requirements and is sensitive to soil pH. Hemp cultivation thrives in various types of fertile soils, such as humus-rich lime soils, alluvial soils, black earth, well-improved peat soils, and loesses. Hemp can thrive in soils contaminated with heavy metals, as it can extract and accumulate cadmium, lead, copper, and zinc without negatively impacting the quality of the crop. Hemp does not possess any specific prerequisites regarding the preceding crop and is competitive with weeds when grown for fiber. Potassium (K) and calcium (Ca) play a crucial role in the cultivation of hemp fiber, while phosphorus is important for hemp grown for seeds. The N:P:K ratio should be 1:7:1.5 for fiber production and 1:8:1 for seed production.

The timing of hemp harvesting is determined by the aim of cultivation. When cultivating hemp specifically for fiber production, it is advisable to harvest the plants at the onset of flowering. This ensures the production of fine and robust fibers that are well-suited for textile manufacturing. Postponing the harvest leads to an increase in both the yield and strength of the fiber but results in stronger lignification that renders it unsuitable for textiles. In cases where hemp is cultivated for both seeds and fiber or solely for seeds, it is recommended to harvest the plants at full maturity. This occurs when the middle seeds in the panicle have reached maturity. The resulting fiber is not appropriate for textile use but can be used for technical applications.

Degumming of Hemp

Enzymatic processes using bacterial laccases and Pectinex 100L were tested for fiber extraction. The laccases remove non-cellulosic components and reduce fiber thinness from 16 Tex (control) to 8.05–9.1 Tex. The mass loss with laccases followed by mechanical upgrading is about twice that of mechanical upgrading alone. Pectinex 100L treatment did not significantly affect fiber thinness, length, or mass loss. The study revealed specific conditions deemed optimal for enzymatic degumming of raw hemp fibers using laccase preparations. These conditions include a temperature of 37 °C, pH level of 4.5, treatment duration of 48 h, and the addition of a dispersing agent called Sultafon UNS, with a quantity of 1 gdm^{-3} and a hydro modulus of 1:25. These findings provide valuable insights for achieving effective enzymatic degumming in hemp fiber processing. Electrolytic degumming of fibers involves both an oxidation process and a reduction process, utilizing hydrochloric acid and sodium hydroxide as electrolytes. The acid is used at a concentration of 0.01 M, while the sodium hydroxide is employed at a concentration of 0.1 M. These processes contribute to the effective removal of gums and impurities from the fibers. The thinness of hemp fiber after electrolysis using these electrolytes ranged from 8.51 to 10.32 Tex, while the control treatment was 16.35 Tex. The total mass loss resulting from both mechanical and electrolytic processing ranged from 18.77 to 21.56%, whereas the control was 12.89–14.19%.

The study revealed that using a 0.1 M concentration of sodium hydroxide in both the oxidation and reduction processes resulted in superior outcomes. The treatment parameters employed were as follows: a temperature range of 20–25 °C, a treatment duration of 48 h, and a voltage of 4.5 V. These specific parameters were found to be effective in achieving favorable results in the electrolytic degumming process.

2.2.3 Applications

Hemp is finding multiple applications in various industries, including textiles, automotive, composites, fiberboard, building construction, and heat-insulating materials [13, 20]. Recent studies have indicated that hemp fibers have the potential for sound insulation and sound absorption [21]. Researchers are also exploring the uses of hemp fiber in musical instruments and the sports industry, given its superior vibration-damping capacity when compared to synthetic fibers [22]. Additionally, hemp fibers are being investigated for use as reinforcements in brake pad applications. With the automotive industry's growing interest in environmentally friendly materials, hemp fibers, and other natural fibers are being widely adopted as reinforcement in polymeric composites for various automotive applications. These composites offer a viable substitute for glass fiber-reinforced composites due to their potential for up to 30% higher strength at equivalent weight. Furthermore, they demonstrate non-brittle fractures upon impact, meeting a critical criterion for interior automotive plastic components [23].

2.3 *Jute*

Jute, derived from the plants Corchorus capsularis and Corchorus olitorius, is one of the most widely used plant-based textile fibers after cotton. While jute was initially thought to originate from China, it is now widely accepted that the most extensively planted species of jute is native to India. This plant has been used for both food and textile fiber for centuries on the Indian subcontinent. The fiber and fabric made from jute were initially known as "pat", and the word "jute" is believed to have originated from the Sanskrit word "yuta" (meaning "fiber"), still used in India as "jhat". The long, lustrous, and pliable nature of jute fiber allows for its transformation into robust, coarse threads. As one of the most economical choices among natural fibers, jute stands out for its cost-effectiveness. These fibers are primarily comprised of lignin, pectin, and cellulose. A solitary jute fiber typically possesses a polygonal shape with rounded corners as shown in Fig. 4a. Figure 4b presents the SEM image of jute fibers, illustrating the multicellular structure of the jute fiber with a smooth surface. The jute plant belongs to the Tiliaceae family, which is part of the Corchorus genus.

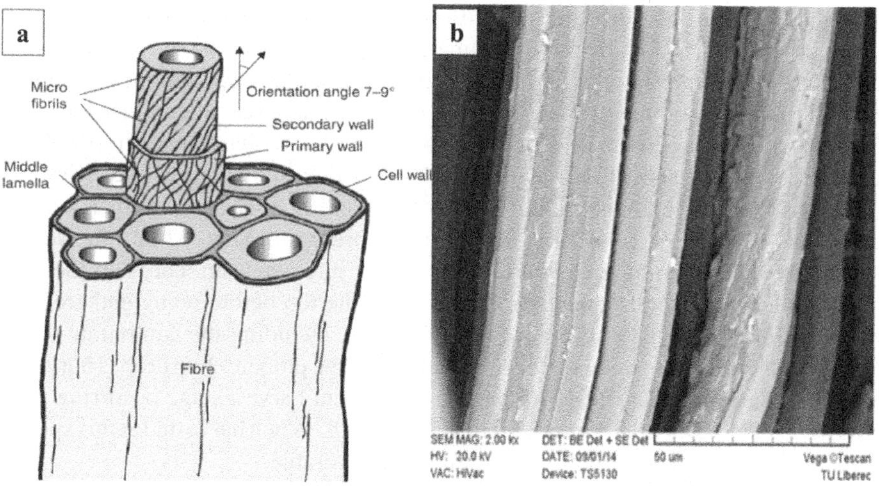

Fig. 4 **a** Schematics of jute fiber morphology [24], **b** SEM image of jute fibers [25]

2.3.1 Economic Importance of Jute

Jute emerged as a commercial commodity during the eighteenth century, finding its initial application in the production of ropes for maritime vessels. Later, it was found that jute could be spun and woven into carpets, sacking bags, and handbags. Jute has been used in packaging and carpet-making since it was first woven into fabric form. Jute's decomposition after harvesting enriches soil quality, which is essential for growing other crops. Jute bags and paper bags are becoming popular alternatives to petrochemical products such as plastic bags, which have become a menace to the environment in some countries. Although the cost of jute bags is higher than that of plastic bags, it is still a small price to pay for saving the environment [26].

Jute products are in demand in India, the USA, Australia, and Europe, and new markets are being explored. Bangladesh, Thailand, Myanmar, Nepal, India, and China are the primary countries responsible for the production of jute and related fibers, collectively contributing to 95% of the global production. Jute cultivation is labor-intensive but requires relatively small quantities of fertilizers and pesticides. Nevertheless, the cultivation of jute competes for land resources with food crops. India is expected to further strengthen its position as a leading global producer of jute, while Bangladesh, China, and Thailand are expected to experience a decline in production.

2.3.2 Production Method

Cultivation

Jute grows well in a humid climate characterized by temperatures ranging from 24 to 38 °C, and humidity between 70 and 90%, with the optimum temperature being around 34 °C. This particular climate is typically observed in regions located between 30° north and south latitudes. Jute cultivation requires a minimum rainfall of 1000 mm, for optimal growth it thrives in fresh, gray alluvial soil that possesses ample depth and receives silt deposits from annual floods Jute comes in two types: capsularis (white) and olitorious (tossa). The capsularis fiber is white, while the olitorious fiber is yellowish reddish, or greyish and is finer and stronger than the capsularis. Capsularis is usually sown from February to May, while olitorious is sown from April to mid-June. During harvesting, jute plants are cut from the base, and their leaves are subsequently removed from the top before being gathered into bundles. Jute stems have several layers of tissues, including the cambium, cortex, epidermis, wide xylem or wood, central pith, and large phloem. Jute crop is harvested at various stages of maturity, with the most prevalent harvest stage occurring when approximately 50% of the plants have developed pods. When harvested at this specific stage, both the yield and quality of the fiber are favorable.

Degumming

The process of extracting jute fiber is known as degumming or retting, where jute stalks or ribbons are submerged in water for a specific duration. The bundles are left in water for 7–10 days, during which organisms in the water and the plant's surface, generally bacteria, act jointly with water to soften and remove mucilaginous substances such as pectins and gums. This procedure facilitates the efficient separation of the fiber from the woody stem and bark. To ensure effective retting, it is recommended to maintain a minimum ratio of 1:20 between the plant material and stagnant water. For proper retting, it is essential to utilize non-saline and clear water with a sufficient volume that enables the jute bundles to float without coming into contact with the bottom. Once the water in the retting tank or ditch becomes contaminated, it is advisable not to reuse it. Retting can be stem retting or ribbon retting, depending on the part of the jute plant used. The duration of retting depends on various factors such as temperature, water movement, age, and size of stems. Generally, the retting process typically spans a duration of 10–30 days, with an optimal temperature of around 30 °C for achieving complete retting within 10–12 days.

Straw Processing

Once the retting process is complete, the jute straw is manually stripped to remove the fibers. This is done by a stripper who lightly taps the root end of the stem with a mallet while holding it in one hand. The fiber is then freed at the bottom of the stalk, and the rest of the fiber is loosened and removed by jerking the stem in the water. Various techniques for hand-stripping retted stems are employed, which vary based on local customs and traditions. The jute fibers are then hung on makeshift hangers to dry, which takes approximately two to three days. The fibers are graded based on their fineness, color, density, and clearness. After grading, the jute is bundled and taken to baling regions where initial grading is done again. After scanning the bundles, the fibers are classified into different grades, which are subsequently stored in designated areas or locations. Machine pressing is employed to compress these graded bundles into bales, and the ropes utilized to tie the bales are crafted from jute waste. Finally, the bales are arranged and stored in the warehouse based on their respective grades, making them readily available for sale.

2.3.3 Applications

Jute is used in producing pulp, paper, geotextiles, traditional materials, home textiles, and composites. Jute represents a renewable energy source that replenishes annually, exhibiting significant biomass production and biodegradability, making it environmentally friendly. Jute also helps increase soil fertility, acts as a hurdle to pests and bugs, provides organic matter and micronutrients to other crops, and is a valuable source of nutrition and medicinal value. Jute can be used instead of wood to make pulp and paper, which will reduce costs and save forest resources. Jute-reinforced composites have several advantages, such as low density, cost, biodegradability, and renewability, and can be used in packaging, automobiles, and construction materials. Jute geo-textiles can be used for vegetation consolidation, soil erosion control, road pavement construction, and agro-mulching materials. Jute has various applications, including yarns, twines, ropes, cordage, fabrics, and handicrafts [27–31].

2.4 *Ramie*

Ramie, a product derived from one or more species of the Boelimeria genus, belongs to the Urticaceae order and is closely related to the stinging nettle genus (Urtica). The Ramie plant is shown in Fig. 5a. The SEM image of Ramie fiber is given in Fig. 5b. Ramie fiber is obtained from the stems of the ramie plant, scientifically known as Boehmeria nivea, and is classified as a bast fiber. Ramie, also known as white ramie (Boehmeria nivea (L.) Gaud., Boehmeria nivea var. tenacissima), is classified chemically as a cellulosic fiber, like cotton and linen. Until recently, ramie was not widely available in the ready-to-wear market, but its presence in garments is gradually increasing. Ramie is often blended with cotton and can be found in a variety of woven and knit fabrics, encompassing a spectrum from delicate linen-like textures to rugged canvas-like materials. However, the adoption of ramie-based materials and fashion goods has been limited to other parts of the world. The primary challenge correlated with ramie fiber is the extraction of gum, which constitutes up to 30% of the fiber's weight. In the presence of gum, the fiber cannot be spun, making the removal of gum an essential step in the production process.

2.4.1 Economic Importance of Ramie

Ramie, a plant fiber used in textiles, has been cultivated in Algiers, China, Congo, India, and Indonesia since ancient times, and was even used in mummy cloths

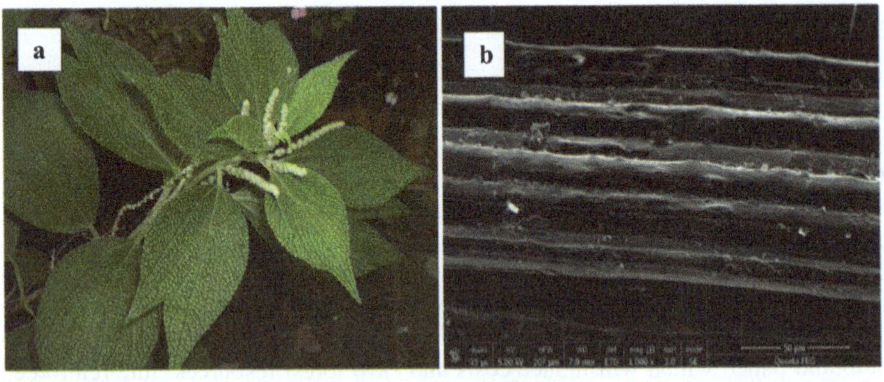

Fig. 5 **a** Ramie plant [32], **b** SEM image of ramie fiber [33]

in Egypt during 5000–3300 BC. The cultivation of ramie can be traced back to 1786 in Bologna, Italy, as documented in historical records. Mechanization of ramie processing began in India in 1869 and France in the late nineteenth century. Nowadays, the main producers of ramie are Thailand, Philippines, South Korea, Brazil, India, China, and Thailand, with Germany, Japan, and France. While ramie has been used in the US since the mid-1980s, legislation was passed in 1986 to eliminate its quota-free status.

2.4.2 Production Method

Cultivation

The ramie plant has unique characteristics that set it apart from other bast fiber crops. It can grow up to 2.4 m and is cultivated from seed, cuttings, layers, or root division. Ramie is known to adapt to a wide range of soil types, although it flourishes best in naturally fertile, moist, and well-drained light loamy soil. It particularly thrives in open-type soils. Ramie prefers high temperatures, high humidity, and an evenly distributed annual rainfall of 1100 cm. Ramie exhibits tolerance to a pH range of 4.3–7.3, although it has a preference for slightly acidic soil conditions. Ramie is classified as a resilient perennial plant and can be harvested as many as six times per year when grown in favorable conditions. Its productive lifespan typically spans from 6 to 20 years, providing an extended period for crop utilization. Chemical or enzymatic treatment is necessary to extract the bast fibers from the bark of the ramie plant, as it contains gums and pectins that need to be processed. The growth of the ramie plant and the strength of its fibers can be impacted by sudden changes in weather conditions.

Boehmeria nivea, a perennial dicot angiosperm, thrives in moist tropical climates with deep soils, making it highly suitable for cultivation in such regions. During cultivation, the Boehmeria nivea plant is maintained without branches by employing apical dominance techniques, which involve the removal of apical buds. Depending on the specific climate and growth conditions, the Boehmeria nivea plant can be harvested multiple times throughout the year, with up to six harvest cycles being possible, with the stems being harvested when they turn brown just before or soon after flowering. To obtain the maximum fiber content, it is important to time the harvest of each stem carefully to avoid harvesting immature or over-mature stems, with the best time being when the female flowers open. To extract the fibers, the outer bark is removed from the stems of the Boehmeria nivea plant, and the fibrous inner bark is then boiled. The harvesting process involves cutting the stems just above the lateral roots or bending the stem to break the core and strip the cortex in its original position. Despite the development of mechanical harvesters, they are not currently utilized on a commercial scale. The harvesting process involves the removal of the fibrous bark from freshly harvested stems, as it becomes increasingly difficult to separate the bark once the plant dries out. Following the decortication process, it is crucial to expedite the drying of the bark ribbons to minimize the risk of bacterial or fungal growth. The harvested stems from both tropical and temperate crops have a dry weight ranging from approximately 3.4 to 4.5 tonnes per hectare per year. A crop yielding 4.5 tonnes can provide approximately 1600 kg of dry-gummed fiber per hectare annually. During degumming, there can be up to a 25% weight loss, resulting in a yield of degummed fiber of about 1200 kg per hectare per year.

Degumming

The raw ramie fiber, obtained through manual scraping, includes a significant amount of non-fibrous materials, such as gums and parenchyma cells (about 30–35%). These non-fibrous materials are hydrophobic and should be eliminated before the fiber's spinning to convert it into yarn form. The gums and cells primarily consist of xylans and arabans, which can readily dissolve in alkaline solutions. The degumming process for ramie fiber typically involves the following steps: boiling the fiber in an alkaline solution, washing and neutralizing, bleaching, and oiling. This procedure can be performed on either dried or undried fiber, although dried fiber is preferred according to Hoefer's research. In the majority of degumming processes, caustic soda is commonly employed to dissolve the remaining gums and pectins. Properly degummed fiber is long and durable with high tensile strength, luster, and absorbency, which increases when moist. In addition to chemical degumming, recent advancements have been made in microbial degumming (retting) techniques as well as the application of ultrasonic vibrations, and enzymatic treatment.

Chemical degumming utilizes hot alkali to dissolve the pectic substances responsible for binding the fibers together. In commercial processes, sodium hydroxide is often used due to its low cost, sometimes in combination with sodium carbonate. The effectiveness and cost of degumming are influenced by factors such as volume, temperature, concentration, and pH of the alkaline solution, degumming interval, and agitation. Degumming can be carried out below or above atmospheric pressure, but yields tend to be lower at high temperatures as hemicelluloses and some cellulose may also be removed. To ensure that the chemically degummed fiber is suitable for spinning, it is important to prevent tangling during the degumming and washing processes, and thorough washing with water is crucial.

In the microbial degumming method, various types of bacteria are used, each comprehending multiple bacterial species that grow in association with each other. The researchers found that isolating and identifying individual organisms failed since they relied on the metabolic products of other organisms in the mixed culture for growth and nutrition. Using mixed bacterial cultures for microbial degumming is a viable choice for chemical degumming, due to achieving high-quality fibers. This process is easy and economical, requiring less alkali and resulting in improved fiber properties such as softness, feel, and luster. Combining microbial and chemical methods is even simpler and more cost-effective.

Straw Processing

The process of degumming results in a fairly white filasse, which may require bleaching if pure white fibers are desired. Bleaching can result in a slight loss of weight and fiber strength and should only be performed when necessary. Degummed fibers can be harsh, stiff, and entangled, so a softening agent such as wax, fat, glycerine, paraffin, oil, soap, or tallow is applied to them before spinning. The fibers are then left to condition for a period of time. To further soften and separate the fibers, they can be guided through a series of paired fluted rollers followed by a pair of smooth rollers, repeating the process if necessary.

2.4.3 Applications

Blends are a popular choice in textile applications, and one of the most common blends consists of 55% ramie and 45% cotton. Incorporating polyester and other synthetic fibers into the blend enhances wrinkle resistance and facilitates easy care and shrinkage control. This addition also helps mitigate the loss of luster while maintaining the characteristic uneven linen texture. Blending ramie with wool has been found to significantly reduce shrinkage compared to pure wool alone. It is employed to prevent miscarriages and facilitate the expulsion of pus. The leaves possess astringent and resolvent properties, making them useful in the treatment of fluxes and wounds.

2.5 Kenaf

Kenaf (Hibiscus cannabinus L.) is a natural fiber derived from the plant belonging to the Malvaceae family. It is cultivated on an annual basis in tropical and sub-tropical regions [34]. This plant has been cultivated for centuries to obtain fiber, particularly in Bangladesh, the United States, Thailand, India, parts of Africa, Southeast Europe, and Brazil, where it is grown throughout the year [35]. This particular species of plant is naturally categorized as a short-day plant, although there are currently photo-insensitive cultivars that have been developed. The kenaf plant and its seeds are given in Fig. 6a. The stem of the kenaf plant comprises two distinct types of fiber: bast fiber and inner core. The bast fiber accounts for approximately 30–40% of the total dry weight of the stalk, while the inner core makes up the remaining 60% [36]. Kenaf fiber is lignocellulosic fiber that has been utilized to produce textiles, fiber board, particle board, reinforcement material for composites, and as fuel [36, 37]. The SEM image of Kenaf fiber is shown in Fig. 6b. The composition of lignocellulosic fibers, such as kenaf bast fiber, is intricate. It consists of elementary fibers that are interconnected through a pectin interface, creating technical fiber bundles. The separation of these bundles is achieved through the partial decomposition of the cell wall, which can be initiated by bacteria or mechanical methods [38].

Kenaf plants yield small seeds, typically ranging in color from brown to black. These seeds are enclosed in seed capsules that measure approximately 1.9–2.5 cm in length and 1.3–1.9 cm in diameter. Kenaf is characterized by its long taproot system, which provides it with drought-tolerant properties. The plant typically has a straight stem with minimal branching and can reach a height of up to 6 m under favorable

Fig. 6 **a** Kenaf plants and seeds [39], **b** SEM image of kenaf fiber [40]

growing conditions. The kenaf stem consists of several layers, including the outer epidermis, a thin layer of bast and wood filling, and the inner part of the stem. These different regions of the stem serve as the sources of bast fibers and core fibers in kenaf. The bast and core fibers make up 40% and 60%, respectively, of the stalk's dry weight. The plant's leaves are lobate and individually stalked, with the degree of lobation depending on the cultivar. The large, bell-shaped flowers of the plant are light yellow or creamy colored, with some cultivars having a deep red or maroon center [41].

2.5.1 Economic Importance of Kenaf

Kenaf (Hibiscus cannabinus L.) is mainly grown in Cuba, India, Thailand, Vietnam, and China. Kenaf is currently generating attention as a potential alternative source of pulp, although its utilization is currently limited in scale. The refined fibers of kenaf exhibit a length of approximately 2.6 mm, resembling high-quality softwood fibers, whereas the core fibers are shorter, measuring around 0.6 mm, similar to hardwood fibers. Kenaf contains approximately 40% cellulose and ten percent lignin which is three times less than southern pine, making it easier and quicker to pulp and bleach with hydrogen peroxide rather than chlorine. In addition to its annual renewability, kenaf's eco-friendliness as a raw material surpasses that of commonly used timber, making it a more sustainable option for various applications.

In the United States, approximately 200 tonnes of kenaf pulp are produced annually, and there are also small-scale mills in countries such as Spain, Thailand, India, and China. Various pulping technologies have been experimented with, but the sulfate (kraft) process is the only one currently used commercially. Other uses of core kenaf fiber include animal bedding, packing material, oil absorbents, soil-fewer potting mixes, drilling mud binder, flower mats, organic filler, decorative fibers, and insulation. Bast fiber can also be blended with cotton and used in textiles. Additionally, kenaf bast fiber can be mixed with plastic for injection molding.

2.5.2 Production Method

Cultivation

Kenaf is a versatile plant that is grown for its fiber and as fodder. It grows best in tropical and sub-tropical regions with temperatures over 20 °C and requires an average of 75–125 mm of water per month. Kenaf has shown adaptability to various soil types and has demonstrated resilience to late-season flooding, low soil fertility, and a broad pH range. It also shows outstanding tolerance to drought environments. While studies have indicated that kenaf can grow without the need for additional fertilizers, it is advisable to maintain a balanced nutrient profile to optimize its growth. Recommended application rates for nitrogen (N), phosphorus (P), and potassium (K) range from approximately 100–130 kg/ha, 35–50 kg/ha, and 110–140 kg/ha, respectively, taking into account the existing nutrient levels in the soil. Kenaf is completely resistant to pests and diseases, although plant parasitic nematodes are a potential problem. Efficient harvesting methods include using machines to pick up whole stalks or using forage choppers to harvest and store the chopped crop.

2.5.3 Applications

The kenaf fiber's cost is significantly lower than synthetic fibers such as E glass, carbon, and S. steel. Kenaf-reinforced polymer composites are easy to process and have potential in high-tech applications like automobile and aerospace components, construction materials, food packaging, and electrical and electronic parts owing to their low cost and lightweight properties. In Brazil and Malaysia, numerous automobile manufacturers have embraced the integration of natural fiber bio composites into their vehicles. By incorporating materials such as kenaf fiber, bio-based polymers, and their composites, these companies can diminish reliance on fossil fuels, mitigate greenhouse gas emissions, and enhance overall environmental conditions within the automotive industry. Furthermore, kenaf paper has several advantages over traditional pine paper, such as being stronger, and brighter, and producing cleaner pages with less bleaching treatment needed. Kenaf fibers find applications in various industries, including the automotive sector, where they are utilized in non-woven mats, fiberboard, textiles, and various commercial applications such as particleboard, sport, furniture, food packaging, animal feeds, medicine, and more.

2.6 Abaca

Abaca, scientifically known as Musa textilis, is an herbaceous plant that is commonly referred to as Manila hemp. It belongs to the Musaceae family. Although it looks like a banana plant, it has distinct properties and uses. Abaca is cultivated and sold successfully in various countries, with numerous durability and quality levels. Originally native to the Philippines, abaca, or Manila hemp, has been introduced to various regions including America, Indonesia, America, and Borneo. The southern part of the Philippines is the native habitat of the abaca plant, where it thrives in rainforests characterized by high humidity. The largest supplier and source of abaca fiber is the Philippines, which is used for making cordage and specialized paper pulp. Cordage manufacturing has been using abaca fiber for a long time, but its use in specialized paper pulp began commercially in the 1930s. Commercial-scale abaca production is concentrated in three regions in the Philippines: Mindanao, Visayas, and the Bicol regions, each supplying different varieties and hybrids. During World War II, production was disrupted by the Japanese army, leading to the exploration of new regions like Ecuador for abaca cultivation. The plant of abaca fibers and its stem is given in Fig. 7a, b.

Abaca, a plant with multiple varieties, can grow over 6 m tall. The SEM images of the cross-section of abaca fiber are given in Fig. 8. Not all varieties are commercially cultivated. The production cycle of abaca is perennial, taking 18–24 months to produce fiber initially, and subsequent harvesting occurs every two to three months.

Fig. 7 **a** Abaca plant, **b** abaca stem [42]

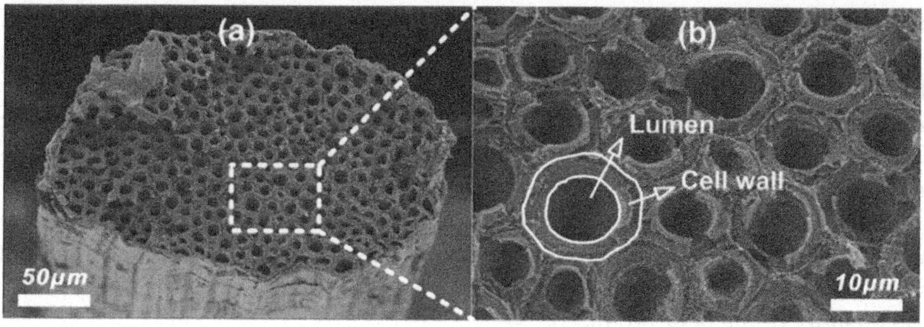

Fig. 8 SEM image of abaca fiber's cross section [43]

Optimal growth conditions include regions with a humid tropical climate, temperatures between 22 and 28 °C, and well-distributed rainfall of 1,800–2,500 mm annually. An altitude ranging from 100 to 140 m above sea level is important. Adequate rainfall and sunlight are crucial for successful abaca production, as excessive sunlight and insufficient rain can harm plant development and reduce productivity. In an ideal environment, abaca plantations can yield fiber for commercial use for 15–20 years.

2.6.1 Economic Importance of Abaca

The Abaca Growers Corporation of Ecuador (CADE) is a major contributor to the national abaca production, accounting for 42% of the total. Despite this, the Philippines continues to maintain its dominant position in global abaca production. In Ecuador, the processing of abaca involves the utilization of specialized machinery to separate the raw material and extract the fiber. Conversely, in the Philippines, the process is predominantly carried out manually, which leads to lower yields and quality of the extracted fiber. The quality of abaca in Ecuador is classified based on color and fiber diameter. Fiber length, which typically ranges from 1.8 to 6.0 m, is also considered in the classification. The grading system consists of five or sometimes six grades, with grade 1 representing white fiber and grade 5 representing dark brown fiber. The quality of the fiber is determined by its diameter, with thinner fibers being considered superior.

2.6.2 Production Method

Cultivation

Abaca is ideally planted at the onset of the winter season, although it can also be cultivated during other seasons provided that the soil retains adequate moisture levels. Abaca plants propagate through a process called suckering, wherein new shoots emerge and grow from the plant's underground roots. When the stem of the abaca plant has fully grown and all its leaves have developed, flower buds begin to appear, signaling that the plant has reached its mature stage and is ready to be harvested. The duration of an abaca plantation, from the time of planting to harvest, can vary and is influenced by various factors such as the characteristics of the property, the specific abaca variety, seed selection, weather conditions, and the extent of maintenance activities. Generally, the time from planting to harvesting is 18–24 months. Harvesting should commence when the flower starts to develop. Harvesting of mature abaca stalks is typically carried out when the flag leaf becomes visible. It is important to avoid harvesting before or after this stage, as it may negatively impact the quality and quantity of the fiber produced.

Fiber Extraction Method

The process of extracting fiber from the plant involves several steps. Firstly, the leaves are removed by defoliating and blunting the mature stems. These stems are then cut at a distance of 10 cm from the ground using a bevel cut that slope outward, preventing stem rot. Next, the surrounding pods are separated from the cut stem, and the stem itself is sliced into strips measuring 5–8 cm in width and 2–4 cm in thickness, with the length varying based on the stem's size. The outer shells of the strips produce a lower quality fiber with a cinnamon or brown color, while the interiors contain higher quality fiber that is white. It is important to complete these steps within 8–12 h after cutting the stem to maintain fiber quality. To facilitate this process, a machine is used, comprising a diesel engine that powers a roller and a shredder equipped with blades. The blades are opened using the other half of the pod or tuxe, which is wrapped around the roller. Once the fiber is obtained, it is necessary to dry it on the same farm due to its high moisture content. The drying time varies depending on atmospheric conditions, ranging from hours to days. Concurrently, a preliminary classification based on fiber color is carried out. After drying, the fiber is stored in a dry and well-ventilated container to ensure its quality.

2.6.3 Applications

Abaca is a versatile plant with diverse applications. Its fibers possess excellent resistance to saltwater, making it a popular choice for manufacturing fishing nets. Additionally, abaca fibers are predominantly utilized in the production of tea bags and meat casings, serving as an alternative to traditional cloth derived from bark. It is also valued as a raw material for machinery filters, diapers, high-quality paper, hospital textiles, napkins, electrical conduction cables, and many other finished products. The fibers extracted from the abaca stalk are used to make clothing, ropes, and paper materials. These plants thrive in shaded and cool environments and share similarities with banana plants.

2.7 *Banana*

Bananas are recognized as one of the earliest cultivated plants globally (see Fig. 9a). The term "banana" is derived from an Arabic word that translates to "finger." They belong to the Musaceae family, which comprises approximately 300 types, although only around 20 varieties are commonly consumed. The fibers used in various applications are extracted from the stems of these plants as shown in Fig. 9b. Utilizing these fibers is particularly important as they are derived from biomass residues, primarily from the pseudo-stems, which would otherwise decompose naturally after fruit processing in agricultural practices. These fibers are readily available, cost-effective, and require minimal energy for extraction, making them highly beneficial from an engineering perspective [44–46].

Banana fibers have been successfully utilized as insulation material in both thermoplastic and thermoset matrices, without incurring additional efforts. The utilization of banana fibers, which serve as a valuable source of fiber, has experienced substantial growth in various countries, contributing to a reduction in dependence on fiber obtained from trees in forested areas. Moreover, utilizing fibers from high-biomass and fast-growing plants aligns with the goal of sourcing renewable and untreated materials. In specific industries that require high phosphorus content, such as certain fiber-based sectors, banana fibers are considered suitable replacements. Given its rapid growth and substantial biomass, the banana plant serves as an excellent example. The SEM image of banana fiber is given in Fig. 10a. Studies on banana fibers have extensively examined their cross-sectional measurements as shown in Fig. 10b, demonstrating that the fibers typically exhibit a circular shape when viewed in cross-section [48–50].

(a) banana plant (b) banana fiber

Fig. 9 **a** Banana plant, **b** extracted banana fibers [47]

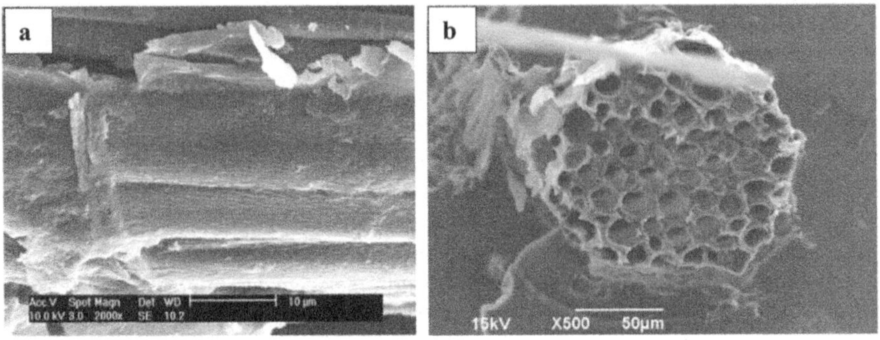

Fig. 10 SEM images of the **a** banana fiber surface and **b** banana fiber cross-section [51]

2.7.1 Economic Importance of Banana

Bananas are considered one of the leading agricultural fruits globally, with a total global production of approximately 119 million tonnes. They are widely consumed in Africa, the Caribbean, Latin America, the Pacific, and Asia, and are enjoyed in various forms such as fresh, fried, cooked, or processed into juice and beer. Many banana cultivars exist worldwide, with some specifically cultivated for direct consumption and others for processing. Distinguishing between bananas and plantains can be challenging for many individuals, as they share similarities in appearance and plant morphology. Consequently, the term "banana" is commonly used to refer to fruit or dessert. Among the top 10 global banana producers, the average production estimates between 2010 and 2020 were approximately 92 million tons, with India being the largest producer, accounting for 29 million tonnes [52, 53].

2.7.2 Production Method

Cultivation

Bananas are crops that can be grown throughout the year, taking approximately 11–12 months to reach maturity. They thrive in tropical regions with well-drained soil and temperatures ranging from 26 to 30 °C. When establishing banana plantations in new fields, it is common practice to use vegetative propagation, specifically by utilizing suckers obtained from well-established colonies. To maintain soil fertility and enhance crop yield, organic fertilizers from livestock waste, nitrogen-based fertilizers, and mulching techniques are often utilized. The pseudo stem of a banana plant is a non-woody structure that resembles a stem but is composed of a soft inner core covered by several sheath layers. As the plant matures, these sheath layers transform into leaves in the upper section of the pseudo stem. The sheath layers of the banana pseudostem are utilized for extracting fibers. Additionally, the peduncle, which extends from the inner core, serves as a supportive structure for the fruit bunch.

When bananas or plantains are harvested, approximately 4 tons of agro-residues are generated per ton of fruit (based on fresh matter). Among these residues, the pseudostem contributes approximately 75%, the peduncle 4%, the leaves 12%, and other parts account for 9% of the total weight. These agricultural residues are typically left to decompose, creating breeding sites for pests, or sometimes ending up in water sources through surface runoff [54–56].

Fiber Extraction

Various methods are employed to extract cellulose-based fibers from lignocellulosic biomass, each presenting its unique set of advantages and disadvantages in terms of output, environmental impact, and fiber quality. The extraction methods can be classified into chemical, manual, mechanical, and biological techniques. Researchers have conducted studies to investigate the utilization of combined techniques to improve the efficiency of the extraction process Among these techniques, chemical and mechanical extraction methods are the most used and well-known methods.

Manual Extraction

A traditional method commonly known as stripping is often practiced in rural cottage setups for extracting cellulosic fibers. This technique involves manual labor and the use of hand tools such as knives. It requires the layer-by-layer separation of the pseudostem (known as tuxying), cutting it into small pieces, and then beating or combing the fibers to remove vegetal matter. After the fibers have been extracted, they undergo a thorough washing process using clean water. Subsequently, they are air-dried and subjected to additional processing stages to prepare them for a wide range of industrial applications [57, 58]. More recently, researchers have also explored the application of this technique for extracting fibers from the peduncle and flower bracts of plants. The quality of the extracted fiber relies on the skills of the personnel involved. However, the low fiber output yield of this method makes it unsuitable for large-scale production.

Mechanical Extraction

Mechanical extraction is a widely adopted technique for commercial extraction of fibers due to its higher fiber output compared to manual extraction. Several decortication machine designs with varying capacities have been developed specifically for this purpose. These machines typically consist of a rotating drum equipped with blades that effectively remove the pulpy matter through a combination of beating and crushing actions. The leaf sheaths are supplied into the machine to extract the fibers, followed by washing and drying of the fibers. However, the mechanical fiber extraction technique may need further treatment to reduce non-cellulosic residue for high-value applications [59, 60].

Extraction via Chemical Treatment

Chemical treatment is a commonly employed approach for extracting banana fibers, using various media such as alkali or acidic solutions, sometimes in combination. The liquor ratio, duration, concentration, and temperature are important factors studied in this process. Alkali treatment, specifically mercerization, is the most widely used method. However, chemical treatment can lead to fiber fibrillation and structural alterations, converting cellulose I to cellulose II. Optimal conditions are necessary to prevent fiber damage. The chemical treatment produces waste chemicals, leading to pollution and requiring costly neutralization. Some studies have explored the use of retting baths to reduce chemical waste, but more research is required to improve chemical recovery and reuse [53].

Biological extraction: Biological treatments have gained attention as eco-friendly alternatives to chemical treatments for extracting banana fibers. These approaches include microbial retting (using fungi and bacteria), enzymatic treatment, and anaerobic digestion. Anaerobic digestion has shown promise with a 40-day duration for optimal results. Various microbial strains and fungi have been explored for their fiber-degrading capabilities. Pond/slow-moving water retting and controlled system tank retting are options for large-scale production, but they require close process control, long durations, and excess water usage. Enzymatic treatments using specific enzymes have also been investigated, with favorable results obtained at specific temperatures and durations. Biological treatments offer advantages such as compatibility with other methods, lower energy obligations, and improved finish of the fiber's surface. However, they require precise control, have long processing times, and involve high enzyme production costs. Further research is needed to optimize and scale up these techniques for efficient and cost-effective fiber degumming [61–64].

2.7.3 Applications

To reduce stiffness in yarn production, banana fibers, which are renowned for their enhanced tensile strength, are frequently combined with other natural fibers in blends. Various extraction approaches have been discussed to judge the appropriateness of banana fibers for textile and yarn processing. Banana fibers offer a potential solution due to their lightweight nature, cost-effectiveness, good shelf life, water repulsion properties, shiny luster, and lower lignin content compared to their cellulosic content. These are currently used in applications such as stationery paper, greaseproof packaging, and currency printing paper. Natural fiber-reinforced composites have gained popularity in various fields including aerospace, leisure, packaging, automotive, biomaterials, construction, sports equipment, and aerospace. These composites offer advantages such as low processing energy requirements, reduced manufacturing costs, abundant availability, and good mechanical properties. Banana fibers find application in numerous industries and products, such as handicrafts (bags, table mats, and baskets), mushroom cultivation substrate, wastewater treatment, sports equipment, food packaging materials, sanitary pads, biorefinery, cordages in the marine sector, and defense helmets [65–67].

2.8 Bamboo

Bamboo, a natural fiber renowned for its low cost and strong properties, has been widely utilized as a structural material in civil construction projects across various countries like China and India. Its toughness, strength, and affordability make it highly desirable. Bamboo possesses several notable advantages, making it a promising alternative to synthetic fibers in fiber-reinforced composites. These advantages include its large aspect ratio, high strength-to-weight ratio, wide availability, biodegradability, and fast renewability. Consequently, the field of composite materials science has shown considerable interest in the exploration of bamboo fiber-reinforced composites. Researchers are actively investigating the potential and properties of these composites. Figure 11 illustrates the distinct honeycomb structure of bamboo, which is composed of vascular bundles containing vessels, phloem, and fibers. These bundles are surrounded by parenchyma cells. These cells possess a lignified wall composed of lignin, cellulose, and hemicellulose, with an intermediate layer between vascular bundles containing a substantial 90% lignin content. The outstanding mechanical properties of bamboo fibers can be attributed to the natural binding component present within them, setting them apart from other types of natural fibers. The utilization of bamboo as a natural reinforcing material offers several key advantages. Firstly, bamboo is known for its rapid growth, making it a highly renewable fiber source. Additionally, the longitudinal alignment of cellulose fibers in bamboo contributes to its exceptional tensile strength, flexural strength, and rigidity. Moreover, bamboo has a high cellulose percentage of around 60%, with relatively low lignin content. Lastly, bamboo exhibits a microfibril angle ranging from 2 to 10°, further enhancing its mechanical properties [68–71].

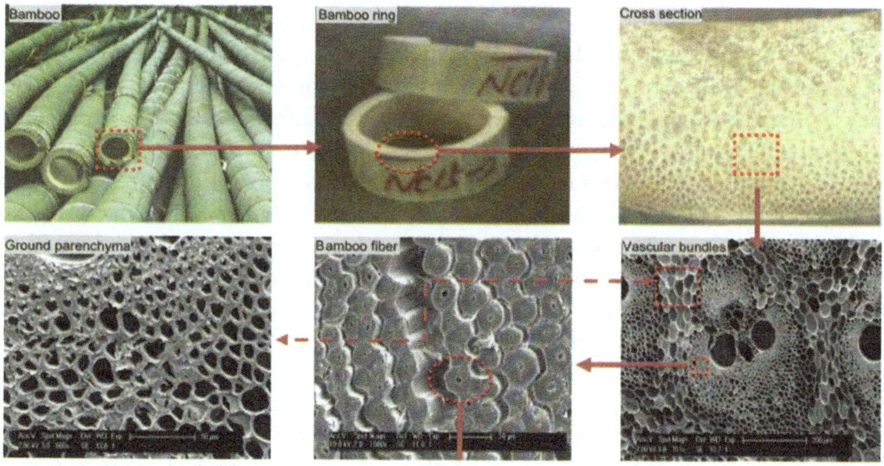

Fig. 11 The morphology and composition of bamboo on different levels [72]

2.8.1 Economic Importance of Bamboo

Bamboo is widely distributed across continents, including Africa, Asia–Pacific, America, North America, and Europe. The Asia–Pacific region is recognized as the primary producer of bamboo worldwide. Countries like Myanmar, Vietnam, Indonesia, India, and China have extensive bamboo plantations. The abundance of bamboo compared to other natural fibers makes it a preferred choice, especially in regions like South America and Asia where it grows naturally without requiring cultivation. Bamboo is a valuable resource for reinforcing composites, especially in regions where forest resources are limited. It offers sustainability benefits as it reaches maturity in about 6–8 months, making it a rapidly renewable resource. Despite being sometimes regarded as a weed due to its rapid growth and widespread presence, bamboo is utilized in small-scale construction and home decor. It holds the potential to be a significant sustainable plant resource that has yet to be extensively utilized [73–75].

2.8.2 Production Method

Bamboo fiber extraction can be achieved through three primary processes: mechanical, chemical, and a combination of mechanical and chemical methods. Chemical extraction methods for bamboo fibers involve various techniques such as alkali/acid retting, chemical retting, and degumming. These processes are utilized to reduce the lignin content of the fibers and enhance their quality. The combined process of bamboo fiber extraction includes techniques such as roller mill (RMT) and compression molding (CMT). These methods are used to extract fibers from bamboo and are commonly employed to enhance the efficiency and quality of the extraction process. When bamboo fibers are extracted manually, they have demonstrated better results with fewer breaks along their length compared to chemical extraction methods. In the mechanical extraction process, the cut bamboo fibers undergo a series of steps. First, they are heated in an autoclave to prepare them for further processing. Then, they are thoroughly washed to remove impurities and ensure their softness. Finally, the fibers are dried, making them ready for extraction. This process helps to enhance the quality and ease of extraction of the bamboo fibers. In the chemical extraction process, specific steps are employed to extract bamboo fibers. Degumming is performed to remove gummy substances present in the fibers. Additionally, the fibers are treated with either sodium hydroxide or trifluoroacetic acid to dissolve the lignin, a complex organic polymer. These chemical treatments effectively eliminate unwanted components and aid in the extraction of high-quality bamboo fibers. Alkali treatment is preferred as it improves the interfacial bond strength in composites. The combined technique of CMT and Roller Milling RMT is utilized for extracting bamboo fibers. This process entails the flattening of bamboo strips by exerting pressure between plates or rollers, followed by immersion in NaOH solution for easy fiber extraction, washing, and drying. Similar methods have been used by Deshpande et al. for bamboo fibers with different diameters [76, 77].

2.8.3 Applications

Bamboo has been widely used in various industries such as handicrafts, agriculture, papermaking, interior furniture, and architecture. The fiber of the Neosinocalamus affinis bamboo species is particularly suitable for textiles due to its crystalline structure, like jute. It has also been recognized as a valuable biofuel-producing plant, with products like firewood, pellets, and charcoal. Bamboo is effectively utilized in the production of wind instruments like flutes and string instruments. Its antimicrobial and hygroscopic properties make it suitable for pharmaceutical applications, including drug preparation and the manufacturing of absorbent pads and doctor's coats. Bamboo fibers and their composites have gained significant popularity in structural applications owing to their exceptional mechanical properties, including high tensile strength and bending strength. These properties make bamboo-based materials well-suited for various structural uses where strength and durability are crucial requirements. The consideration of environmental impact has become a crucial factor in material selection, leading to the use of natural fibers like bamboo in wind turbine blades. Bamboo composites are also popular in the sports industry for products like surfboards and baseball equipment, thanks to their moisture permeability and waterproof surface [78, 79].

3 Sustainable Synthetic Fibers

3.1 Alginate

Seaweed extracts have been used for food and medicine since 3000 BC, with alginate being utilized in various industries like food, pharmaceuticals, and textiles. In 1880, British scientist Stanford discovered and named "algin," a substance found in seaweed after extracting iodine and salts. Further research led to the realization that it was an acid, resulting in the renaming of the substance as "alginic acid." Brown algae are the primary source of commercial alginates, mainly originating in the cell wall and intercellular regions. Three types of brown algae, laminaria, microcystis, and ascophyllum, are abundant enough for commercial extraction. Commercially available high-viscosity alginate typically possesses approximately 150,000 molecular weights, while ordinary alginate has an average molecular weight of 15,000. Laminariales, the prime and highly complex algae, consist of a frond, stem, and roots, requiring firm anchoring due to their exposure to turbulent environments. Harvesting brown seaweed is relatively easy as they grow in shallow waters [80–82].

Alginate is a linear polysaccharide composed of mannuronic acid (M block) and guluronic acid (G block) linked by 1,4-linkages that can be extracted from brown algae or certain bacteria. The distribution and proportion of M and G blocks impact the physicochemical properties of alginate, with M-rich alginate displaying flexibility and better biocompatibility, while G-rich alginate has a rigid structure. Alginate contains carboxyl and hydroxyl groups along its backbone, allowing for chemical modification and cross-linking as shown in Fig. 12. Cross-linking can be achieved by divalent or trivalent metal cations like calcium ions, forming an egg-box structure as shown in Fig. 13 [83–86].

Polymeric fibers that are useful for textile applications typically have long-chain molecules arranged in a parallel manner along the fiber's length. These fibers exhibit cohesion and strength due to lateral forces that hold the molecules together. It is also beneficial for the fibers to have some degree of molecular movement, which allows for extensibility, moisture absorption, and the uptake of chemicals like dyes. Alginic acid, a linear polymer with reactive groups along its length, possesses these characteristics and can potentially be used as a fiber-forming material. The unique functionality of alginate chains enables the reinforcement of lateral forces within the fiber by creating cross-links between active groups using appropriate agents. The SEM images of alginate fiber are given in Fig. 14. The smooth surface of alginate fiber is shown in Fig. 14a. The cross-sectional perspective of the alginate fiber is depicted in Fig. 14b, revealing the internal pores. The use of reagents based on formaldehyde in cross-linking can also enhance the water resistance of the fiber by reducing the hydroxyl content within the system. In the pursuit of obtaining textile-grade fibers, scientists and researchers have made efforts to spin fibers using different types of alginate salts [87].

The orientation and crystallinity of alginates, such as calcium alginate, as fibers can be enhanced through stretching. However, compared to cellulose and other natural fibers, alginate has a lower orientation and crystallinity even after stretching.

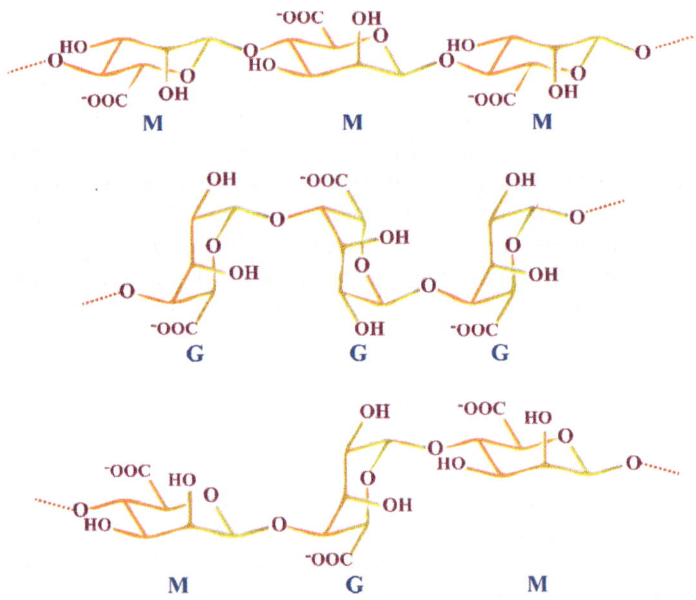

Fig. 12 Schematics for chain configuration and MGM block division of molecular chain of alginate

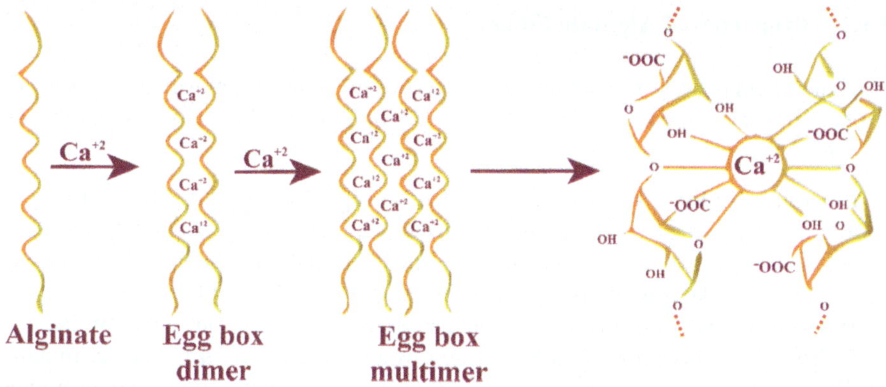

Fig. 13 Egg box multimers of alginate polymer chains by crosslinker (calcium ion)

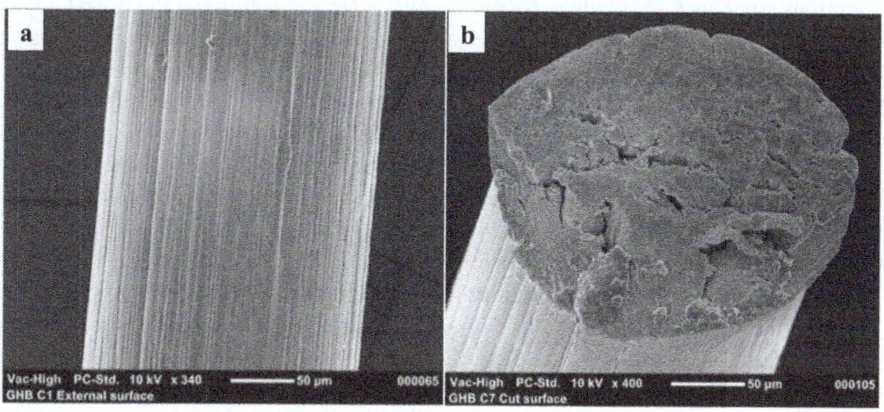

Fig. 14 **a** SEM image of alginate fiber's surface, **b** SEM image of the cut surface of alginate fiber [88]

This is because the chains of alginate undergo cross-linking with metal ions during the chemical coagulation process in the spinneret. While this cross-linking is important for gel formation, it reduces molecular rearrangements and decreases the effectiveness of normal stretching compared to the traditional system of wet-spun fiber. Typically, a maximum stretch of around 30% can be achieved before fracture. However, there is a possibility for improvement by applying tension during the drying process of the filaments. This allows the filaments to contract in a specific direction, facilitating the straightening of molecular chains and potentially enhancing the overall quality of the fibers. Elevated temperatures, combined with the presence of water that acts as a plasticizer, contribute to the breaking and reformation of bonds, further enhancing the fiber's properties [89–91].

3.1.1 Properties of Alginate Fibers

Alginate fibers possess several unique properties that make them valuable in various applications. Here are some detailed properties of alginate fibers along with relevant sources:

- **Biocompatibility**: Alginate fibers are biocompatible, meaning they are well-tolerated by living organisms and do not cause adverse reactions. Due to this characteristic, they find wide applicability in medical and biomedical fields, including applications like drug delivery, tissue engineering, and wound dressings.
- **Hydrophilicity**: Alginate fibers have high water absorption capacity and are hydrophilic. This property allows them to absorb and retain moisture, making them useful in wound dressings to create a moist environment that promotes healing.
- **Gel-Forming Ability**: Alginate fibers can form gels when exposed to divalent cations, such as Ca^{+2}. This gel-forming characteristic is utilized in several applications, including controlled release in drug delivery systems and immobilization of cells or enzymes.
- **Mechanical Strength**: Alginate fibers possess good mechanical strength, allowing them to withstand stretching and deformation. This property is beneficial in applications where the fibers need to provide structural support or endure mechanical stress, such as in tissue engineering scaffolds or reinforced composite materials.
- **Biodegradability**: Alginate fibers are biodegradable, meaning they can be broken down by natural processes over time. This property is advantageous in applications where the fibers are intended to degrade and be replaced by new tissue, such as in tissue engineering or wound healing.
- **Rheological Properties**: Alginate fibers exhibit interesting rheological properties, including shear thinning behavior and high viscoelasticity. These properties make alginate solutions suitable for 3D bioprinting and other manufacturing processes where precise control of flow and gelation is required [92–99].

3.1.2 Production Method

Alginate-based materials can be fabricated using various methods, such as ion cross-linking, wet spinning, freeze-drying, microfluidic spinning, and immersive rotary/centrifugal jet spinning techniques. The wet spinning technique is commonly used to produce alginate fibers. In this technique, a spinning solution containing alginate is extruded via a spinneret into a coagulation bath comprising a calcium salt solution as shown in Fig. 15 [98, 99]. This bath induces ion cross-linking, leading to the formation of the primary fiber. To improve the mechanical properties of the fiber, draft rollers are utilized during the coagulation bath to achieve an appropriate drawing ratio. In addition, researchers have explored the combination of calcium ions (Ca^{2+}) and other metallic ions such as zinc (Zn^{2+}), barium (Ba^{2+}), copper (Cu^{2+}), aluminum (Al^{3+}), and others to create a coagulation bath with multiple metal ions. Different

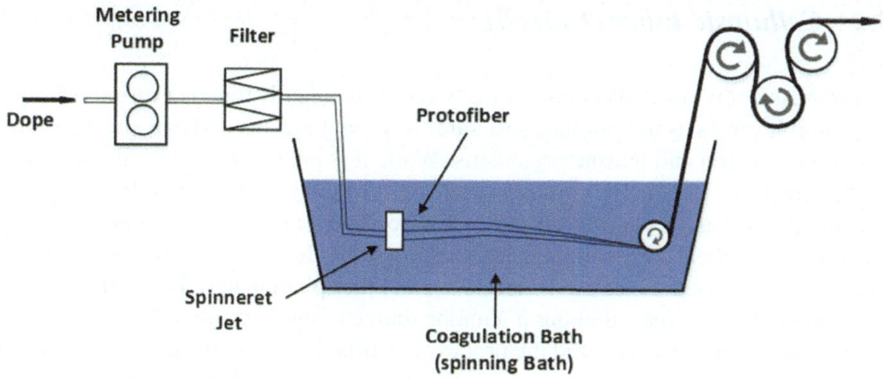

Fig. 15 Schematic of the wet spinning process [101]

metal ions interact with alginate molecules in unique ways, resulting in variations in the rate of fiber formation [100].

3.1.3 Applications

Alginate wound dressings, including hydrogels, fibers, nonwoven fabrics, and freeze-dried scaffolds, offer several advantages such as promoting a moist wound environment, facilitating wound healing, and reducing the risk of bacterial infections. To enhance their functionality, alginate dressings have been combined with therapeutic drugs, growth factors, metal ions, and polymers, resulting in improved antibacterial properties, angiogenesis stimulation, and wound healing promotion. The incorporation of multifunctional additives in alginate dressings shows promising progress and represents a future development trend [102–104]. Calcium alginate-based fibers are utilized as tissue engineering scaffolds, offering advantages such as fast nutrient transport and controlled fiber orientation for cell alignment. Challenges in using alginate fibers include the need for reducing fiber diameter for better cell integration, which can be addressed by electrospinning and the addition of other polymers. Alginate fibers are also employed for drug delivery and can be modified by substituting calcium with other physiological bivalent cations like Zn^{2+}, Mg^{2+}, and Cu^{2+} to enhance tissue repair and angiogenesis [105–108]. Researchers have utilized the active groups of alginate chains to cross-link with polymers, resulting in highly stretchable fiber strain sensors for innovative electronics. Alginate fibers also show potential for applications in soft actuators, thermosensitivity, pH-sensitivity sensors, and smart wearable devices [109–111]. Alginate fibers are being explored as environmentally sustainable flame-retardant materials for textiles. While alginate fibers show low carbon emissions and improved flame retardancy, the presence of minor afterglow limits their fireproof abilities. Researchers are addressing this issue by blending alginate fibers with synthetic fibers such as polyester to reduce afterglow and improve overall fire resistance [112, 113].

3.2 Cellulosic Fiber (Lyocell)

Cellulose is a polymer that occurs naturally and is abundantly found in various sources in nature. It serves as the primary structural material in plants and can also be found in certain bacteria and marine organisms. While it is present in cotton fibers almost in its pure form, it is more commonly mixed with other substances. For instance, dry wood consists of varying proportions of cellulose, lignin, and hemicelluloses. Cellulose has been used for thousands of years, with one of its earliest applications being firewood. Over time, cellulose and other plant materials transform peat, coal, oil, and natural gas, making it a major source of energy, either directly or indirectly. Early humans discovered the process of extracting fibers from plant stems and twisting them together to create a string. This technique can still be observed today by extracting fibers from wild nettles. In the context of current concerns about climate change, cellulose is recognized as a means of sequestering carbon dioxide. Cellulose is synthesized through enzymatic reactions within living cells. In plants, sunlight converts carbon dioxide from the air and water from the ground into glucose, which is then used for cellulose biosynthesis. Glucose undergoes a condensation polymerization reaction, where water molecules are eliminated, leading to the formation of cellulose [114, 115].

Glucose ($C_6H_{12}O_6$) molecule consists of a ring structure having five –CH functional groups, four –OH and one –CH_2OH pendant functional groups, and one oxygen atom. When two adjacent –OH groups undergo dehydration, a disaccharide like sucrose is created. In the case of cellulose formation in plant cells, glucose molecules join enzyme complexes and sequentially combine, releasing water molecules, to create long polymer chains. Figure 16 depicts the chemical formula of cellulose, showcasing its geometric shape and a simplified representation [114].

Lyocell is a revolutionary cellulosic fiber created through a solvent-spinning method. The motivation behind its development stemmed from the demand for an eco-friendly process that utilizes renewable resources such as raw materials. Lyocell, a textile fabric known for its comfort and exceptional physical performance, was initially developed in 1984 and entered commercial production in 1988. It offers a versatile range of attractive fabric options. Lyocell stands apart from other cellulosic fibers, like viscose, due to its unique production method known as direct solvent spinning, which entails dissolving cellulose directly in an organic solvent without the formation of any intermediate compound. Despite this distinction, it shares the benefits of being a cellulosic fiber, including full biodegradability and excellent absorbency. It is known for its exceptional strength in both wet and dry conditions, making it a suitable choice for blending with fibers such as cotton, linen, and wool, resulting in effective fiber combinations. During wet abrasion, lyocell fibers exhibit fibrillation, where small fiber-like structures peel away from the main body while remaining attached. This fibrillation behavior is exploited to create various appealing fabric aesthetics [6]. The SEM image of lyocell fiber is shown in Fig. 17 revealing the smooth fibrous surface.

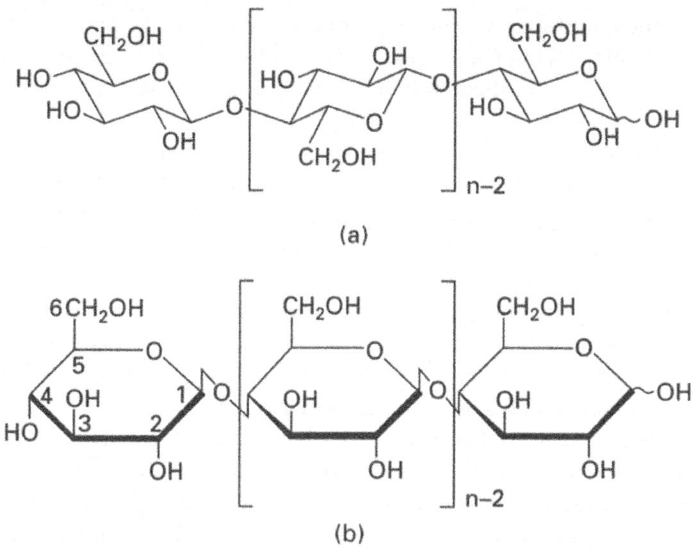

Fig. 16 Chemical formula of cellulose

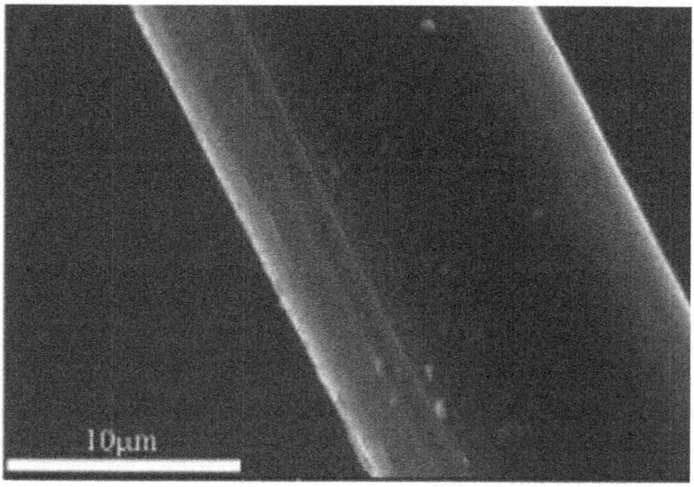

Fig. 17 SEM image of lyocell fiber's surface [116]

3.2.1 Properties

Lyocell fibers possess several general properties that contribute to their popularity in various applications. Here are some general properties of lyocell fibers along with relevant sources:

- **Softness and Comfort**: Lyocell fibers are known for their exceptional softness and smooth texture, providing a luxurious and comfortable feel against the skin. This property makes lyocell fibers desirable for applications such as apparel, bedding, and home textiles.
- **Moisture Absorption**: Lyocell fibers have excellent moisture absorption properties, allowing them to absorb moisture efficiently and release it quickly. This moisture management property helps to keep the skin dry and comfortable, making lyocell fibers suitable for activewear, sportswear, and other moisture-regulating textiles.
- **Breathability**: Lyocell fibers exhibit excellent breathability, facilitating the circulation of air through the fabric. This characteristic assist in regulating the body's temperature and prevent the accumulation of moisture, making lyocell fibers suitable for applications in hot and humid climates.
- **Environmental Sustainability**: Lyocell fibers are produced from renewable raw materials, typically sourced from sustainably managed forests. The manufacturing process of lyocell fibers involves a closed-loop system with minimal chemical waste and water consumption, making them environmentally friendly and biodegradable.
- **Strength and Durability**: Lyocell fibers exhibit good tensile strength and durability, allowing them to withstand regular use and laundering without significant loss of strength. This property contributes to the longevity and durability of lyocell-based products.
- **Color Retention**: Lyocell fibers have excellent color retention properties, meaning they can retain their vibrant colors even after repeated washing and exposure to sunlight. This property makes lyocell fibers suitable for applications where colorfastness is important, such as in fashion and home textiles [117, 118].

3.2.2 Production Method

The process begins by pulling the pulp from the reels and passing it through a shredder, which cuts it into small pieces. These pieces are then mixed with an amine oxide solvent. The shredded pulp is mixed with a solution containing 76–78% amine oxide in water in vessels. Mixing takes place in a ploughshare mixer at temperatures ranging from 70 to 90 °C. The mixer is equipped with high-speed refiners to effectively break down the pulp and promote the wetting of the solvent. The resultant slurry exhibits a consistency similar to the dough, with the pulp fibers swollen due to the presence of the solvent. Afterward, the premix is transferred to a storage hopper equipped with agitation, where it is accurately measured for the subsequent stage of the process. The premix is subjected to vacuum heating to remove water and achieve a dark, clear, amber-colored cellulosic solution. In general, the solutions used in this process contain cellulose concentrations ranging from 10 to 18%. The evaporation of water from the premix to achieve the desired solution is achieved through the utilization of a wiped thin film evaporator. To prevent overheating, the evaporator operates under a vacuum, which lowers the evaporation of water temperature (around 90–120 °C) [119].

The solution exiting the extruder undergoes pumping through a series of specialized pumps in the transport system. Due to the high viscosity of the solution, it is necessary to pump it at elevated pressures, which can reach up to 180 bar. To ensure safety, strategically placed bursting discs are installed at various points in the plant to release pressure in case of an exothermic reaction. The bursting discs installed in the plant are carefully designed to prevent flow dead spots and ensure the efficient release of excess pressure. When activated, they divert the pressure to disentrainment pots, where solid degradation products are separated, and the gases are safely vented into the atmosphere. Before the spinning process, it is crucial to eliminate different impurities from the solution. The solution is then extruded and passed through an air gap, eventually entering a spin bath. The spin bath consists of a diluted amine oxide solution. Within each jet, numerous tiny holes facilitate the extrusion of the solution into individual fibers. After the fibers traverse the air gap, they are drawn downward through the spin bath, where the cellulose undergoes regeneration within the diluted solvent. These fibers, commonly referred to as "tow," are then subjected to additional drawing and stretching in the air gap. Traction units or godets are often employed to facilitate this process [120, 121].

3.2.3 Applications

Lyocell fibers are commonly used in the production of high-quality and comfortable clothing items. They are known for their softness, breathability, and excellent moisture-wicking properties. Lyocell fabrics drape well, have a smooth texture, and are often used in garments such as dresses, shirts, skirts, and activewear. These fibers also find applications in various home textile products, including bed linens, curtains, upholstery, and towels. The nonwoven fabrics of lyocell fibers are commonly found in products such as wet wipes, medical dressings, feminine hygiene products, and filtration materials [118].

3.3 Polylactic Acid

Poly (lactic acid) (PLA) is a compostable thermoplastic polyester derived from renewable resources such as corn. Initially, PLA found limited applications in the field of biomedical products, recent developments have allowed for large-scale production for packaging and fiber applications. The use of polyester fibers, particularly poly(ethylene terephthalate) (PET), is widespread in the textile industry but poses environmental challenges due to their reliance on fossil fuel resources and non-biodegradability. On the other hand, PLA fibers are obtained from sustainable agricultural sources, are 100% compostable, and have the potential to reduce carbon dioxide levels in their life cycle. The recognition by regulatory bodies such as the

EU Commission further validates these fibers as a new and sustainable approach to producing synthetic fibers with performance characteristics [122, 123].

3.3.1 Properties

Polylactic acid (PLA) is a polymer known for its biodegradability and biocompatibility, making it a highly regarded material due to its distinct characteristics. Here are some key characteristics of PLA along with relevant sources:

- **Biodegradability**: PLA is a biodegradable polymer that is sourced from renewable resources like corn starch or sugarcane. It can undergo microbial degradation in composting conditions, making it an environmentally friendly material.
- **Biocompatibility**: PLA exhibits excellent biocompatibility, which means it is well-tolerated by living organisms and does not cause harmful effects. The biodegradable nature of PLA makes it well-suited for a range of medical and biomedical applications, such as drug delivery systems, tissue engineering scaffolds, and implants.
- **Mechanical Strength**: PLA possesses good mechanical strength, including tensile strength and stiffness, allowing it to withstand certain structural and load-bearing applications. The mechanical characteristics of PLA can be tailored by adjusting its molecular weight, crystallinity, and processing conditions.
- **Thermal Stability**: PLA demonstrates moderate thermal stability, characterized by a glass transition temperature (Tg) typically falling between 55 and 65 °C and a melting temperature (Tm) ranging from 150 to 180 °C. These values may vary depending on factors such as the molecular weight and stereochemistry of the polymer. The versatile thermal properties of PLA allow for its processing through various techniques, such as injection molding and extrusion.
- **Transparency and Clarity**: PLA has good transparency, like traditional plastics, making it appropriate for applications that require clarity, such as packaging and disposable items. The transparency of PLA can be improved by optimizing processing conditions and reducing the presence of impurities.
- **Barrier Properties**: PLA exhibits moderate barrier properties against gases, such as O_2 and CO_2. However, it has higher gas permeability compared to some petroleum-based polymers. To tailor the barrier properties of PLA for specific applications, it is possible to modify or blend PLA with other materials. This enables the enhancement of its barrier performance, expanding its potential use in various industries and packaging applications [124–127].

3.3.2 Production Method

Conventional synthetic polymers heavily rely on non-renewable fossil fuel reserves for their production, which is unsustainable due to the limited availability and long regeneration times of these resources. In contrast, the manufacturing of poly (lactic acid) (PLA) utilizes a monomer obtained from annually renewable crops. Through the process of photosynthesis, plants convert carbon dioxide and water into starch, which can be extracted and enzymatically hydrolyzed to produce fermentable sugars like glucose. These natural sugars are then fermented to produce lactic acid, which serves as the precursor for PLA production. This renewable and environmentally friendly process reduces dependence on fossil fuels and promotes the utilization of sustainable resources for polymer manufacturing [128].

PLA Synthesis

Lactic acid can undergo polymerization to produce polylactic acid (PLA) through two main methods: direct polycondensation and ring-opening polymerization (ROP). Many researchers have worked on these methods, leading to the development of various techniques, catalysts, and solutions for synthesizing PLA. The direct polycondensation process was initially developed in 1932, involving high temperatures and high vacuum conditions. The condensation reaction of lactic acid is reversible and generates water, which can negatively impact the quality of PLA. Hence, it is essential to shift the balance of dehydration towards esterification to produce premium PLA characterized by a high weight-average molecular weight (Mw). Various techniques such as the utilization of solvents, catalysts, and branching agents have been implemented in the condensation process [129, 130].

The polycondensation method of producing PLA resin faces challenges in effectively removing water and impurities, resulting in PLA with low to moderate molecular weight (Mw). As a result, the industry has primarily determined the ROP method. ROP involves the conversion of lactide, a cyclic intermediate dimer of lactic acid, into PLA. This approach offers the advantage of producing PLA with high Mw under moderate reaction conditions. The ROP of lactide typically takes place at higher temperatures without the use of solvents, and it can generate different stereopolymers depending on the specific lactides used. Due to the presence of two forms of lactic acid, the conversion of lactic acid to lactide results in the formation of three distinct stereoisomers: D-lactide, L-lactide, and meso-lactide [131, 132].

PLA Fiber Synthesis

The manufacturing of PLA filament from PLA polymer resin has developed into a well-established industry. Generally, PLLA (poly (L-lactic acid)) polymer resin is commonly employed to produce PLA fiber. PLA filament production can be accomplished using melt-spinning technology, similar to other melt-spun fibers such as polypropylene fiber. Additionally, solution spinning methods, including both dry spinning and wet spinning, can also be employed for PLA filament manufacturing. At present, the majority of commercially accessible PLA fibers are manufactured using the melt-spinning method, primarily due to their ability to achieve high-speed production. Additionally, melt spinning is advantageous as it is a solvent-free process, contributing to its cost-effectiveness and environmentally friendly nature. However, the PLA polymer is prone to degradation in the melt state. When aiming for minimal degradation during the manufacturing process, solution spinning can be utilized. Both dry spinning and wet spinning methods necessitate the use of specific solvents that can be toxic and pose difficulties in terms of recycling. The PLA fiber produced through solution spinning is not directly suitable for applications in medical products and healthcare textiles, which are important sectors for PLA utilization [133–135].

During the wet spinning process, PLA is first dissolved in an appropriate solvent, such as chloroform, to form the solution for spinning. The concentration of PLA in the solution is determined by considering the solubility and molecular weight of the PLA polymer, typically ranging from 6 to 12 wt%, as well as the pressure limitations of solution spinning. The solution that has been prepared is subsequently extruded into a coagulation bath using a spinneret that is immersed in the bath. The coagulation bath comprises a non-solvent solution, such as toluene. Multiple coagulation baths may be used in PLA spinning. As the PLA fibers are immersed in the coagulation bath, their appearance and structure are established. Afterward, the filaments undergo drying and are gathered. In laboratory settings, the production speed for wet spinning can vary from 25 to 35 cm/min [135].

In the dry-spinning process, the PLA solution is extruded into a thermally insulated chamber, where it comes into contact with hot air. Inside this chamber, the solvent gradually evaporates, leading to the formation of PLA fiber. Previous research has explored the utilization of both effective solvents like chloroform and less effective solvents such as toluene for dry-spinning PLA. Moreover, combining these solvents has demonstrated improvements in the tensile properties of PLA fiber. Various factors, including the choice of solvent for dissolving PLA, spinning temperature, winding speed, hot-drawing, extruding speed, and the incorporation of additives, can all exert an influence on surface morphology and characteristics of the dry-spun PLA fiber [136–138].

3.3.3 Applications

At present, PLA-based materials find significant usage in three main markets: biomedical (the primary market), textiles, and packaging, with a particular emphasis on short-term applications in the food industry. Examples of manufactured products in these markets include injection-molded cups, forks, spoons, thermoformed trays, blow-molded bottles, paper coatings, fibers for the textile industry, films, sutures, and various molded articles [139].

3.4 Polyhydroxy Alkanoates

Bacterial polyesters, specifically poly(hydroxyalkanoates) (PHAs) with poly(hydroxybutyrate) (PHB) as the initial homologue, belong to a fascinating yet contentious category of biodegradable polymers. These materials offer numerous advantages, including manufacturing from renewable resources, biocompatibility, rapid biodegradability, and excellent mechanical characteristics, positioning them as promising polymers for the future. However, their wider application is impeded by several significant drawbacks. Some notable characteristics of PLA include its high vulnerability to thermal degradation, difficulties in processing arising from thermal instability, low melt elasticity, material brittleness leading to low toughness (which worsens over time due to a phenomenon called physical aging), and relatively high cost. These factors contribute to limited production volumes and a less-than-desirable range of applications [140].

PHAs, hydroxy alkanoate polyesters, are synthesized by bacteria from various sources including renewable resources, organic acids, fossil resources, and wastes. Different species of bacteria produce PHAs, with the number and size of the inclusions depending on the monomer's carboxyl and hydroxyl groups. PHAs are synthesized by aerobic and anaerobic bacteria, photosynthetic bacteria, Gram-positive and Gram-negative bacteria, as well as archaea. The most commonly used PHA is poly(3-hydroxybutyrate), but other monomers have been discovered over time. Initially, it was believed that poly(3-hydroxybutyrate) was the only constituent, but subsequent research revealed the presence of other monomers. Currently, there are approximately 150 different monomers that makeup PHAs [141, 142].

PHAs have different properties depending on their composition, whether they are homopolyesters or copolyesters. Under aerobic conditions, these biodegradable materials can decompose into carbon dioxide and water. Under anaerobic conditions, they can decompose into carbon dioxide and methane. Microorganisms in nature can degrade PHAs using PHA depolymerases. Solvent extraction is the recommended method for obtaining pure PHA polymer from biomass. PHAs possess desirable properties for medical applications, including biocompatibility, piezoelectricity (which aids in bone growth and wound healing), and a wide range chemical of and physical properties due to their diverse chemical structures [143, 144].

3.4.1 Properties

Polyhydroxyalkanoates (PHA) are a type of biodegradable polymers that are synthesized by various microorganisms as internal carbon and energy storage compounds. PHA has garnered considerable interest because of its promising potential as a sustainable substitute for traditional plastics. Here are some general properties of polyhydroxyalkanoates:

- **Biodegradability**: PHAs are biodegradable polymers, meaning they can be broken down by microorganisms into safe products such as CO_2 and water. This property makes them environmentally friendly and reduces their impact on waste accumulation.
- **Biocompatibility**: PHAs are considered biocompatible materials, meaning they are generally non-toxic and do not elicit adverse reactions when in contact with living tissues. This property makes them suitable for various biomedical applications, including drug delivery systems and tissue engineering scaffolds.
- **Versatile Chemical Structure**: The chemical structures of PHAs can vary significantly due to the wide array of monomers that can be incorporated into their polymer chains. The monomers can vary in terms of their carbon chain length,

functional groups, and degree of saturation, allowing for the development of PHA polymers with different physical and chemical properties.

- **Mechanical Properties**: PHAs can exhibit a wide range of mechanical characteristics, including high tensile strength, flexibility, and elasticity.
- **Thermal Stability**: PHAs typically exhibit favorable thermal stability, characterized by melting temperatures from 40 to 180 °C. The thermal properties of PHAs can be modified by copolymerization or blending with other polymers, expanding their potential applications in various industries.
- **Processing Methods**: Conventional polymer processing techniques like injection molding, extrusion, and film blowing can be employed for the processing of PHAs. This makes them compatible with existing manufacturing infrastructure, facilitating their integration into commercial production processes [145–149].

3.4.2 Production Method

Fiber-forming polymers can be developed using different fabrication methods such as melt spinning, cold or hot drawing, and gel spinning. These methods have been applied to the production of PHA fibers, although research on PHA fibers is relatively limited compared to other synthetic polymers. Extensive research has been conducted on the spinning and drawing processes for PHA fibers. It is crucial to examine the variations between these processes and the resulting fiber properties as reported by various researchers [150, 151].

The ability to successfully spin and draw fibers is crucial for the further development and industrial production of PHA fibers. The spinning line to synthesize the PHA fibers consisted of various components, including a spinning pump, an extruder, two winders, and heated godets. It allowed for production speeds ranging from 2000 to 6000 m/min, which is comparable to the production speed of synthetic commercial fibers. Special interest was given to thoroughly drying the PHA powder before spinning it to minimize hydrolytic degradation. Despite the careful removal of moisture, the viscometric molecular weight of PHB (polyhydroxybutyrate) decreased from 540,000 (for the original powder) to 175,000 after the spinning process. The observed decrease in molecular weight is most likely attributed to thermal chain scission, considering that the moisture content of the dried pellets was merely 0.01%. The viscometric molecular weight of the PHA dropped during spinning, indicating thermal chain scission. However, the mechanical characteristics of the resulting fibers are satisfactory. The research paper provides detailed information on the procedure, allowing for an estimation of the impact of several preparation requirements on the characteristics of the fibers [152].

The second method to synthesize the PHA fibers is the gel spinning method which starts by dissolving PHA in a suitable liquid, with 1,2-dichloromethane recommended as the optimal solvent. A concentrated solution of PHA is typically prepared with a high concentration, usually around 20 wt%, for PHA with a molecular weight of approximately 300,000 g/mol. The solution is then used to create a solid gel by evaporating some of the solvents, resulting in a polymer concentration of approximately

30 wt%. The gel can be extruded at around 170 °C. The subsequent processing of the extruded gel occurs in three stages. During the pre-conditioning stage, the fiber is wound around a drum at a controlled speed. The ideal pre-conditioning draw ratio is determined by comparing the Tex values of the extruded and drawn fibers. The second stage entails a continuous hot drawing process between two rollers set at 120 °C, with a total draw ratio of approximately 10. Subsequently, the fibers undergo stretching at room temperature to reach 180% of their original length after the second stage. This is followed by fixation and annealing at 150 °C for 1 h. The "as-spun" fibers obtained in the first stage can undergo necking, either at room temperature or above, with necking beginning at a strain of approximately 6–7%. The ability to draw the fibers is influenced by the drawing temperature, with the highest draw ratio (around 5) observed at 120 °C. While the initial drawing after spinning requires immediate action, the pre-conditioned fibers, once drawn, can be stored for several months without losing their capability to undergo further drawing in the second stage. Interestingly, the material that underwent drawing demonstrated elastic behavior like rubber. The drawing process may introduce variations to the brittle PHB bulk, like the effects observed in cold-rolled polyester. However, the impact of drawing on PHB is more significant, likely due to a higher degree of chain orientation [153].

3.4.3 Applications

PHA finds potential use in tissue engineering, agricultural foils, surgical sutures, and packaging for food storage. With its good barrier properties, like PVC and PET, PHA can contribute to addressing environmental pollution issues as a biodegradable plastic in the packaging industry. In medical applications, PHA demonstrates compatibility with mammalian blood and tissues, rendering it suitable for use in surgical implants, wound healing, and blood vessel repair. In the field of pharmacology, PHA can find applications as microcapsules for therapy and as packaging materials for cells and tablets. Additionally, PHA finds applications in deep drawing articles for the food industry, such as sanitary goods, disposable cups, fast foods, foils, flowerpots, bottles, agricultural foils, fishnets, and textile fibers. PHA shows promise as a substitute for PP, PE, and even PET in specific applications, given its appealing properties.

3.5 Poly(Caprolactone)-Based Fibers

Polycaprolactone (PCL) fibers, a type of synthetic polymer fibers, have garnered considerable interest across diverse industries because of their distinct properties and wide-ranging applications. PCL is a biodegradable polyester that can be produced from renewable sources like plant-based feedstocks or petroleum-based raw materials. It possesses excellent processability and can easily melt-spun into fibers with different diameters and lengths. One of the key advantages of PCL fibers is their biodegradability. PCL fibers can undergo degradation through the action of naturally existing microorganisms in the environment, making them an eco-friendly alternative to traditional synthetic fibers. This characteristic makes PCL fibers particularly suitable for applications where biodegradability is desired, such as in medical implants, tissue engineering, and controlled drug delivery systems. PCL fibers also exhibit good mechanical properties, including flexibility and elongation, which make them suitable in textile fields such as nonwoven fabrics, sutures, and wound dressings. Additionally, to enhance their properties and broaden their application possibilities, PCL fibers can be combined with various materials, including natural fibers or other synthetic polymers, through blending processes. Overall, PCL fibers offer a promising solution in the field of sustainable materials, combining the advantages of synthetic polymers with biodegradability. Their versatility and biocompatibility make them a preferred choice in various industries, from healthcare to textiles. PCL fibers are expected to find even more innovative applications, contributing to a greener and more sustainable future [154].

3.5.1 Properties

Poly(ε-caprolactone) (PCL) fibers are a type of synthetic polymer fibers that are commonly used in various applications, including textiles, medical devices, tissue engineering, and filtration systems. Here are some properties of PCL fibers:

Biodegradability: PCL fibers are biodegradable, meaning they can be broken down by natural processes over time. This property makes them suitable for applications where controlled degradation is desired, such as in biomedical scaffolds or environmentally friendly textiles.

Mechanical Strength: PCL fibers exhibit favorable mechanical strength characteristics, including high tensile strength and elongation. This property allows them to withstand external forces and makes them suitable for applications requiring durability and flexibility, such as in textile fabrics and tissue engineering scaffolds.

Thermal Stability: PCL fibers have a relatively low melting point, typically around 60–65 °C. This characteristic makes them thermally processable using techniques such as melt spinning, electrospinning, or solution spinning. However, it also means that PCL fibers may soften or deform at higher temperatures, limiting their use in fields that need high-temperature resistance.

Hydrophobicity: PCL fibers are hydrophobic, meaning they repel water and are not easily wetted. This property can be advantageous in certain applications, such as in filtration systems, where water or moisture resistance is required. However, it can also limit their compatibility with aqueous environments or applications that require water absorption.

Biocompatibility: PCL fibers are recognized as biocompatible materials, indicating that they are generally well-tolerated by living tissues and typically do not provoke significant immune or inflammatory reactions. This property makes them suitable for different biomedical areas.

Processability: PCL fibers can be easily processed into numerous shapes, including microfibers and nanofibers, using techniques like electrospinning. This versatility in processing methods allows for the manufacturing of PCL fibers with different surface morphologies, surface structures, and porosity, expanding their range of applications [155–158].

3.5.2 Production Method

PCL, a synthetic polymer, is produced through the ring-opening polymerization of ε-caprolactone. The study of PCL dates back to the 1930s. In recent times, an extensive review has been carried out, encompassing various catalysts employed for the ring-opening polymerization of caprolactone. To aid the polymerization process, catalysts such as stannous octoate are employed, while low molecular weight alcohols are used to regulate the molecular weight of the resultant polymer. The polymerization of PCL is influenced by several mechanisms, including coordination, radical, anionic, and cationic processes. Each mechanism has consequences for the molecular weight, molecular weight distribution, end group composition, and chemical structure of the copolymers. The number average molecular weight of PCL samples commonly ranges from 3,000 to 80,000 g/mol, and the molecular weight can be classified based on its distribution. As mentioned earlier, PCL exhibits excellent biocompatibility, making it highly suitable for long-term implantable systems in biomedical applications. Additionally, the potential applications of PCL in fiber form have been extensively investigated, including its use in drug delivery systems, absorbable sutures, and, more recently, as 3D scaffolds for tissue engineering. To fabricate PCL fibers, various methods such as melt spinning, solution spinning, and electrospinning can be employed, each offering unique characteristics and influencing the properties of the resulting fibers [159–161].

Melt Spinning

Melt spinning is a widely employed approach for synthesizing PCL fibers. To ensure successful melt spinning, it is recommended to employ a vertical spinning direction with a small gap between the die and the cooling bath. Effective cooling using ice water at temperatures ranging from 5 to 10 °C is vital. The temperature during spinning should be carefully maintained at around 85–90 °C. Although fibers with uniform diameters can still be achieved at higher temperatures (approximately 120 °C), there may be indications of capillary instability. The diameter of the resulting fibers typically ranges from 0.49 to 0.91 mm, which is influenced by various factors including take-up rate, extrusion rate, ram speed, and the ratio of take-up to extrusion rate. In certain cases, melt spinning of PCL can be conducted with the incorporation of additives like N-(3,4-dimethoxycinnamoyl)-anthranilic acid, a drug known for suppressing fibroblast hyperplasia. This enables the production of PCL fibers with drug-incorporated properties [162–164].

The method for PCL fiber preparation, known as gravity spinning, was performed by following the steps. The PCL polymer was dissolved in acetone to form a solution with a PCL concentration ranging from 6 to 20 wt%. The polymer solution was transferred into a container and then passed through a spinneret located at the vessel's bottom, allowing it to flow out. The polymer solution naturally descended into a solvent due to the force of gravity, specifically methanol, resulting in the formation of fibers. The "as-spun" fibers were collected by winding them onto a mandrel at a variable speed. It was observed that no fiber formation occurred when the concentration of the polymer solution was 5% or lower. In the concentration range of 6–20%, the production rate of the fibers ranged from 2.5 to 0.9 m/min, resulting in fiber diameters between 0.19 and 0.15 mm. It was observed that the production rate and fiber diameter decreased as the solution concentration increased. The resulting fibers exhibited a round shape with a rough, porous surface.

3.5.3 Applications

PCL fibers have found numerous uses in various areas due to their properties. One significant application of PCL fibers is in tissue engineering. These scaffolds have found extensive applications in tissue regeneration and repair. The biocompatibility and biodegradability of PCL fibers make them suitable for promoting cell adhesion and growth. They provide structural support to the growing tissue, allowing for proper tissue regeneration. PCL fibers can be fabricated into intricate three-dimensional structures, mimicking the natural extracellular matrix, and facilitating the formation of new tissues. PCL fibers are also utilized in drug delivery systems, where their capability to encapsulate and release drugs in a controlled manner adds value to pharmaceutical applications. The porous structure of PCL fibers allows for high drug-loading capacity, while their biodegradability ensures gradual release over an extended period. PCL fibers can be engineered to achieve specific release kinetics, enabling targeted and sustained drug delivery. The versatility of PCL fibers in drug delivery systems has opened up possibilities for personalized medicine and improved therapeutic outcomes [165, 166].

3.6 Silk

Silk, renowned for its luxurious texture and versatility, serves as a crucial and sustainable raw material for many products. It is highly sought after in the world of haute couture for its opulent feel and graceful draping. Different countries boast indigenous varieties of wild silk, but the production and promotion of silk are predominantly dominated by China, particularly with a species called Bombyx mori. Silk filament represents a unique and significant class of natural structural proteins that play diverse roles in nature. It serves as a protective membrane during development, shielding organisms from environmental challenges, and acts as a strong web for capturing insects or serving as a lifeline for spiders. The silk proteins are synthesized in specialized gland cells, stored in the gland's lumen, and later spun into filaments. The chemical, physical, and biological properties of silk have sparked considerable interest, leading to extensive research in this field. Silk is composed of polymers derived from amino acids, with peptide bonds linking the 20 different building blocks. Both interchain and intrachain interactions are crucial for the structural functionality of fibrous proteins. The remarkable structure and versatility of silk have made it a fascinating subject of study, revealing its potential applications in various domains [167, 168].

The majority of the world's silk supply is sourced from mulberry silk, which is derived from silkworms (Bombyx mori) that feed on mulberry bushes. Mulberry silk accounts for about 90% of the silk supply, but it requires special care to preserve its smooth texture. However, the extraction of silk filament from silkworms involves killing them in the cocoon stage, which is an ethical concern. Tussah silk, produced by tussah silkworms, has a distinct light golden to dark brown color due to their consumption of tannin-rich leaves. It is commonly used for couches, jackets, and sweaters. Muga silk, characterized by its golden-brown hue and lustrous texture, is produced by the semi-domesticated silkworm species Antheraea assamensis, much like tussah silk. Eri silk, also called "peace silk," is made from the domesticated silkworm Philosamia rinini, and its caterpillars are allowed to complete their life cycle, making it an ethical alternative. It is a durable material used in clothing and furnishings but can be heavy to wash. Mussel silk, also known as sea silk, is produced by mussels found on seabeds and is different from silk produced by silkworms. It is rare and expensive due to sourcing challenges and pollution. Spider silk, like mussel silk, was historically used by ancient cultures. Spider silk production is difficult as spiders cannot be bred like silkworms. However, its exceptional durability makes it valuable for applications such as telescopes, wear-resistant clothing, and bulletproof vests [167–169].

3.6.1 Properties

Silk fibers are natural fibers produced by certain insect larvae, such as silkworms, as well as spiders. Silk is known for its exceptional strength, lustrous appearance, and softness. Here are some properties of silk fibers:

Strength: Silk fibers have remarkable tensile strength, making them one of the strongest natural fibers. They possess high resistance to breaking or tearing, allowing silk fabrics to withstand stress and strain. The strength of silk is attributed to its molecular structure, which consists of interlocking protein chains held together by hydrogen bonds.

Softness and Smoothness: Silk fibers have a smooth and soft texture, which contributes to their luxurious feel. The smoothness of silk arises from the regular alignment of the protein chains, resulting in a uniform surface. This property gives silk fabrics a distinctive silky touch and enhances their comfort against the skin.

Luster: Silk fibers exhibit a natural luster or shine, often described as having a pearl-like sheen. The luster is a result of the way silk fibers reflect light due to their fine structure and the presence of triangular prism-like microstructures on their surface. This unique property contributes to the aesthetic appeal of silk fabrics.

Lightweight: Silk fibers are lightweight, which makes silk fabrics comfortable to wear. Despite their lightness, silk fibers provide insulation, helping to regulate body temperature by trapping air and keeping the body warm in cold weather while allowing breathability in warm conditions.

Biocompatibility: Silk fibers are biocompatible and compatible with human skin. They are hypoallergenic and less likely to cause irritation or allergic reactions. These properties make silk fabrics suitable for sensitive skin and medical applications, including wound dressings and sutures.

Moisture Absorption: Silk can absorb and release moisture, making it a breathable fabric that helps regulate body temperature and prevent the build-up of perspiration. This property contributes to the comfort of silk garments in different climates and conditions.

Durability: Silk fibers possess good durability and resistance to deformation. With proper care, silk fabrics can retain their shape and quality over time, allowing them to be long-lasting and withstand repeated use and washing.

It's worth noting that silk fibers can have variations in properties depending on the specific type of silk, such as mulberry silk or wild silk, and the processing methods employed [170–173].

3.6.2 Production Method

The process of silk production starts with silk moths depositing their eggs on specifically prepared paper surfaces. The eggs hatch and the caterpillars start feeding on mulberry leaves. Over 35 days and four moultings, the caterpillars grow significantly in weight. Silk moths are responsive to external factors such as noise, temperature changes, and odors. Once fully grown, the caterpillar forms a pupa and begins spinning a cocoon using its head in a specific pattern. The silkworm generates silk by combining a silk fibroin protein and a sticky substance called sericin, both of which are produced in its two salivary glands. The resulting mixture is then extruded through an aperture in the lips. Upon contact with air, the liquid solidifies into a solid filament. During about 2–3 days, the silkworm spins approximately 1.5 km of filament, completely enveloping itself within the cocoon. Once the metamorphosis is finished, the moth secretes alkali-based fluids that help dissolve a portion of the cocoon, allowing it to create an opening and emerge.

Before the emergence of the moths, the cocoons are subjected to heat treatment, which serves the purpose of exterminating the silkworms, making it easier to extract the silk and facilitate the breeding of the subsequent generation. The life cycle for the synthesis of silk is given in Fig. 18.

To extract silk filaments, the cocoons are immersed in boiling water, causing the sericin to soften and enabling the unwinding process. In commercial applications, multiple silk filaments, typically ranging from three to ten, are combined and spun together to achieve the desired strength and quality. Upon emerging from the cocoon, the blind, flightless, and toothless moth engages in immediate mating and lays approximately 500 eggs within the initial few days before ultimately perishing. While the

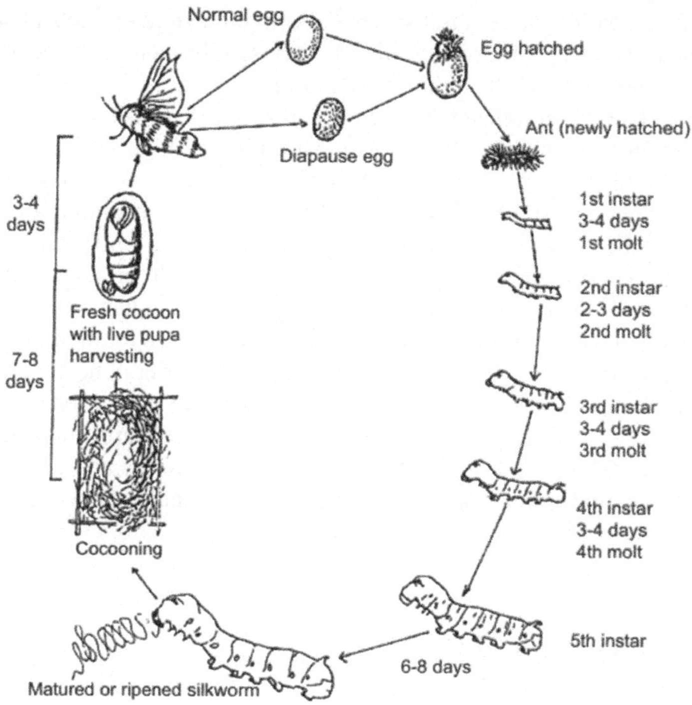

Fig. 18 Production cycle of silk fibers [5]

extraction of silk involves killing the silkworms, both the cultivation of silk and the life cycle of wild silkworms are sustainable processes. The handloom production of silk fabric has a negligible energy footprint, which enhances the sustainability aspect of silk production. The ethical treatment of silkworms is one of the key sustainability considerations in silk production. In the process of obtaining domesticated silks, it is common practice to boil the intact cocoons. However, a small number of silk moths are preserved for breeding purposes to sustain the production cycle [167, 168, 174].

3.6.3 Applications

Silk fibers have many applications owing to their exceptional properties and luxurious feel. Here are some common applications of silk fibers.

Silk is highly valued in the textile industry for its smoothness, softness, and lustrous appearance. It is commonly used to produce high-quality fabrics for clothing such as dresses, blouses, scarves, and ties. The lightweight and breathable nature of silk makes it comfortable to wear in various climates. It is also favored for its draping quality, making it ideal for elegant and flowing garments. These are also used in the production of luxurious home furnishing items such as bedding, curtains, upholstery, and decorative textiles. The natural sheen and smooth texture of silk add an elegant touch to interior spaces. Silk fabrics are often chosen for their aesthetic appeal and ability to enhance the overall ambiance of a room.

Silk fibers have been utilized in medical and surgical fields for various applications. Silk sutures, known for their biocompatibility and high tensile strength, are commonly used in wound closure. Silk-based biomaterials have also been explored for tissue engineering and regenerative medicine purposes, including the development of scaffolds, wound dressings, and drug delivery systems. Silk is a sought-after ingredient in the cosmetics and personal care industry. Silk proteins and peptides derived from silk fibers are incorporated into skincare products, hair care products, and cosmetics. These silk-based formulations are known for their moisturizing, smoothing, and anti-aging properties, contributing to the overall health and appearance of the skin and hair. Art and Craft: Silk fibers are used in various art and craft applications. Silk painting, embroidery, and other textile arts often employ silk as a canvas or thread due to its ability to hold vibrant colors and create intricate designs. The natural sheen and smoothness of silk enhance the visual appeal of artistic creations [175, 176].

4 Sustainable Natural Dyes

4.1 Introduction

Natural dyes have been used for centuries as a means of coloring fabrics, fibers, and other materials. Unlike synthetic dyes that are chemically derived, these organic dyestuffs are derived from various natural sources such as plants, insects, minerals, and even certain animals, microbes and fungi. With growing concerns about the environmental impact of synthetic dyes, there has been a renewed interest in natural dyes due to their eco-friendliness and potential health benefits. Unlike synthetic dyes, which were introduced in the late nineteenth century, natural dyes offer distinct advantages along with some limitations [177, 178].

4.1.1 Advantages of Natural Dyes

- **Environmental Sustainability**: Natural dyes are eco-friendly as they are derived from renewable sources and biodegradable materials. Their production involves lower energy consumption and generates fewer pollutants, making them a more sustainable choice compared to synthetic dyes, which often contain toxic chemicals. The application of natural dyes also produces less harmful effluent. The use of metal mordants for the improvement of exhaustion and fastness properties of natural dyes may cause serious environmental pollution and must be avoided. The use of bio-mordants such as tannin or tannin-rich pant materials as well as chitosan etc. has been considered as a promising approach to solve this problem. So, unlike synthetic dyes, if applied without toxic heavy metal salts, natural dyes do not contribute to environmental pollution and contamination of water bodies [179–181].

- **Biodegradability**: Natural dyes are biodegradable and do not cause harm to the environment. When disposed of, they easily break down into harmless components, whereas synthetic dyes can persist in the environment for extended periods, polluting water bodies and causing ecological damage [181].
- **Health Benefits**: Natural dyes are generally considered safe and hypoallergenic. Unlike synthetic dyes that often contain harmful chemicals, most of common natural dyes pose no significant health risks to users or workers involved in the dyeing process. This makes them a preferable option, especially for individuals with sensitive skin or allergies. Also, some natural dyes add antibacterial, antifungal, anti-odor, or UV protection properties to the dyed textiles [182, 183].
- **Unique aesthetics and cultural and historical significance**: Natural dyes offer a unique color palette that cannot be replicated by synthetic dyes. Each natural dye source exhibits distinctive shades, textures, and variations, giving the finished product an organic and visually appealing appearance. This uniqueness adds value to textiles and works as a differentiating factor in the market. Many traditional cultures have rich histories of dyeing fabrics and creating natural pigments. Using natural dyes helps to preserve and promote these cultural practices, which can be valuable for heritage and art preservation [184–186].

4.1.2 Disadvantages of Natural Dyes

Besides the various advantages mentioned above, natural dyes have some disadvantages that should be taken into account when dyeing textiles with this type of colorants.

- **Limited color range and reproducibility**: One significant limitation of natural dyes is the limited range of colors they can produce. Unlike synthetic dyes, which provide an almost infinite array of color options, natural dyes typically offer a more subdued and earthy color palette. Achieving bright or vibrant hues may require complex dyeing procedures or the use of additional mordants. The inherent variations in the proportion of chemical constituents within natural materials, caused by factors such as maturity, variety, and agroclimatic conditions like soil type and region, contribute to the difficulty in reproducing shades with natural dyes. As a result, it is difficult to consistently achieve the same shade using a specific natural dye in every dyeing process and precise optimization and trials are essential [187, 188].
- **Limited Colorfastness**: Natural dyes often exhibit lower colorfastness compared to synthetic dyes. They may fade or change in color over time due to factors such as exposure to light, washing, or prolonged use. To enhance colorfastness, specialized dyeing techniques or the application of natural mordants may be

necessary, adding complexity to the dyeing process. The exploration of new dye sources can definitely contribute to expanding the range of natural dyes with better colorfastness properties [189, 190].

- **Cost and Availability**: Natural dyes can be relatively expensive, especially when compared to synthetic dyes that are mass-produced. Sourcing and extracting natural dye materials also requires significant efforts and resources. Natural dyeing processes may require longer dyeing time compared with synthetic dyes. Moreover, some natural dye sources may be region-specific or seasonal, limiting their availability on a larger scale [185, 191].

While natural dyes possess several advantages, they also come with a few inherent limitations. Nonetheless, their eco-friendliness, biodegradability, unique aesthetics, and health benefits make them an appealing choice for individuals and industries striving for more sustainable and natural approaches to dyeing.

4.2 Classification of Natural Dyes

Natural dyes can be classified based on source, chemical structure, or application method. Here we will discuss the classification based on the source. The classification of natural dyes based on sources categorizes them according to the type of plant or organism from which they are derived. This classification system allows researchers, artists, and textile industries to identify and utilize natural dyes in a more efficient and organized manner. By understanding the source of a dye, one can have a better understanding of its properties, characteristics, and limitations.

4.2.1 Plant Dyes

One of the primary sources of natural dyes is plant-based materials. Plant dyes can be extracted from various parts of plants, including leaves, flowers, stems, roots, and barks. Some common plant-based dyes include indigo, obtained from the leaves of the Indigofera plant; turmeric, extracted from the roots of the turmeric plant; and madder, derived from the roots of the Rubia tinctorum plant. Each plant-based dye has a distinct composition and chemical structure, leading to variations in color intensity, shade, and durability. There are many agricultural wastes which contain considerable dye content and can be employed for the natural dyeing of textiles [192].

4.2.2 Animal Dyes

Another significant source of natural dyes is derived from insects or animal materials. This category includes dyes obtained from insects such as cochineal, which is obtained from the dried bodies of female insect (Dactylopius coccus), and lac dye, derived from the resinous secretion of the lac insect (Laccifer lacca). These insect-based dyes are known for their vibrant red and purple hues, and they have been used for centuries in various cultures. Tyrian purple or royal purple was obtained from the mucus secreted by the hypobranchial gland of the mollusk Muricidae. The process of extracting this dye was labor-intensive, which made it highly valuable and associated with wealth and royalty [192].

4.2.3 Mineral Pigments

Furthermore, natural colorants can also come from minerals and earth pigments. The most notable example in this category is ochre, a yellow–brown pigment derived from iron oxide. Ochre has been used for thousands of years, and its availability in different natural shades makes it a versatile and widely used dye in various applications [192].

4.2.4 Microbial Dyes

In recent years, microbial natural dyes have emerged as a promising alternative, offering a new dimension to the world of natural dyes. Microbial natural dyes are derived from various microorganisms, including bacteria, fungi, and algae. These microorganisms possess the ability to produce a wide range of pigments, which can be extracted and used as natural dyes. The utilization of microorganisms for dye production not only enhances the repertoire of available natural colorants but also provides a sustainable source of dyes [193].

One of the major advantages of microbial natural dyes is the ability to achieve a wide range of colors. Different strains of microorganisms yield distinct colors due to variations in their pigment production. The use of microbial natural dyes opens up endless possibilities for creating vibrant and unique shades, including those that are not easily obtained from traditional plant-based dyes. This versatility allows for more creativity and customization in dyeing applications. Furthermore, microbial natural dyes offer improved light and wash fastness compared to some plant-based dyes. The use of these dyes ensures that the colors do not fade quickly or bleed during washing, providing a more durable and long-lasting coloration on textiles. This attribute is particularly advantageous for the textile industry, where colorfastness is an essential requirement. Moreover, microbial natural dyes have shown potential in various other applications beyond textiles. They can be used in the food industry as natural food colorants, providing a safe and healthy alternative to synthetic dyes. Additionally, microbial natural dyes have been explored in the field of medicine for their potential antimicrobial and antioxidant activities, thus offering additional benefits beyond coloration [194–196].

The production of microbial natural dyes holds promise for a sustainable and eco-friendly future. Unlike some traditional natural dyes derived from plants, which often require large amounts of land, water, and energy for cultivation and extraction, microorganisms can be cultivated under controlled conditions, minimizing resource consumption and environmental impact. Furthermore, microbial dye production can be easily scaled up. Overall, the classification of natural dyes based on sources provides a useful framework for understanding the origins and properties of these dyes. Whether derived from plants, insects, minerals, or organisms, each dye source offers distinctive characteristics that contribute to the vibrant and diverse world of natural dyes [197–199].

Lichens are unique organisms that consist of a fungus and green or blue-green algae. Lichens have adapted to diverse and sometimes extreme ecological conditions. Lichens have been found to contain valuable compounds that have commercial applications, such as antimicrobial agents, dyeing agents, and ingredients used in spices and perfumes. These compounds, known as lichen acids, are especially important in the production of dyes that can color natural fibers. Lichens have a long history of being used as a source of dyes, with purple orchils being particularly significant during ancient Greek and Roman times. There are various methods for extracting dyes from lichens, including the cow urine method, ammonia fermentation method, boiling water method, and dimethyl sulphoxide extraction method. In addition to coloring fibers, lichen dyes also impart a characteristic odor. The growth rate of lichens is very slow, making them unsuitable for commercial exploitation. But recent research has found that the fungal part of lichens can be cultured to produce secondary metabolites that can create colored compounds [199].

Recently, algae have been considered as a new source of natural dye for textiles. Blue-green algae, also known as cyanobacteria, are considered an environmental nuisance due to their rapid growth rate which can potentially harm aquatic ecosystems. However, recycling the blue pigment produced by these algae into textile dyes presents a unique opportunity for sustainable production of blue colors. The phycobiliproteins, such as phycocyanin, found in some species of cyanobacteria have shown potential as natural food colorants with promising applications in various industries. Baek et al. applied the phycocyanin obtained from blue-green algae for dyeing of cotton and silk fibers and obtained the best results on silk using metal mordants [197, 200].

4.3 Some Common Natural Dyes

Several natural dyes have been investigated in the literature and the dyeing of various textile fibers with different natural dyes has been studied and optimized. Natural dyes are mostly applied for dyeing of wool and silk and the exhaustion and fastness properties of all natural dyes are not satisfactory. Here a number of the most commonly used natural dyes are presented.

4.3.1 Madder (Rubia tinctorum)

Madder is a plant that has been used for centuries as a natural dye for coloring fabrics, particularly red clothing. It belongs to the Rubiaceae family of plants and is native to India but also grows in other parts of Asia and Europe. The root of the plant contains alizarin, purpurin and munjistin which are responsible for its red color. In ancient times, madder was highly valued due to its ability to produce bright and long-lasting colors on fabric. Madder roots are harvested from the ground during autumn or early winter when they have developed their highest concentration of alizarin content. They are then cleaned, dried, and powdered and stored for later use. Madder has been used in various ways throughout history including dyeing leather, paper making, and even medicinal purposes. However, with the advent of synthetic dyes, the use of natural dyes like madder declined significantly. Today, it is still used by some artisans who prefer the unique color quality and sustainable nature of natural dyes over chemical alternatives. Madder continues to be an important part of traditional crafts such as carpet-making in Iran and Turkey where its vibrant red hues are highly valued [201–203].

4.3.2 Weld (Reseda luteola)

Reseda luteola is a plant species of the family Resedaceae that has been used for centuries as a source of natural dye for coloring textiles. The flowers and leaves of weld contains luteolin, a flavonoid (flavone) which is responsible for its yellow color. This makes reseda luteola an excellent choice for creating bright golden yellows, buttercup yellows, and mustard yellows in various shades depending on the mordant used during the dye process. Reseda luteola is also known for its medicinal properties such as being an anti-inflammatory, antispasmodic, and sedative which makes it useful in traditional medicine. In textile dyeing, reseda luteola has been used since ancient times to color fabrics like wool, silk, cotton, linen, and even leather. The process involves harvesting the leaves during peak flowering season when they contain the highest levels of pigments. They can be collected manually or mechanically using a machine then dried before use. To extract the pigment, the leaves are boiled with water for several hours until the desired shade is achieved. The liquid is strained out and applied directly onto the fabric or mixed with other mordants such as alum or tin to fix the color [177, 202].

4.3.3 Cochineal

Cochineal (C.I. Natural Red 4) is a natural red dye that has been used for centuries to color fabrics, food, cosmetics, and medicines. It comes from the female cochineal insect (Dactylopius coccus) which lives on cacti in Mexico and South America. This tiny scale insect feeds on the pulp of prickly pear cactus plants and produces carminic acid, which gives it its distinctive red color. Cochineal is known for its bright hue, lightfastness, and durability. It has excellent fastness to light, meaning that it does not easily fade or lose its color even after repeated washing or exposure to sunlight. Its vibrant color makes it ideal for use in fashion designers' collections, especially for creating bold statement pieces like dresses, jackets, and accessories. In addition, cochineal is also a natural dye, which means it is considered eco-friendly and safe for people with allergies or sensitivities to synthetic dyes [204–206].

The process of extracting the dye involves collecting the insects from cacti plants, crushing them into a paste, and then soaking the mixture in water to release the carminic acid. The liquid can be used as is or further processed through boiling and filtering to create different shades of red. Cochineal has been used historically in traditional textile production methods such as handloom weaving but nowadays, it can also be incorporated into modern manufacturing processes like printing and dyeing. It blends well with other natural dyes for creating unique color combinations that are both vibrant and sustainable [204, 206, 207].

4.3.4 Indigo

Indigo is a natural dye extracted from the leaves of the plant Indigofera tinctoria, which grows primarily in tropical regions around the world including India, Africa, and Asia. The process of extracting indigo involves soaking the leaves in water, then fermenting them to create a vat or solution that can be used to dye fabric. This process results in a rich blue hue that was highly valued by ancient cultures throughout history due to its rarity and beauty. Indigo has been used for thousands of years in textile production. Indigo has also been used in modern times as a natural alternative to synthetic dyes due to its eco-friendly properties. It can be found in clothing made from organic cotton or hemp, as well as in home decor items such as bedding and throws. The color produced by indigo is unique and rich, adding depth to any material it touches, making it highly sought after for both fashion and interior design purposes. Additionally, the process of extracting indigo involves minimal environmental impact compared to synthetic dye production methods, making it an attractive option for those seeking sustainable alternatives.

Indigo dyeing is a traditional process used for coloring fabrics with blue hues. It involves three main stages—reduction, exhaustion, and oxidation. In the first stage called reduction, the indigo molecules are converted into a leuco form that can dissolve in water and diffuse the fibers. This step takes place at temperatures between 50 and 60 °C. The second stage of the process is called exhaustion where the dye penetrates the fibers. Finally, during the third stage of oxidation, the indigo molecules which diffused the fibers react with oxygen in the air (or an oxidizing agent) and become permanent blue pigment on the fiber. These steps can be repeated multiple times for darker shades of blue. Indigo dyeing has been used since ancient times and continues to be popular due to its unique color and durability [208, 209].

4.4 Application of Natural Dyes on Natural and Synthetic Textile Fibers

Considering the exhaustion and fastness properties, natural dyes exhibit the best results on natural protein fibers such as wool and silk. However, they can be applied on cotton as well as synthetic fibers such as nylon, polyester, and acrylic. As the substantivity of natural dyes toward the fibers is lower compared to the synthetic dyes, the use of auxiliaries such as metal mordants are usually necessary before the dyeing. Alum (aluminum potassium sulfate), copper sulfate, sodium dichromate, stannous chloride, and ferrous sulfate are among the common salts used for mordanting of textiles in natural dyeing process. Mordanting is normally done at boil before, after, or simultaneous with the dyeing process. Since the metals such as chromium and stannous are considered as toxic, mordanting with metal salts is not recommended for the eco-friendly natural dyeing. In this regard, several attempts have been made to replace the metal mordants with sustainable auxiliaries or bio mordants such as chitosan, tannin (extracted from plant materials), etc. to maintain the sustainability of natural dyeing while obtaining high exhaustion and acceptable fastness properties [189, 207, 210, 211].

The dyeing of wool, silk, cotton, nylon, and acrylic can be done at boil or below it, while polyester dyeing may need temperatures above boil (up to 130 °C). It should be noted that some natural dyes (such as anthocyanins) are unstable at elevated temperatures and the suitable temperature should be determined for each dye-fiber system individually [201, 212, 213].

4.5 Future Trends

As the demand for sustainable textiles grows, future trends in the industry focus on developing advanced biodegradable fibers that offer improved performance, and durability, and align with circular economy principles. This involves exploring new sources of natural fibers derived from bio-based and renewable raw materials, such as agricultural waste, algae, fungi, and bio-based synthetic fibers. Additionally, the incorporation of recycled materials like polyester, nylon, and cotton into textile production continues to gain momentum, driven by advancements in recycling technologies, enhanced quality of recycled fibers, and increased availability of recycled raw materials. Moreover, the integration of nanotechnology and smart materials into sustainable raw materials opens possibilities for the development of smart and functional fibers with properties like self-cleaning, moisture-wicking, and temperature regulation. Simultaneously, innovations in utilizing waste streams such as coffee grounds, food waste, and discarded textiles offer opportunities to convert these materials into high-quality textile fibers through efficient and scalable processes.

The development of sustainable synthetic fibers will continue to progress, focusing on bio-based alternatives to conventional synthetic fibers. The exploration of new polymers, improved manufacturing processes, and the incorporation of renewable feedstocks in synthetic fiber production. Circular design principles will become more integrated into the textile industry, emphasizing product life extension, repairability, and recyclability. Designing textiles for disassembly, implementing take-back programs, and developing innovative business models that prioritize the circularity of textile products. Increasing consumer awareness and demand for sustainable textiles will shape future trends in raw material selection. Consumers are becoming more conscious of the environmental and social impact of their purchases, driving the demand for sustainable and ethically produced textiles. Educating consumers, promoting sustainable fashion initiatives, and creating market incentives for sustainable raw materials. Collaboration among various stakeholders, including manufacturers, researchers, brands, and policymakers, will continue to drive innovation and accelerate the adoption of sustainable raw materials. It involves fostering industry partnerships, research collaborations, and knowledge sharing to advance sustainable practices and address common challenges.

By embracing these future trends, the textile industry can move towards a more sustainable and circular approach to raw material selection, contributing to a more environmentally and socially responsible supply chain.

References

1. Azam F, Ahmad S (2020) Fibers for agro textiles. pp 151–168
2. Lo CKY, Yeung ACL, Cheng TCE (2012) The impact of environmental management systems on financial performance in fashion and textiles industries. Int J Prod Econ 135:561–567. https://doi.org/10.1016/j.ijpe.2011.05.010
3. Fresner J (1998) Starting continuous improvement with a cleaner production assessment in an Austrian textile mill. J Clean Prod 6:85–91. https://doi.org/10.1016/S0959-6526(97)00049-8
4. Anandjiwala RD (2007) Textiles for sustainable development. Nova Publishers
5. Muthu SS (2017) Sustainable fibres and textiles. Woodhead Publishing
6. Blackburn R (2005) Biodegradable and sustainable fibres. Taylor & Francis US
7. Gedik G, Avinc O (2020) Hemp fiber as a sustainable raw material source for textile industry: can we use its potential for more eco-friendly production? pp 87–109
8. Kaplan DL (1998) Introduction to biopolymers from renewable resources. In: Kaplan DL (ed) Biopolymers from renewable resources. Springer Berlin Heidelberg, Berlin, Heidelberg, pp 1–29
9. Vink ETH, Rábago KR, Glassner DA, Gruber PR (2003) Applications of life cycle assessment to NatureWorks^{TM} polylactide (PLA) production. Polym Degrad Stab 80:403–419. https://doi.org/10.1016/S0141-3910(02)00372-5
10. Debnath S (2016) Unexplored vegetable fibre in green fashion. In: Muthu SS, Gardetti MA (eds) Green fashion: volume 2. Springer Singapore, Singapore, pp 1–19
11. Lörcks J (1998) Properties and applications of compostable starch-based plastic material. Polym Degrad Stab 59:245–249. https://doi.org/10.1016/S0141-3910(97)00168-7
12. Heller K, Byczynska M (2004) Ontogenesis of fibre flax (Linum usitatissimum L.). In: CD proceedings of the conference “bast fibrous plants for healthy life”. pp 24–28

13. Jones D, Brischke C (2017) 3—Nonwood bio-based materials. In: Performance of bio-based building materials. Woodhead Publishing, pp 97–186
14. Zeng X, Mooney SJ, Sturrock CJ (2015) Assessing the effect of fibre extraction processes on the strength of flax fibre reinforcement. Compos Part A Appl Sci Manuf 70:1–7. https://doi.org/10.1016/j.compositesa.2014.12.004
15. Charlet K, Jernot JP, Eve S et al (2010) Multi-scale morphological characterisation of flax: from the stem to the fibrils. Carbohydr Polym 82:54–61. https://doi.org/10.1016/j.carbpol.2010.04.022
16. Shekhar Sharma HS, van Sumere CF (1990) The biology and processing of flax. M Publications, Belfast
17. Ranalli P, Venturi G (2004) Hemp as a raw material for industrial applications. Euphytica 140:1–6. https://doi.org/10.1007/s10681-004-4749-8
18. Chen Y, Sun L, Negulescu I et al (2007) Comparative study of hemp fiber for nonwoven composites. J Ind Hemp 12:27–45. https://doi.org/10.1300/J237v12n01_04
19. Pietak A, Korte S, Tan E et al (2007) Atomic force microscopy characterization of the surface wettability of natural fibres. Appl Surf Sci 253:3627–3635. https://doi.org/10.1016/j.apsusc.2006.07.082
20. Netinger Grubeša I, Marković B, Gojević A, Brdarić J (2018) Effect of hemp fibers on fire resistance of concrete. Constr Build Mater 184:473–484. https://doi.org/10.1016/j.conbuildmat.2018.07.014
21. Santoni A, Bonfiglio P, Fausti P et al (2019) Improving the sound absorption performance of sustainable thermal insulation materials: natural hemp fibres. Appl Acoust 150:279–289. https://doi.org/10.1016/j.apacoust.2019.02.022
22. Teixeira FP, Gomes O da FM, Silva F de A (2019) Degradation mechanisms of curaua, hemp, and sisal fibers exposed to elevated temperatures. BioResources 14:1494–1511. https://doi.org/10.15376/biores.14.1.1494-1511
23. Mohanty AK, Misra M, Drzal LT (2002) Sustainable bio-composites from renewable resources: opportunities and challenges in the green materials world. J Polym Environ 10:19–26. https://doi.org/10.1023/A:1021013921916
24. Roy S, Lutfar LB (2012) 3—Bast fibres: jute. In: Kozłowski RM (ed) Handbook of natural fibres. Woodhead Publishing series in textiles. Woodhead Publishing, pp 24–46
25. Militký J, Jabbar A (2015) Comparative evaluation of fiber treatments on the creep behavior of jute/green epoxy composites. Compos Part B Eng 80:361–368. https://doi.org/10.1016/j.compositesb.2015.06.014
26. Dempsey JM (1963) Long vegetable fiber development in South Vietnam and other Asian countries, 1957–1962
27. Feng T, Qin J, Shao Y et al (2019) Size-controlled transparent jute fiber for replacing transparent wood in industry production area. Coatings 9
28. Aziz A, Shah SS, Kashem A (2020) Preparation and utilization of jute-derived carbon: a short review. Chem Rec 20:1074–1098. https://doi.org/10.1002/tcr.202000071
29. Aranci K, Uzun M, Su S et al (2020) 3D propolis-sodium alginate scaffolds: influence on structural parameters, release mechanisms, cell cytotoxicity and antibacterial activity. Molecules 25
30. Hofstätter T, Pedersen DB, Tosello G, Hansen HN (2017) State-of-the-art of fiber-reinforced polymers in additive manufacturing technologies. J Reinf Plast Compos 36:1061–1073. https://doi.org/10.1177/0731684417695648
31. Filho WL, Salvia AL, Bonoli A et al (2021) An assessment of attitudes towards plastics and bioplastics in Europe. Sci Total Environ 755:142732. https://doi.org/10.1016/j.scitotenv.2020.142732
32. Debnath S (2017) 3—Sustainable production of bast fibres. In: Muthu SS (ed) Sustainable fibres and textiles. The Textile Institute book series. Woodhead Publishing, pp 69–85
33. Marinho NP, de Cademartori PHG, Nisgoski S et al (2020) Feasibility of ramie fibers as raw material for the isolation of nanofibrillated cellulose. Carbohydr Polym 230:115579. https://doi.org/10.1016/j.carbpol.2019.115579

34. Kaldor AF, Karlgren C, Verwest H (1990) Kenaf—a fast growing fiber source for papermaking. Tappi J 73:205–208
35. Shi J, Shi SQ, Barnes HM et al (2011) Kenaf bast fibers—part I: hermetical alkali digestion. Int J Polym Sci 2011:212047. https://doi.org/10.1155/2011/212047
36. Abdul Khalil HPS, Yusra AFI, Bhat AH, Jawaid M (2010) Cell wall ultrastructure, anatomy, lignin distribution, and chemical composition of Malaysian cultivated kenaf fiber. Ind Crops Prod 31:113–121. https://doi.org/10.1016/j.indcrop.2009.09.008
37. Buzarovska A, Bogoeva-Gaceva G, Grozdanov A et al (2007) Crystallization behavior of poly(hydroxybytyrate-co-valerate) in model and bulk PHBV/kenaf fiber composites. J Mater Sci 42:6501–6509. https://doi.org/10.1007/s10853-007-1527-8
38. Amel BA, Paridah MT, Sudin R et al (2013) Effect of fiber extraction methods on some properties of kenaf bast fiber. Ind Crops Prod 46:117–123. https://doi.org/10.1016/j.indcrop.2012.12.015
39. Mariod AA, Saeed Mirghani ME, Hussein I (2017) Chapter 9—Hibiscus cannabinus L. Kenaf. In: Mariod AA, Saeed Mirghani ME, Hussein I (eds) Unconventional oilseeds and oil sources. Academic Press, pp 45–51
40. Anuar NIS, Zakaria S, Gan S et al (2019) Comparison of the morphological and mechanical properties of oil Palm EFB fibres and kenaf fibres in nonwoven reinforced composites. Ind Crops Prod 127:55–65. https://doi.org/10.1016/j.indcrop.2018.09.056
41. Webber III, Charles L, Bhardwaj HL, Bledsoe VK et al (2002) Kenaf production: fiber, feed, and seed. Trends New Crop New Uses 13:327–339
42. Kozłowski R, Mackiewicz-Talarczyk M (2012) Handbook of natural fibres. Elsevier
43. Liu K, Zhang X, Takagi H et al (2014) Effect of chemical treatments on transverse thermal conductivity of unidirectional abaca fiber/epoxy composite. Compos Part A Appl Sci Manuf 66:227–236. https://doi.org/10.1016/j.compositesa.2014.07.018
44. El-Meligy MG, Mohamed SH, Mahani RM (2010) Study mechanical, swelling and dielectric properties of prehydrolysed banana fiber—waste polyurethane foam composites. Carbohydr Polym 80:366–372. https://doi.org/10.1016/j.carbpol.2009.11.034
45. Ramesh M (2018) 9—Hemp, jute, banana, kenaf, ramie, sisal fibers. In: Bunsell AR (ed) Handbook of properties of textile and technical fibres. The Textile Institute book series, 2nd edn. Woodhead Publishing, pp 301–325
46. Liu H, Wu Q, Zhang Q (2009) Preparation and properties of banana fiber-reinforced composites based on high density polyethylene (HDPE)/nylon-6 blends. Bioresour Technol 100:6088–6097. https://doi.org/10.1016/j.biortech.2009.05.076
47. Saxena T, Chawla VK (2021) Banana leaf fiber-based green composite: an explicit review report. Mater Today Proc 46:6618–6624. https://doi.org/10.1016/j.matpr.2021.04.099
48. Brindha D, Vinodhini S, Alarmelumangai K, Malathy NS (2012) Physico-chemical properties of fibers from banana varieties after scouring. Indian J Fundam Appl Life Sci 2:217–221
49. Rao KMM, Rao KM (2007) Extraction and tensile properties of natural fibers: vakka, date and bamboo. Compos Struct 77:288–295. https://doi.org/10.1016/j.compstruct.2005.07.023
50. Saikia CN, Goswami T, Ali F (1997) Evaluation of pulp and paper making characteristics of certain fast growing plants. Wood Sci Technol 31:467–475. https://doi.org/10.1007/BF00702569
51. Deepa C, Ramesh M (2022) Banana fibers, their composites and applications. In: Plant fibers, their composites, and applications. Elsevier, pp 161–180
52. Scott GJ (2021) A review of root, tuber and banana crops in developing countries: past, present and future. Int J Food Sci Technol 56:1093–1114. https://doi.org/10.1111/ijfs.14778
53. Akatwijuka O, Gepreel MA-H, Abdel-Mawgood A et al (2022) Overview of banana cellulosic fibers: agro-biomass potential, fiber extraction, properties, and sustainable applications. Biomass Convers Biorefinery. https://doi.org/10.1007/s13399-022-02819-0
54. Guerrero AB, Aguado PL, Sánchez J, Curt MD (2016) GIS-based assessment of banana residual biomass potential for ethanol production and power generation: a case study. Waste Biomass Valorization 7:405–415. https://doi.org/10.1007/s12649-015-9455-3

55. Fernandes ERK, Marangoni C, Souza O, Sellin N (2013) Thermochemical characterization of banana leaves as a potential energy source. Energy Convers Manag 75:603–608. https://doi.org/10.1016/j.enconman.2013.08.008
56. Meya AI, Ndakidemi PA, Mtei KM et al (2020) Optimizing soil fertility management strategies to enhance banana production in volcanic soils of the northern highlands, Tanzania. Agronomy 10
57. Manimaran P, Pillai GP, Vignesh V, Prithiviraj M (2020) Characterization of natural cellulosic fibers from Nendran Banana Peduncle plants. Int J Biol Macromol 162:1807–1815. https://doi.org/10.1016/j.ijbiomac.2020.08.111
58. Fasake V, Dashora K (2022) Characterization of raw and anaerobic digested cattle dung fibers: a sustainable source of non-wood material. Biomass Convers Biorefinery. https://doi.org/10.1007/s13399-022-02487-0
59. Paramasivam SK, Panneerselvam D, Sundaram D et al (2022) Extraction, characterization and enzymatic degumming of banana fiber. J Nat Fibers 19:1333–1342. https://doi.org/10.1080/15440478.2020.1764456
60. Xu S, Xiong C, Tan W, Zhang Y (2015) BPSFs mechanical extraction. BioResources 10:3724–3735
61. Kumar KP, Sekaran ASJ (2014) Some natural fibers used in polymer composites and their extraction processes: a review. J Reinf Plast Compos 33:1879–1892. https://doi.org/10.1177/0731684414548612
62. Kohli P, Gupta R (2019) Application of calcium alginate immobilized and crude pectin lyase from Bacillus cereus in degumming of plant fibres. Biocatal Biotransformation 37:341–348. https://doi.org/10.1080/10242422.2018.1564745
63. Nagarajan KJ, Ramanujam NR, Sanjay MR et al (2021) A comprehensive review on cellulose nanocrystals and cellulose nanofibers: pretreatment, preparation, and characterization. Polym Compos 42:1588–1630. https://doi.org/10.1002/pc.25929
64. Chen S, Zhang X, Singh D et al (2010) Biological pretreatment of lignocellulosics: potential, progress and challenges. Biofuels 1:177–199. https://doi.org/10.4155/bfs.09.13
65. Jacob N, Niladevi KN, Anisha GS, Prema P (2008) Hydrolysis of pectin: an enzymatic approach and its application in banana fiber processing. Microbiol Res 163:538–544. https://doi.org/10.1016/j.micres.2006.07.016
66. Manimaran A, Vatsala TM (2007) Biobleaching of banana fibre pulp using Bacillus subtilis C 01 xylanase produced from wheat bran under solid-state cultivation. J Ind Microbiol Biotechnol 34:745–749. https://doi.org/10.1007/s10295-007-0248-y
67. Deepa C, Ramesh M (2022) Chapter 7—Banana fibers, their composites and applications. In: Mavinkere Rangappa S, Parameswaranpillai J, Siengchin S et al (eds) The Textile Institute book series. Woodhead Publishing, pp 161–180
68. Huang J-K, Young W-B (2019) The mechanical, hygral, and interfacial strength of continuous bamboo fiber reinforced epoxy composites. Compos Part B Eng 166:272–283. https://doi.org/10.1016/j.compositesb.2018.12.013
69. Yu Y, Wang H, Lu F et al (2014) Bamboo fibers for composite applications: a mechanical and morphological investigation. J Mater Sci 49:2559–2566. https://doi.org/10.1007/s10853-013-7951-z
70. Amada S, Untao S (2001) Fracture properties of bamboo. Compos Part B Eng 32:451–459. https://doi.org/10.1016/S1359-8368(01)00022-1
71. Yang Z, Peng H, Wang W, Liu T (2010) Crystallization behavior of poly(ε-caprolactone)/layered double hydroxide nanocomposites. J Appl Polym Sci 116:2658–2667. https://doi.org/10.1002/app.31787
72. Wang G, Chen F (2017) Development of bamboo fiber-based composites. In: Advanced high strength natural fibre composites in construction. Elsevier, pp 235–255
73. Okubo K, Fujii T (2013) Improvement of interfacial adhesion in bamboo polymer composite enhanced with microfibrillated cellulose. In: Polymer composites. pp 317–329
74. Suhaily SS, Khalil HPSA, Nadirah WOW, Jawaid M (2013) Bamboo based biocomposites material, design and applications. In: Mastai Y (ed), chap 19. IntechOpen, Rijeka

75. Bystriakova N, Kapos V, Lysenko I, Stapleton CMA (2003) Distribution and conservation status of forest bamboo biodiversity in the Asia-Pacific Region. Biodivers Conserv 12:1833–1841. https://doi.org/10.1023/A:1024139813651
76. Deshpande AP, Bhaskar Rao M, Lakshmana Rao C (2000) Extraction of bamboo fibers and their use as reinforcement in polymeric composites. J Appl Polym Sci 76:83–92. https://doi.org/10.1002/(SICI)1097-4628(20000404)76:1<83::AID-APP11>3.0.CO;2-L
77. Zakikhani P, Zahari R, Sultan MTH, Majid DL (2014) Extraction and preparation of bamboo fibre-reinforced composites. Mater Des 63:820–828. https://doi.org/10.1016/j.matdes.2014.06.058
78. Imadi SR, Mahmood I, Kazi AG (2014) Bamboo fiber processing, properties, and applications. In: Hakeem KR, Jawaid M, Rashid U (eds) Biomass and bioenergy: processing and properties. Springer International Publishing, Cham, pp 27–46
79. Yueping W, Ge W, Haitao C et al (2009) Structures of bamboo fiber for textiles. Text Res J 80:334–343. https://doi.org/10.1177/0040517509337633
80. Frei EVA, Preston R (1961) Variants in the structural polysaccharides of algal cell walls. Nature 192:939–943
81. Haug A, Smidsrød O (1967) Strontium–calcium selectivity of alginates. Nature 215:757. https://doi.org/10.1038/215757a0
82. Evans LV., Holligan MS (1972) Correlated light and electron microscope studies on brown algae I. Localization of alginic acid and sulphated polysaccharides in Dictyota. New Phytol 71:1161–1172. https://doi.org/10.1111/j.1469-8137.1972.tb01994.x
83. Xu Y-J, Qu L-Y, Liu Y, Zhu P (2021) An overview of alginates as flame-retardant materials: pyrolysis behaviors, flame retardancy, and applications. Carbohydr Polym 260:117827. https://doi.org/10.1016/j.carbpol.2021.117827
84. Zhang X, Wang L, Weng L, Deng B (2020) Strontium ion substituted alginate-based hydrogel fibers and its coordination binding model. J Appl Polym Sci 137:48571. https://doi.org/10.1002/app.48571
85. Benslima A, Sellimi S, Hamdi M et al (2021) The brown seaweed Cystoseira schiffneri as a source of sodium alginate: chemical and structural characterization, and antioxidant activities. Food Biosci 40:100873. https://doi.org/10.1016/j.fbio.2020.100873
86. Bierhalz ACK, da Silva MA, Braga MEM et al (2014) Effect of calcium and/or barium crosslinking on the physical and antimicrobial properties of natamycin-loaded alginate films. LWT - Food Sci Technol 57:494–501. https://doi.org/10.1016/j.lwt.2014.02.021
87. Zhang X, Wang X, Fan W et al (2022) Fabrication, property and application of calcium alginate fiber: a review. Polymers (Basel) 14:3227. https://doi.org/10.3390/polym14153227
88. Yeung RA, Kennedy RA (2019) A comparison of selected physico-chemical properties of calcium alginate fibers produced using two different types of sodium alginate. J Mech Behav Biomed Mater 90:155–164. https://doi.org/10.1016/j.jmbbm.2018.10.011
89. Morton WE, Hearle JWS (2008) 5—Fibre density. In: Morton WE, Hearle JWS (eds) Physical properties of textile fibres. Woodhead Publishing series in textiles, 4th edn. Woodhead Publishing, pp 163–167
90. Morton WE, Hearle JWS (2008) 1—An introduction to fibre structure. In: Morton WE, Hearle JWS (eds) Physical properties of textile fibres. Woodhead Publishing series in textiles, 4th edn. Woodhead Publishing, pp 1–81
91. Morton WE, Hearle JWS (2008) 4—Fibre length. In: Morton WE, Hearle JWS(eds) Physical properties of textile fibres. Woodhead Publishing series in textiles, 4th edn. Woodhead Publishing, pp 134–162
92. Deng X, Gould M, Ali MA (2022) A review of current advancements for wound healing: biomaterial applications and medical devices. J Biomed Mater Res Part B Appl Biomater 110:2542–2573. https://doi.org/10.1002/jbm.b.35086
93. Tomić SL, Babić Radić MM, Vuković JS et al (2023) Alginate-based hydrogels and scaffolds for biomedical applications. Mar Drugs 21:177. https://doi.org/10.3390/md21030177
94. Farshidfar N, Iravani S, Varma RS (2023) Alginate-based biomaterials in tissue engineering and regenerative medicine. Mar Drugs 21:189. https://doi.org/10.3390/md21030189

95. Ahmad F, Mushtaq B, Ahmad S et al (2023) A novel composite of hemp fiber and alginate hydrogel for wound dressings. J Polym Environ 31:2294–2305. https://doi.org/10.1007/s10924-023-02756-7
96. Ahmad F, Mushtaq B, Butt FA et al (2021) Preparation and characterization of wool fiber reinforced nonwoven alginate hydrogel for wound dressing. Cellulose 28:7941–7951. https://doi.org/10.1007/s10570-021-04043-x
97. Azam F, Ahmad F, Ulker Z et al (2022) The role and applications of aerogels in textiles. Adv Mater Sci Eng 2022:1–22. https://doi.org/10.1155/2022/2407769
98. Azam F, Ahmad F, Ahmad S et al (2023) Synthesis and characterization of natural fibers reinforced alginate hydrogel fibers loaded with diclofenac sodium for wound dressings. Int J Biol Macromol 241:124623. https://doi.org/10.1016/j.ijbiomac.2023.124623
99. Azam F, Ahmad F, Ahmad S et al (2022) Preparation and characterization of alginate hydrogel fibers reinforced by cotton for biomedical applications. Polymers (Basel) 14:4707. https://doi.org/10.3390/polym14214707
100. Zhang X, Weng L, Liu Q et al (2019) Facile fabrication and characterization on alginate microfibres with grooved structure via microfluidic spinning. R Soc Open Sci 6:181928. https://doi.org/10.1098/rsos.181928
101. Khayyam H, Jazar RN, Nunna S et al (2020) PAN precursor fabrication, applications and thermal stabilization process in carbon fiber production: experimental and mathematical modelling. Prog Mater Sci 107:100575. https://doi.org/10.1016/j.pmatsci.2019.100575
102. Hopper GP, Deakin AH, Crane EO, Clarke JV (2012) Enhancing patient recovery following lower limb arthroplasty with a modern wound dressing: a prospective, comparative audit. J Wound Care 21:200–203. https://doi.org/10.12968/jowc.2012.21.4.200
103. Moreira Filho RNF, Vasconcelos NF, Andrade FK et al (2020) Papain immobilized on alginate membrane for wound dressing application. Colloids Surf B Biointerfaces 194:111222. https://doi.org/10.1016/j.colsurfb.2020.111222
104. Suzuki Y, Tanihara M, Nishimura Y et al (1999) In vivo evaluation of a novel alginate dressing. J Biomed Mater Res 48:522–527. https://doi.org/10.1002/(SICI)1097-4636(1999)48:4<522::AID-JBM18>3.0.CO;2-O
105. Ashton RS, Banerjee A, Punyani S et al (2007) Scaffolds based on degradable alginate hydrogels and poly(lactide-co-glycolide) microspheres for stem cell culture. Biomaterials 28:5518–5525. https://doi.org/10.1016/j.biomaterials.2007.08.038
106. Cao H, Liu T, Chew SY (2009) The application of nanofibrous scaffolds in neural tissue engineering. Adv Drug Deliv Rev 61:1055–1064. https://doi.org/10.1016/j.addr.2009.07.009
107. Chae T, Yang H, Leung V et al (2013) Novel biomimetic hydroxyapatite/alginate nanocomposite fibrous scaffolds for bone tissue regeneration. J Mater Sci Mater Med 24:1885–1894. https://doi.org/10.1007/s10856-013-4957-7
108. Roopavath UK, Soni R, Mahanta U et al (2019) 3D printable SiO_2 nanoparticle ink for patient specific bone regeneration. RSC Adv 9:23832–23842. https://doi.org/10.1039/C9RA03641E
109. Kim D, Ahn S, Yoon J (2019) Highly stretchable strain sensors comprising double network hydrogels fabricated by microfluidic devices. Adv Mater Technol 4:1800739. https://doi.org/10.1002/admt.201800739
110. Zou X, Shang S, Liu J et al (2021) Facile fabrication of temperature triggered thermochromic core-sheath alginate microfibers from microfluidic spinning. Fibers Polym 22:1535–1542. https://doi.org/10.1007/s12221-021-0778-3
111. Cui L, Hu J, Wang W et al (2020) Smart pH response flexible sensor based on calcium alginate fibers incorporated with natural dye for wound healing monitoring. Cellulose 27:6367–6381. https://doi.org/10.1007/s10570-020-03219-1
112. Liu Y, Zhao X-R, Peng Y-L et al (2016) Effect of reactive time on flame retardancy and thermal degradation behavior of bio-based zinc alginate film. Polym Degrad Stab 127:20–31. https://doi.org/10.1016/j.polymdegradstab.2015.12.024
113. Zhang C-J, Liu Y, Cui L et al (2016) Bio-based calcium alginate nonwoven fabrics: flame retardant and thermal degradation properties. J Anal Appl Pyrolysis 122:13–23. https://doi.org/10.1016/j.jaap.2016.10.030

114. Kraessig H (1987) In: Nevell TP, Zeronian SH (eds) Cellulose chemistry and its applications. Halsted Press, John Wiley, New York, 1985, 552 pp. J Polym Sci Part C Polym Lett 25:87–88. https://doi.org/10.1002/pol.1987.140250212
115. Kumar Gupta P, Sai Raghunath S, Venkatesh Prasanna D et al (2019) An update on overview of cellulose, its structure and applications. In: Cellulose. IntechOpen
116. Zhang H, Yang G, Edgar KJ et al (2021) Preparation and properties of dual-wavelength excitable fluorescent Lyocell fibers and their applications in papermaking. Carbohydr Polym 261:117861. https://doi.org/10.1016/j.carbpol.2021.117861
117. Chen J (2015) Synthetic textile fibers. In: Textiles and fashion. Elsevier, pp 79–95
118. Jiang X, Bai Y, Chen X, Liu W (2020) A review on raw materials, commercial production and properties of lyocell fiber. J Bioresour Bioprod 5:16–25. https://doi.org/10.1016/j.jobab.2020.03.002
119. Mortimer SA, Peguy A, Ball RC (1996) Influence of the physical process parameters on the structure formation of lyocell fibres. Cell Chem Technol 251–266
120. Mortimer SA, Péguy A (1995) Spinning of fibres through the N-methylmorpholine-N-oxide process. In: Cellulose and cellulose derivatives. Elsevier, pp 561–567
121. Mortimer SA, Péguy AA (1996) Methods for reducing the tendency of Lyocell fibers to fibrillate. J Appl Polym Sci 60:305–316. https://doi.org/10.1002/(SICI)1097-4628(19960418)60:3%3c305::AID-APP3%3e3.0.CO;2-P
122. Tsuji H, Ikada Y (1998) Blends of aliphatic polyesters. II. Hydrolysis of solution-cast blends from poly(L-lactide) and poly(E-caprolactone) in phosphate-buffered solution. J Appl Polym Sci 67:405–415. https://doi.org/10.1002/(SICI)1097-4628(19980118)67:3%3c405::AID-APP3%3e3.0.CO;2-Q
123. Vert M, Schwarch G, Coudane J (1995) Present and future of PLA polymers. J Macromol Sci Part A 32:787–796. https://doi.org/10.1080/10601329508010289
124. Vayshbeyn LI, Mastalygina EE, Olkhov AA, Podzorova MV (2023) Poly(lactic acid)-based blends: a comprehensive review. Appl Sci 13:5148. https://doi.org/10.3390/app13085148
125. Castañeda-Rodríguez S, González-Torres M, Ribas-Aparicio RM et al (2023) Recent advances in modified poly (lactic acid) as tissue engineering materials. J Biol Eng 17:21. https://doi.org/10.1186/s13036-023-00338-8
126. Ruz-Cruz MA, Herrera-Franco PJ, Flores-Johnson EA et al (2022) Thermal and mechanical properties of PLA-based multiscale cellulosic biocomposites. J Mater Res Technol 18:485–495. https://doi.org/10.1016/j.jmrt.2022.02.072
127. Singha S, Hedenqvist MS (2020) A review on barrier properties of poly(lactic acid)/clay nanocomposites. Polymers (Basel) 12:1095. https://doi.org/10.3390/polym12051095
128. Lunt J (1998) Large-scale production, properties and commercial applications of polylactic acid polymers. Polym Degrad Stab 59:145–152. https://doi.org/10.1016/S0141-3910(97)00148-1
129. Carothers WH, Dorough GL, van Natta FJ (1932) Studies of polymerization and ring formation. X. The reversible polymerization of six-membered cyclic esters. J Am Chem Soc 54:761–772. https://doi.org/10.1021/ja01341a046
130. Avinc O, Khoddami A (2009) Overview of poly(lactic acid) (PLA) fibre. Fibre Chem 41:391–401. https://doi.org/10.1007/s10692-010-9213-z
131. Maharana T, Mohanty B, Negi YS (2009) Melt–solid polycondensation of lactic acid and its biodegradability. Prog Polym Sci 34:99–124. https://doi.org/10.1016/j.progpolymsci.2008.10.001
132. Kricheldorf HR, Boettcher C, Tönnes K-U (1992) Polylactones: 23. Polymerization of racemic and mesod, l-lactide with various organotin catalysts—stereochemical aspects. Polymer (Guildf) 33:2817–2824. https://doi.org/10.1016/0032-3861(92)90459-A
133. Lim L-T, Auras R, Rubino M (2008) Processing technologies for poly(lactic acid). Prog Polym Sci 33:820–852. https://doi.org/10.1016/j.progpolymsci.2008.05.004
134. Agrawal AK, Bhalla R (2003) Advances in the production of poly(lactic acid) fibers. A review. J Macromol Sci Part C Polym Rev 43:479–503. https://doi.org/10.1081/MC-120025975

135. Tsuji H, Ikada Y, Hyon S-H et al (1994) Stereocomplex formation between enantiomeric poly(lactic acid). VIII. Complex fibers spun from mixed solution of poly(D-lactic acid) and poly(L-lactic acid). J Appl Polym Sci 51:337–344. https://doi.org/10.1002/app.1994.070510216
136. Horáček I, Kalíšek V (1994) Polylactide. III. Fiber preparation by spinning in precipitant vapor. J Appl Polym Sci 54:1767–1771. https://doi.org/10.1002/app.1994.070541119
137. Horáček I, Kalíšek V (1994) Polylactide. I. Continuous dry spinning–hot drawing preparation of fibers. J Appl Polym Sci 54:1751–1757. https://doi.org/10.1002/app.1994.070541117
138. Leenslag JW, Pennings AJ (1987) High-strength poly(l-lactide) fibres by a dry-spinning/hot-drawing process. Polymer (Guildf) 28:1695–1702. https://doi.org/10.1016/0032-3861(87)90012-7
139. Avérous L (2008) Polylactic acid: synthesis, properties and applications. In: Monomers, polymers and composites from renewable resources. Elsevier, pp 433–450
140. Sudesh K, Abe H, Doi Y (2000) Synthesis, structure and properties of polyhydroxyalkanoates: biological polyesters. Prog Polym Sci 25:1503–1555. https://doi.org/10.1016/S0079-6700(00)00035-6
141. Campos E, Branquinho J, Carreira AS et al (2013) Designing polymeric microparticles for biomedical and industrial applications. Eur Polym J 49:2005–2021. https://doi.org/10.1016/j.eurpolymj.2013.04.033
142. Keshavarz T, Roy I (2010) Polyhydroxyalkanoates: bioplastics with a green agenda. Curr Opin Microbiol 13:321–326. https://doi.org/10.1016/j.mib.2010.02.006
143. Tan G-Y, Chen C-L, Li L et al (2014) Start a research on biopolymer polyhydroxyalkanoate (PHA): a review. Polymers (Basel) 6:706–754. https://doi.org/10.3390/polym6030706
144. Wang Y, Yin J, Chen G-Q (2014) Polyhydroxyalkanoates, challenges and opportunities. Curr Opin Biotechnol 30:59–65. https://doi.org/10.1016/j.copbio.2014.06.001
145. Chen G-Q, Patel MK (2012) Plastics derived from biological sources: present and future: a technical and environmental review. Chem Rev 112:2082–2099. https://doi.org/10.1021/cr200162d
146. Sudesh K, Iwata T (2008) Sustainability of biobased and biodegradable plastics. CLEAN - Soil, Air, Water 36:433–442. https://doi.org/10.1002/clen.200700183
147. Philip S, Keshavarz T, Roy I (2007) Polyhydroxyalkanoates: biodegradable polymers with a range of applications. J Chem Technol Biotechnol 82:233–247. https://doi.org/10.1002/jctb.1667
148. Rujnić-Sokele M, Pilipović A (2017) Challenges and opportunities of biodegradable plastics: a mini review. Waste Manag Res J Sustain Circ Econ 35:132–140. https://doi.org/10.1177/0734242X16683272
149. Koller M, Salerno A, Dias M et al (2010) Biotechnological polymer synthesis. Food Technol Biotechnol 48:255–269
150. Chodák I (1998) High modulus polyethylene fibres: preparation, properties and modification by crosslinking. Prog Polym Sci 23:1409–1442. https://doi.org/10.1016/S0079-6700(98)00006-9
151. Chodák I, Blackburn RS (2009) Sustainable synthetic fibres: the case of poly(hydroxyalkanoates) (PHA) and other fibres. In: Sustainable textiles. Elsevier, pp 88–112
152. Schmack G, Jehnichen D, Vogel R, Tändler B (2000) Biodegradable fibers of poly(3-hydroxybutyrate) produced by high-speed melt spinning and spin drawing. J Polym Sci Part B Polym Phys 38:2841–2850. https://doi.org/10.1002/1099-0488(20001101)38:21%3C2841::AID-POLB130%3E3.0.CO;2-%23
153. Gordeyev SA, Nekrasov YP, Shilton SJ (2001) Processing of gel-spun poly(β-hydroxybutyrate) fibers. J Appl Polym Sci 81:2260–2264. https://doi.org/10.1002/app.1665
154. Charuchinda A, Molloy R, Siripitayananon J et al (2003) Factors influencing the small-scale melt spinning of poly(ε-caprolactone) monofilament fibres. Polym Int 52:1175–1181. https://doi.org/10.1002/pi.1234

155. Lee KY, Mooney DJ (2001) Hydrogels for tissue engineering. Chem Rev 101:1869–1880. https://doi.org/10.1021/cr000108x
156. Tsigkou O, Pomerantseva I, Spencer JA et al (2010) Engineered vascularized bone grafts. Proc Natl Acad Sci 107:3311–3316. https://doi.org/10.1073/pnas.0905445107
157. Huang Z-M, Zhang Y-Z, Kotaki M, Ramakrishna S (2003) A review on polymer nanofibers by electrospinning and their applications in nanocomposites. Compos Sci Technol 63:2223–2253. https://doi.org/10.1016/S0266-3538(03)00178-7
158. Nair LS, Laurencin CT (2007) Biodegradable polymers as biomaterials. Prog Polym Sci 32:762–798. https://doi.org/10.1016/j.progpolymsci.2007.05.017
159. Woodruff MA, Hutmacher DW (2010) The return of a forgotten polymer—polycaprolactone in the 21st century. Prog Polym Sci 35:1217–1256. https://doi.org/10.1016/j.progpolymsci.2010.04.002
160. Okada M (2002) Chemical syntheses of biodegradable polymers. Prog Polym Sci 27:87–133. https://doi.org/10.1016/S0079-6700(01)00039-9
161. Hayashi T (1994) Biodegradable polymers for biomedical uses. Prog Polym Sci 19:663–702. https://doi.org/10.1016/0079-6700(94)90030-2
162. Krishnanand K, Deopura BL, Gupta B (2013) Determination of intrinsic birefringence values of polycaprolactone filaments. Polym Int 62:49–53. https://doi.org/10.1002/pi.4304
163. Meng Q, Hu J (2008) Study on poly(ε-caprolactone)-based shape memory copolymer fiber prepared by bulk polymerization and melt spinning. Polym Adv Technol 19:131–136. https://doi.org/10.1002/pat.985
164. Hayashi T, Nakayama K, Mochizuki M, Masuda T (2002) Studies on biodegradable poly(hexano-6-lactone) fibers. Part 3. Enzymatic degradation in vitro (IUPAC Technical Report). Pure Appl Chem 74:869–880. https://doi.org/10.1351/pac200274050869
165. Kim GH (2008) Electrospun PCL nanofibers with anisotropic mechanical properties as a biomedical scaffold. Biomed Mater 3:025010. https://doi.org/10.1088/1748-6041/3/2/025010
166. Ghasemi-Mobarakeh L, Prabhakaran MP, Morshed M et al (2008) Electrospun poly(ε-caprolactone)/gelatin nanofibrous scaffolds for nerve tissue engineering. Biomaterials 29:4532–4539. https://doi.org/10.1016/j.biomaterials.2008.08.007
167. Mishra S (2000) A text book of fibre science and technology. New Age International (P) Ltd, India
168. Karthik T, Rathinamoorthy R (2017) Sustainable silk production. In: Sustainable fibres and textiles. Elsevier, pp 135–170
169. Rheinberg L (1991) The romance of silk: A review of sericulture and the silk industry. Text Prog 21:1–43. https://doi.org/10.1080/00405169108688854
170. Bini E, Knight DP, Kaplan DL (2004) Mapping domain structures in silks from insects and spiders related to protein assembly. J Mol Biol 335:27–40. https://doi.org/10.1016/j.jmb.2003.10.043
171. Cao L, Qiu X, Jiao Q et al (2021) Polysaccharides and proteins-based nanogenerator for energy harvesting and sensing: a review. Int J Biol Macromol 173:225–243. https://doi.org/10.1016/j.ijbiomac.2021.01.109
172. Jasmine S, Mandal BB (2014) Types and properties of non-mulberry silk biomaterials for tissue engineering applications. In: Silk biomaterials for tissue engineering and regenerative medicine. Elsevier, pp 275–298
173. Perez-Rigueiro J, Viney C, Llorca J, Elices M (2000) Mechanical properties of single-brin silkworm silk. J Appl Polym Sci 75:1270–1277. https://doi.org/10.1002/(SICI)1097-4628(20000307)75:10%3c1270::AID-APP8%3e3.0.CO;2-C
174. Franck RR (2001) Silk, mohair, cashmere and other luxury fibres. Elsevier
175. Kundu SC (2014) Silk biomaterials for tissue engineering and regenerative medicine. Elsevier
176. Kozlowski RM, Mackiewicz-Talarczyk M (2020) Handbook of natural fibres: volume 1: types, properties and factors affecting breeding and cultivation. Woodhead Publishing
177. Safapour S, Mazhar M, Abedinpour S (2023) Color shade extension of Reseda luteola L. natural colorant on wool textiles via binary combination of metal salts: colorimetric and fastness studies. Fibers Polym. https://doi.org/10.1007/s12221-023-00296-4

178. Haji A, Naebe M (2020) Cleaner dyeing of textiles using plasma treatment and natural dyes: a review. J Clean Prod 265:121866. https://doi.org/10.1016/j.jclepro.2020.121866
179. Shahmoradi Ghaheh F, Haji A, Daneshvar E (2023) Sustainable dyeing process for nylon 6 fabrics by Rhubarb flower using different bio-mordants. Sustainability 15:9232. https://doi.org/10.3390/su15129232
180. Elsahida K, Fauzi AM, Sailah I, Siregar IZ (2019) Sustainability of the use of natural dyes in the textile industry. In: IOP conference series: earth and environmental science. IOP Publishing, p 12065
181. Lin L, Jiang T, Xiao L et al (2022) Sustainable fashion: eco-friendly dyeing of wool fiber with novel mixtures of biodegradable natural dyes. Sci Rep 12:21040. https://doi.org/10.1038/s41598-022-25495-6
182. Fang J, Meng C, Zhang G (2022) Agricultural waste of Ipomoea batatas leaves as a source of natural dye for green coloration and bio-functional finishing for textile fabrics. Ind Crops Prod 177:114440. https://doi.org/10.1016/j.indcrop.2021.114440
183. Repon MR, Islam T, Islam T et al (2023) Cleaner pathway for developing bioactive textile materials using natural dyes: a review. Environ Sci Pollut Res 30:48793–48823. https://doi.org/10.1007/s11356-023-26131-0
184. Do KL, Su M, Zhao F (2022) From historical dye to bio-colourant: processing, identification in historical textiles and potential applications of anthraquinone-based morindone. Dye Pigment 205:110482. https://doi.org/10.1016/j.dyepig.2022.110482
185. Saxena S, Raja ASM (2014) Natural dyes: sources, chemistry, application and sustainability issues. In: Muthu SS (ed) Roadmap to sustainable textiles and clothing. Springer Singapore, pp 37–80
186. Melo MJ, Nabais P, Vieira M et al (2023) Between past and future: advanced studies of ancient colours to safeguard cultural heritage and new sustainable applications. Dye Pigment 208:110815. https://doi.org/10.1016/j.dyepig.2022.110815
187. Bechtold T, Manian A (2023) Technical aspects and requirements to produce natural colorants—processes and product standardization. In: Handbook of natural colorants. pp 573–589
188. Omerogullari Basyigit Z, Eyupoglu C, Eyupoglu S, Merdan N (2023) Investigation and feed-forward neural network-based estimation of dyeing properties of air plasma treated wool fabric dyed with natural dye obtained from Hibiscus sabdariffa. Color Technol 139:441–453. https://doi.org/10.1111/cote.12665
189. Shahmoradi Ghaheh F, Moghaddam MK, Tehrani M (2021) Comparison of the effect of metal mordants and bio-mordants on the colorimetric and antibacterial properties of natural dyes on cotton fabric. Color Technol 137:689–698. https://doi.org/10.1111/cote.12569
190. Mandal S, Venkatramani J (2023) A review of plant-based natural dyes in leather application with a special focus on color fastness characteristics. Environ Sci Pollut Res 30:48769–48777. https://doi.org/10.1007/s11356-023-26281-1
191. Dulo B, Phan K, Githaiga J et al (2021) Natural quinone dyes: a review on structure, extraction techniques, analysis and application potential. Waste Biomass Valorization. https://doi.org/10.1007/s12649-021-01443-9
192. Mansour R (2018) Natural dyes and pigments: extraction and applications. In: Handbook of renewable materials for coloration and finishing. Wiley, pp 75–102
193. Singh K, Kumar P, Singh NV (2020) Natural dyes: an emerging ecofriendly solution for textile industries. Pollut Res 39:S87–S94
194. Almoulki T, Akkaya E Algae-based dyes for the textile industry. In: 6th Eurasia waste management symposium
195. El-Khatib EM, Ali NF, El-Mohamedy RSR (2016) Enhancing dyeing of wool fibers with colorant pigment extracted from green algae. J Chem Pharm Res 8:614–619
196. Ferrándiz M, Moldovan S, Mira E et al (2016) Phycobiliproteins—new natural dyes from algae as a sustainable method. Vlakna a Text 23:56–61
197. Baek N, Zhang X, Lou J, Fan X (2022) Dyeing fabrics with a colorant extracted from blue-green algae. AATCC J Res 9:223–230. https://doi.org/10.1177/24723444221103673

198. Dufossé L (2023) Natural colorants from cyanobacteria and algae. In: Handbook of natural colorants. pp 417–437
199. Mir RA, Adeel S, Azeem M et al (2019) Green algae, Cladophora glomerata L.–based natural colorants: dyeing optimization and mordanting for textile processing. J Appl Phycol 31:2541–2546. https://doi.org/10.1007/s10811-018-1717-6
200. Moldovan S, Ferrandiz M, Bonet MA (2017) Natural cotton printing with red macroalgae biomass of Gracilaria gracilis and Gracilaria cornea. Ann Univ Oradea Fascicle Text Leatherwork 20:61–66
201. Haji A, Vadood M (2023) Prediction of color coordinates of cotton fabric dyed with binary mixtures of madder and weld natural dyes using artificial intelligence. Fibers Polym 24:1759–1769. https://doi.org/10.1007/s12221-023-00184-x
202. Willemen H, van den Meijdenberg GJP, van Beek TA, Derksen GCH (2019) Comparison of madder (Rubia tinctorum L.) and weld (Reseda luteola L.) total extracts and their individual dye compounds with regard to their dyeing behaviour, colour, and stability towards light. Color Technol 135:40–47. https://doi.org/10.1111/cote.12384
203. Blackburn RS (2017) Natural dyes in madder (Rubia spp.) and their extraction and analysis in historical textiles. Color Technol 133:449–462. https://doi.org/10.1111/cote.12308
204. Čorak I, Brlek I, Sutlović A, Tarbuk A (2022) Natural dyeing of modified cotton fabric with cochineal dye. Molecules 27:1100
205. Sajed T, Haji A, Mehrizi MK, Nasiri Boroumand M (2018) Modification of wool protein fiber with plasma and dendrimer: effects on dyeing with cochineal. Int J Biol Macromol 107:642–653. https://doi.org/10.1016/j.ijbiomac.2017.09.038
206. Karadag R (2022) Cotton dyeing with cochineal by just in time extraction, mordanting, dyeing, and fixing method in the textile industry. J Nat Fibers 1–11. https://doi.org/10.1080/15440478.2022.2108184
207. Haji A, Ashraf S, Nasiriboroumand M, Lievens C (2020) Environmentally friendly surface treatment of wool fiber with plasma and chitosan for improved coloration with cochineal and safflower natural dyes. Fibers Polym 21:743–750. https://doi.org/10.1007/s12221-020-9587-3
208. Chavan RB (2015) Indigo dye and reduction techniques. In: Paul R (ed) Denim. Woodhead Publishing, pp 37–67
209. Mocquard J, Le Lamer A-C, Fabre P-L et al (2022) Indigo dyeing from Isatis tinctoria L.: from medieval to modern use. Dye Pigment 207:110675. https://doi.org/10.1016/j.dyepig.2022.110675
210. Adeel S, Azeem M, Habib N et al (2023) Sustainable application of microwave assisted extracted tea based tannin natural dye for chemical and bio-mordanted wool fabric. J Nat Fibers 20:2136322. https://doi.org/10.1080/15440478.2022.2136322
211. Haji A (2017) Improved natural dyeing of cotton by plasma treatment and chitosan coating; optimization by response surface methodology. Cellul Chem Technol 51:975–982
212. Vadood M, Haji A (2022) A hybrid artificial intelligence model to predict the color coordinates of polyester fabric dyed with madder natural dye. Expert Syst Appl 193:116514. https://doi.org/10.1016/j.eswa.2022.116514
213. Elmaaty TA, Abouelenin S, Elsisi H, Okubayashi S (2022) Eco-friendly approach for dyeing synthetic fabrics with natural dyes using electron beam irradiation. Fibers Polym 23:759–767. https://doi.org/10.1007/s12221-022-3144-1

Sustainable Production Practices in Textiles

Shahood uz Zaman, Muhammad Umair, and Amjed Javid

Abstract Sustainability and circularity are the major issues the world is facing nowadays. Industrial revolutions impacted the natural ecosystem, and we are facing some unexpected weather conditions in various areas of the world. As the population of the world increased, industrial production also fastened to meet the requirements. This industry caused the drainage of natural resources with a speed that is exceptionally faster than expected. Additionally, this industrialization increased the natural temperature of the earth and caused a major source of carbon footprints. If it continues, natural resources will be finished soon which is a disaster for living hood in Earth. Sustainable production development in industries including textile is a need of the day if we want to avoid uncertain disasters in the coming future. It will not only reduce the resource drainage but also reduce greenhouse gas emissions. This chapter explains the sustainable practices that should be adopted in the textile industry to reduce the environmental impact. All major textile sectors including spinning, weaving, textile processing, and garment manufacturing are explained in this chapter with some possible sustainable practices that should be adopted in the industries to reduce the environmental impact caused by the textile industry.

1 Sustainable Practices in Spinning

Spinning is the initial stage of converting textile fibers into usable textile products. Textile spinning is the process of yarn manufacturing and prepared yarn is further used for manufacturing fabric and other textile products. Currently focus is being shifted towards sustainable industrial development and textile is one of the sectors that are considered harmful to the environmental ecosystem. Textiles are claimed to produce 5–10% of global greenhouse emissions. The use of water and other chemicals in textile processes is challenging for the achievement of sustainability goals. Water pollution, disposal of chemically contaminated water, and textile waste is causing harmful impact on environmental ecosystems [1, 2]. These issues are triggered by

S. Zaman (✉) · M. Umair · A. Javid
School of Engineering and Technology, National Textile University, Faisalabad 37610, Pakistan
e-mail: shahood.zaman@ntu.edu.pk

S. R. Batool et al. (eds.), *Circularity in Textiles*, Textile Science and Clothing Technology, https://doi.org/10.1007/978-3-031-49479-6_5

the increasing demand for textile products. With the increasing population, modern technologies are focused on higher production without compromising sustainable practices. The recent exceptional growth of the world economy also surged the consumption behavior in the population [3, 4]. Overall, consumption per capita has increased but this is not the case for every individual. The spending and consumption gap has also broadened over time and should be addressed in terms of sustainable practices [5, 6]. With time, solid waste disposal developed into a serious issue in urban cities and countries leading toward the development of sustainable initiatives. However, there is a need for a balance between the current production requirements and sustainability expectations for subsequent generations. The primary changing elements for producing a sustainable textile have been associated with sustainable and environment-friendly materials. On the other hand, reduced waste, recycling, and reusing products are also getting interest. Ethical and sustainable production practices also play a vital role in claiming sustainable and eco-friendly products [6–9].

Spinning is a vital part of the textile manufacturing process and a source of major energy and capital consumption. However, the profit margin of this sector is limited to a small portion of net profit. Hence, implementing the new methods or modifying the current ones to adopt sustainability is quite challenging in this sector. Current production activities not only fulfilled the increased consumer demands, but they also compromised the global environment. Air and water pollution damages the ozone layer which ultimately forecasts the upcoming future [10, 11]. To balance industrial production and environmental safety, new methodologies should be adopted in the textile sectors, especially in the spinning industry. However, profit margins in the spinning sectors are much more limited, as compared to other sectors, which creates hesitations in further investments. It is also considered that cleaning the pollution afterward is more costly than limiting it during the process. Hence more serious efforts should be adopted during the textile spinning process instead of reducing the pollution after it is produced [12, 13].

Environmental issues related to the textile industry start from the cultivation of fibers and the usage of fertilizers during the production process. The preparation of synthetic fibers also caused the release of various chemicals that are dangerous to health. However, the current chapter is focused on sustainable practices in the industrial processes only. The following pages discuss the various possibilities that can reduce the environmental damages to claim sustainable growth [7, 14].

1.1 Energy Consumption

Energy conservation is important in the spinning sector of the textile industry. It is one of the major energy-consuming sectors in the textile industry. Hence one of the major environmental pollution sources. It is claimed that the textile sector creates 1.2 billion carbon emissions yearly and it is projected to increase by 49% in 2030.

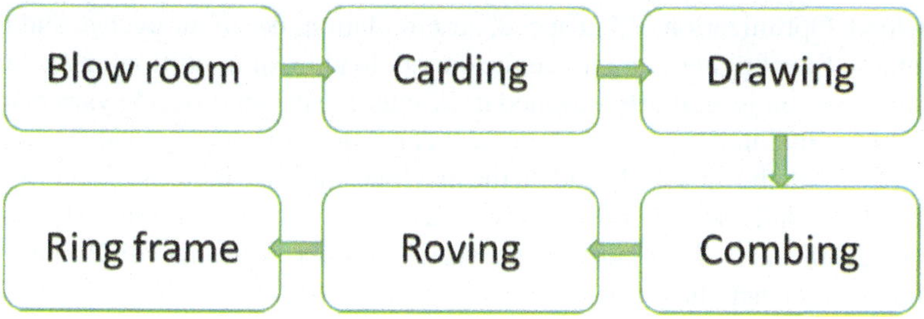

Fig. 1 The production line in the ring-spinning process

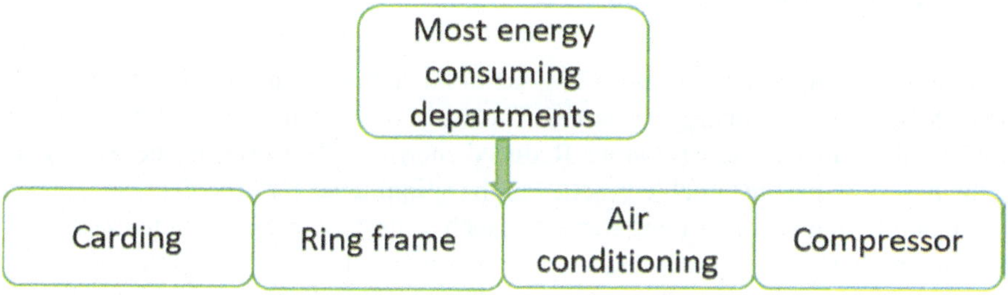

Fig. 2 Most energy-consuming areas in the spinning process

One of the major portions comes from energy consumption and it should be limited if we want to reduce the carbon footprints [14, 15].

Figure 1 explains the production line for the ring-spinning process. In the spinning industry, most energy is consumed in air-conditioning, compressors, lighting, and production machinery lines (Fig. 2). Ring spinning is considered one of the major techniques in yarn manufacturing, which is why used as a reference for the discussion.

If we discuss the production machinery line, card and ring frames are one of the most energy-consuming areas of the yarn manufacturing assembly line. Energy-efficient machines and optimizing the process can reduce energy consumption in the spinning industries. The use of modern equipment with reduced energy consumption can minimize this issue. Energy-saving motors can be replaced in existing card machines. Speed adjustment and waste control can also reduce the energy load on the card machines. Similarly, modification in ring frame gearing and belt/tape system can reduce the machine energy consumption. Periodic machine maintenance and cleaning of machine waste suction pipes can reduce the load on the motors. The use of a modified gearing head with reduced friction between metal parts is also one option to reduce energy consumption. Nowadays metallic gearing heads are also being replaced with specially designed materials to reduce friction and energy.

Air conditioning is also one of the energy-consuming factors in the textile. Various departments have their temperature ranges that should be maintained for efficient working. However, the modernization of the air conditioning system can reduce the energy load. Optimization of fan speed, air circulation, no. of air cycles, and proper adjustment of duct plates can reduce the energy load in air conditioning systems.

Compressed air is regularly supplied in various departments during yarn manufacturing. Proper maintenance for compressors can reduce the energy load for compressors. Leakages in the air supply and on the machine may also increase the compressor working. Especially in the auto winding department air is continuously supplied on each spindle and if all nozzles are not in proper working conditions, it can dramatically enhance the air usage.

1.2 Renewable Energy Sources

As one of the major energy-consuming sectors in textile, spinning industries should proceed toward green energy resources. The usage of solar and wind energy is one of the best alternatives where possible. Reduced energy with renewable energy sources is the best possible sustainable practice in the spinning sector [16, 17].

Implementing energy-saving measures such as using energy-efficient machinery, optimizing production processes, and utilizing renewable energy sources like solar, or wind power can reduce the carbon footprint of textile spinning. Upgrading to modern, energy-efficient spinning machines and optimizing their operation can result in significant energy savings.

1.3 Recycling of Leftover and Waste

The spinning sector produces waste at various places during the yarn manufacturing process. Some waste is usable and others are non-recyclable [5, 18]. Figure 3 explains the waste at various departments in the spinning.

Spinning process waste may be divided into two parts. One that may be reused directly in the blow room and, a second that needs to be processed before it can be

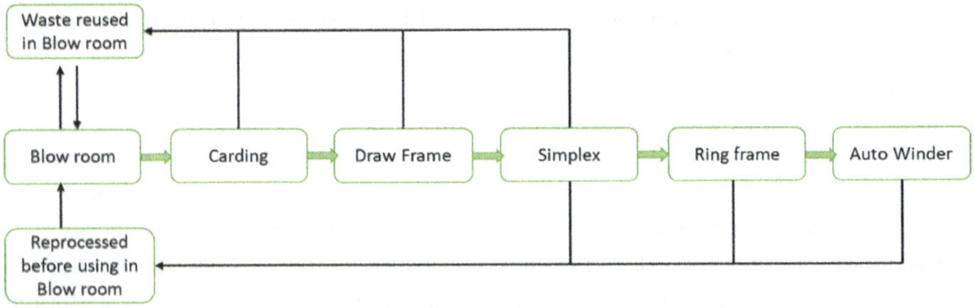

Fig. 3 Waste flow chart and its possible reuse in the ring-spinning process

used in the process again. During the process waste is separated at various points throughout the yarn manufacturing line. There is also unwanted waste that is added due to machine or worker faults. For example, in the draw frame, a sliver is discarded when the machine is stopped or during the can-changing procedure. Similarly, roving is wasted in the roving frame as well as in the ring frame during the yarn preparation. These wastages are consuming material and energy during its production and should be minimized. Almost all industries are reusing these unwanted wastages in blowroom mixing areas. However, its percentage depends on the yarn quality being developed.

Hard waste in the auto winding section is also one of the major issues in the spinning industry. Yarn is processed through multiple machinery lines and finally reaches the auto-winding section. Any yarn lost here is directly causing the loss of material and energy consumed during its manufacturing. This waste is produced by multiple factors including a greater number of cuttings and mechanical faults in machinery. Auto winder quality setting should be adjusted to avoid extra wastage and if there is an issue in yarn quality, it should be verified from previous stages instead of wasting the produced yarn at the final stage which will cause the loss of energy, material, and manpower.

1.4 Socioeconomic Factors in the Industry

The textile sector and especially yarn manufacturing evolves a large amount of manpower starting from fiber mixing until the dispatch of yarn. These workers should be trained to get optimal performance through sustainable processes. Industries should also work for the benefit of workers including health and food facilities. Sound cancellation devices should be available for the workers. Similarly, floor atmosphere and safety measures should be available for the better safety of the workers. Hygienic water and food supply should be available for the workers in a green atmosphere. Other socioeconomic factors should also be addressed for the betterment of atmosphere and social culture [13, 19, 20].

1.5 Packaging Material

Spinning industries are usually designed as stand-alone units or integrated units with other textile processes. Packaging material is one major cost in the spinning industry. Study claims that about 50% of virgin paper and plastic is used for packaging purposes [21]. The use of paper cones and polyethylene bags for packing should be discouraged. In integrated units, reusable cones are adopted nowadays, however in standalone units, single-use paper cones and packing material are still in practice. Single-use packing material should be minimized and alternate solutions including pallet packing with reusable cones should be adopted to obtain a sustainable level of production. The use of recycled material, ecofriendly, biodegradable material, to prepare the packaging, is another solution to obtain sustainable textiles practices goal. Hydrocarbon-free jute sacking is another alternate solution of polythene-based packing materials [22–24].

1.6 Sustainable Storage Equipment in the Spinning Process

During the spinning process, various storage materials are used to store the carded, combed sliver, roving, and yarn. Similarly, ring bobbins and roving bobbins are being used in the yarn manufacturing process. The use of sustainable and recycled material to produce these storage components should be adopted to achieve sustainable spinning practices. Similarly, recycling these used bobbins and cans is another challenge to reduce carbon emissions [25].

2 Sustainable Production Practices in Weaving

In recent decades, the global textile industry has undergone a profound transformation, driven by increasing consumer awareness and concern for the environment. Fabric manufacturers are under growing pressure to adopt sustainable production practices that minimize their ecological footprint, reduce resource consumption, and address social and ethical issues throughout the supply chain. This chapter explores the key sustainable production practices in fabric manufacturing and their significance in building a more responsible and environmentally conscious industry. Initially discuss the fabric manufacturing process, Fabric manufacturing encompasses several key stages, beginning with winding. In this initial step, spinning packages containing continuous yarn strands are unraveled and rewound onto larger spools or cones. This process not only alters the yarn package size but also eliminates any imperfections in the yarn [26]. Figure 4 shows the Process flow of fabric manufacturing.

The 1st step is the warping process in which the warp beam is created during the warping phase. Here, the yarn from the previous step is wound onto a specialized

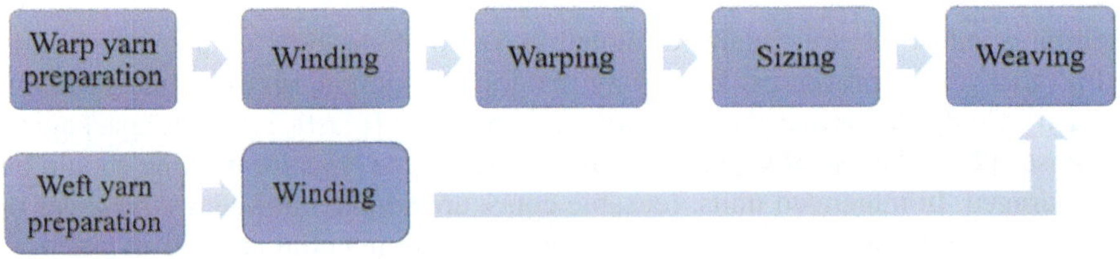

Fig. 4 Process flow of fabric manufacturing process

drum, where multiple yarns are arranged in parallel. Subsequently, these parallel yarns are wound onto a warp beam, forming a cylindrical structure. Sizing, or slashing, is the next vital stage. During sizing, a protective coating, often composed of starch or synthetic substances, is applied to the warp yarns. This coating serves to strengthen and lubricate the yarn, reducing friction and minimizing breakage during the weaving process. The final step, weaving, involves interlacing the warp and weft yarns to produce the fabric. The sized warp beam is installed on a weaving loom, and the weft yarn is woven through the warp threads. Various weaving techniques, such as shuttle-based or shuttleless methods, are employed to create distinct fabric textures and designs. Post-weaving, additional processes like dyeing, printing, finishing, and quality control may be carried out to meet specific requirements for color, texture, strength, and appearance. The fabric is then rolled onto bolts or rolls, ready for distribution and use in diverse applications, ranging from apparel and upholstery to industrial and technical textiles. Each of these manufacturing steps is crucial in determining the quality and characteristics of the final fabric product. The choice of materials, machinery, and techniques varies based on the intended purpose and characteristics of the fabric, ensuring a wide array of fabrics tailored to different needs [27].

2.1 Importance of Sustainability in Fabric Manufacturing

The textile industry is a major global player, but its impact on the environment has raised growing concerns. Unsustainable practices within fabric manufacturing have far-reaching consequences. Here's how they affect the environment [28].

2.1.1 Resource Depletion

Traditional fabric manufacturing relies heavily on non-renewable resources such as petroleum for synthetic fibers and vast amounts of water for processing natural fibers like cotton. The overuse of these resources depletes them and places a strain on ecosystems.

2.1.2 Chemical Pollution

The widespread utilization of sizing agents in warp yarn processing may result in environmental concerns. These agents can contain harmful chemicals such as (Formaldehyde or Fluorocarbon) which, when released into water systems, have the potential to pollute aquatic habitats and pose risks to human health.

2.1.3 Wastewater Generation

Sizing materials, which are commonly applied to warp yarns, often contain chemicals, and during the weaving process, an excess of these sizing agents can leach into the wastewater, potentially carrying toxic or harmful substances with them. Additionally, the rinsing or washing of fabric, either during or after weaving to eliminate impurities, contributes to this wastewater stream. If not adequately treated, this wastewater can contaminate water sources and disrupt ecosystems.

2.1.4 Energy Consumption

Traditional textile factories consume significant amounts of energy, often sourced from fossil fuels. This not only contributes to greenhouse gas emissions but also exacerbates the world's energy crisis.

2.1.5 Waste Generation

Fabric manufacturing can produce vast quantities of waste, from defective fabric to discarded-size baths and chemicals. These waste products, if not managed properly, can have a detrimental impact on landfills and the environment [29].

Considering these environmental concerns, the fabric manufacturing industry is increasingly recognizing the urgency of adopting sustainable practices. By choosing eco-friendly materials, implementing energy-efficient processes, reducing waste, and addressing ethical concerns, fabric manufacturers can not only reduce their ecological footprint but also meet the growing demand from consumers for responsibly produced textiles. Some sustainable practices and their significance in forging a more environmentally conscious and ethical fabric manufacturing industry are discussed below [1].

2.2 Energy Consumption

Fabric manufacturing is an energy-intensive process that involves several stages, each contributing to the overall energy consumption. Here is an overview of energy consumption in the fabric manufacturing process [30]. Figure 5 shows the highly energy-consuming departments during the fabric manufacturing process.

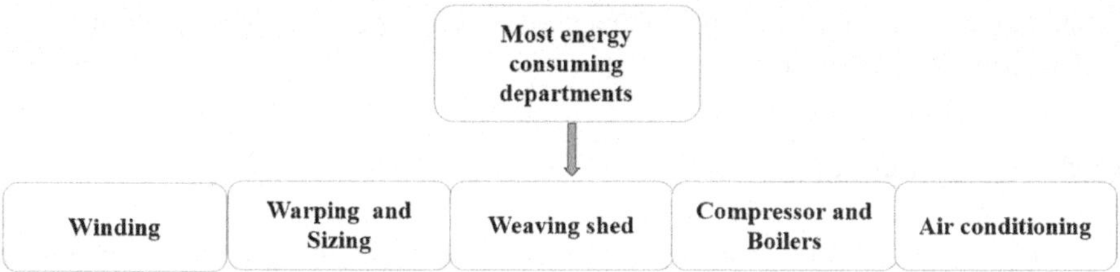

Fig. 5 High energy-consuming departments

2.2.1 Warping and Sizing

- Energy use: Warping is the process of creating the warp beam with parallel yarns. Sizing follows, where a protective coating is applied to the warp yarns to reduce friction. Both stages require energy to operate machinery.
- Energy-efficiency measures: Implementing energy-efficient machines, recycling size materials, and optimizing processes can help reduce energy consumption in these stages.

2.2.2 Weaving Shed

- Energy use: Weaving is a crucial step where the fabric is produced by interlacing the warp and weft yarns. Energy, often in the form of electricity, is needed to power weaving machines.
- Energy-efficiency measures: Upgrading to modern, energy-efficient looms and optimizing weaving processes can significantly lower energy consumption.

2.2.3 Quality Control and Packaging

- Energy use: Quality control processes, as well as packaging and transportation, also contribute to energy consumption in the form of electricity, fuel, etc.
- Energy-efficiency measures: The implementation of effective quality control methods and sustainable packaging techniques can serve to curtail energy consumption within these domains.

2.2.4 Compressor

- Energy use: Compressors play a vital role in supplying compressed air, widely used in various weaving processes, such as powering air jet looms and pneumatic devices. These compressors depend on electricity to generate compressed air.
- Energy-efficiency measures: To optimize energy usage in compressors, regular maintenance, and servicing are necessary to prevent air leaks and ensure peak performance. Additionally, the selection of energy-efficient compressor models

and the implementation of load management systems can contribute significantly to energy conservation.

2.2.5 Air Conditioning

- Energy use: Air conditioning systems hold a crucial role in regulating temperature and humidity levels within weaving facilities, fostering a comfortable and productive working environment. These systems consume electricity for the cooling and dehumidification of the air.
- Energy-efficiency measures: To improve energy efficiency in air conditioning, implementing practices such as the utilization of programmable thermostats, regular filter and coil maintenance, and investment in energy-efficient HVAC equipment can effectively mitigate energy consumption while upholding a comfortable indoor atmosphere for employees.

2.2.6 Overall Plant Operations

- Energy Use: Beyond the specific manufacturing processes, energy is also consumed in plant operations, including lighting, HVAC systems, and general machinery.
- Energy-Efficiency Measures: Retrofitting facilities with energy-efficient lighting, insulation, and HVAC systems can reduce the overall energy footprint.

By focusing on these stages and adopting energy-efficient machinery, optimizing production processes, recycling materials, and investing in renewable energy sources, textile manufacturers can reduce their energy consumption and environmental impact while producing quality fabrics [31].

2.3 *Energy Utilizing Sources During Fabric Manufacturing*

Types of energy utilized during fabric manufacturing, including various energy sources and their applications [32].

2.3.1 Electricity

- Usage: Electricity is a primary energy source in fabric manufacturing. It powers most machinery, including looms, warping machines, and quality control equipment.
- Energy-Efficiency Measures: Companies can adopt energy-efficient motors and lighting systems, as well as control systems that optimize electricity usage.

2.3.2 Compressed Air

- Usage: Compressed air is commonly used to drive pneumatic systems in weaving machines, air jet looms, and other textile equipment.
- Energy-Efficiency Measures: Regular maintenance of compressed air systems and the use of efficient compressors can help minimize energy wastage.

2.3.3 Water

- Usage: Water is a critical resource in fabric manufacturing, particularly in processes like wet processing and water jet weaving.
- Energy Implication: While water itself is not an energy source, its heating and pumping for processes like washing can be energy intensive.
- Energy-Efficiency Measures: Implementing closed-loop water systems, recycling water, and optimizing water usage can reduce associated energy consumption.

2.3.4 Natural Gas

- Usage: Some textile processes, especially those involving high-temperature operations, may rely on natural gas for heating.
- Energy-Efficiency Measures: Upgrading to more energy-efficient burners and improving insulation can help minimize natural gas consumption.

2.3.5 Steam

- Usage: Steam is used in various fabric finishing processes, such as calendaring and setting of fabrics.
- Energy-Efficiency Measures: Efficient steam generation and distribution systems can reduce energy loss.

By considering the types of energy sources utilized and adopting energy-efficient technologies and practices specific to each source, textile manufacturers can reduce their overall energy consumption, lower operational costs, and lessen their environmental impact. Energy management and sustainability are critical aspects of modern fabric manufacturing [33, 34].

2.4 Enhancing Sustainability in Fabric Manufacturing Processes

2.4.1 Sustainable Material Selection

One of the fundamental steps towards sustainability in fabric manufacturing is the choice of materials. Fabrics made from organic fibers, such as organic cotton and hemp, are grown without synthetic pesticides or genetically modified organisms, reducing environmental impact. Additionally, the use of recycled fibers, often derived from post-consumer textile waste, conserves resources and reduces the need for virgin materials. These sustainable choices not only reduce the environmental footprint but also cater to the growing demand for eco-friendly products [35, 36].

2.4.2 Renewable Energy Sources

Fabric manufacturers are increasingly turning to renewable energy sources, such as solar and wind power, to reduce their reliance on fossil fuels. By investing in clean energy solutions, companies can significantly decrease their carbon emissions and contribute to a more sustainable energy landscape.

2.4.3 Energy-Efficient Equipment

The adoption of energy-efficient machinery and production processes can substantially lower energy consumption. Innovations like heat recovery systems, LED lighting, and automated controls can result in significant cost savings while reducing the environmental impact.

2.5 Waste Reduction and Recycling

2.5.1 Zero-Waste Manufacturing

Zero-waste manufacturing during the weaving stage is a critical component of sustainable textile production. This approach minimizes waste by precisely calculating yarn requirements, optimizing loom efficiency, and implementing strict quality control. Recycling and reusing fabric, weaving pattern optimization, lean manufacturing, supplier collaboration, efficient yarn splicing, and finishing further reduce waste. Employee training and energy-efficient practices enhance both sustainability and production efficiency.

2.5.2 Textile Recycling

Retaining the recycling of post-production textile waste holds substantial potential for diminishing the environmental impact of the textile industry. Fabric producers can establish partnerships with recycling facilities to guarantee the transformation of end-of-life textiles into fresh materials, thereby lessening the reliance on virgin resources. Additionally, it is advisable to repurpose the yarns that remain unused, either from winding or during the weaving process, further contributing to sustainability efforts.

2.5.3 Net Zero in Weaving

Working towards achieving net-zero emissions during the weaving phase of textile manufacturing is a pivotal step in fostering a sustainable and environmentally responsible industry. This objective entails diminishing the carbon footprint linked to weaving processes to a level where any remaining emissions are counterbalanced using various methods, such as reforestation or the adoption of carbon capture technologies. To attain this goal, weaving facilities can implement several strategies, including the transition to energy-efficient equipment, the incorporation of renewable energy sources, the adoption of effective waste management practices, and the optimization of transportation and supply chain logistics. Collaborative efforts across the textile industry and with suppliers are essential for sharing best practices and driving innovation toward achieving net-zero emissions during weaving operations. By prioritizing sustainability and embracing these measures, the weaving stage can significantly contribute to a future where the textile industry achieves net-zero emissions, aligning with broader global initiatives to combat climate change and reduce carbon emissions.

2.6 Ethical and Fair Labor Practices

2.6.1 Fair Wages and Working Conditions

Sustainability in fabric manufacturing goes beyond environmental concerns; it extends to social responsibility. Fabric manufacturers should uphold fair labor practices, ensuring that workers are paid fair wages and provided with safe working conditions. Certifications like Fair Trade can help verify and promote ethical production.

2.6.2 Supply Chain Transparency

Transparency in the supply chain is essential for identifying and addressing potential ethical issues. Fabric manufacturers should trace the origins of their materials and ensure that their suppliers adhere to ethical and environmental standards. Blockchain and other technologies have the potential to improve transparency.

2.7 *Sustainable Packaging and Transportation*

2.7.1 Eco-Friendly Packaging

Eco-Friendly Packaging Incorporating sustainable production practices should also encompass the packaging of fabric products. Manufacturers can opt for biodegradable or recyclable packaging materials while minimizing excess packaging to reduce waste.

2.7.2 Sustainable Transportation

Selecting environmentally friendly transportation options, such as electric vehicles or streamlining shipments to reduce the carbon footprint, can contribute significantly to sustainable fabric manufacturing by lessening the environmental impact of distribution.

2.8 *Consumer Education and Engagement*

2.8.1 Labeling and Certification

It's vital to educate consumers about sustainable fabric choices. Clear labeling and certification systems, such as OEKO-TEX and GOTS (Global Organic Textile Standard), can aid consumers in making well-informed decisions when purchasing textiles.

2.8.2 Sustainable Fashion Movements

The sustainable fashion movement is gaining momentum, urging consumers to buy fewer, higher-quality items and support brands that prioritize sustainability. Fabric manufacturers can align themselves with these movements to drive positive change in the industry [37].

3 Sustainable Production Practices in Textile Processing

Sustainability refers to the use of global resources in such a way that future generations can meet their needs. It deals with maintaining life systems in a productive, efficient, and safe manner so that minimal damage to society, the environment, and human beings may occur. The concept of a comfortable and extra-conscious lifestyle has directed human beings to always seek better options for products, facilities, processes, and structures. To fulfill these requirements, science, and technology have played a vital role. Incredible progress can be observed in terms of new/improved products and processes. On the other side of the picture, one of the outcomes of these advancements is the depletion of resources and the pollution of the environment. Product manufacturing is always accompanied by the consumption of natural raw materials and other resources along with land.

The concept of sustainability has forced individuals and organizations to restructure their business and manufacturing strategies towards the introduction of sustainable practices and processes. Owing to the scarcity of natural resources, enhanced polluted environment, economic imbalance, and severe health issues, sustainable manufacturing is the need of the recent era. Among textile manufacturing, textile wet processing is the most pollution-generated section where a large quantity of water, chemicals, auxiliaries, and energy is utilized to run the different processes emitting the pollutants into disposal, rain, atmosphere, and throughout the whole supply chain as well. Textile wet processing comprises the various substrate modification processes to make them more attractive, induce functionality, higher performance, and improved aesthetics. Textile wet processes include pretreatment (desizing, scouring, bleaching, mercerizing), coloration (dyeing and printing), finishing (properties induction and modification), and garments wet processing. All these processes pose a heavy load on the environment if performed conventionally. Sustainable processes are a dire need of the planet to ensure the conservancy of resources preserving the capacity of regeneration.

3.1 Origins of Unsustainability in Textile Processing

The textile manufacturing industry is considered the second most pollutant industry in the globe. The increasing demand for high fashion articles and their quick turnover leads to a heavy impact on resources, the environment, and society. The textile industry relies predominantly on natural fibers that need intensive treatments to add value. The treatments are comprised of extensive use of chemicals and colorants to generate the required characteristics. The pollutants produced from the treatments not only poison the waterways, aquatic life, and environment but also affect the health of interactants. The waste either in the form of chemicals/products and end of end-of-life textiles mostly goes to landfills that can pose hazardous effects on the environment. The energy requirements of the treatment processes are filled through fossil fuels used in the operation of machinery and transport which leads to greenhouse gases and carbon emissions. The detrimental effects of conventional unsustainable practices originate from the consumption of huge energy, massive quantities of water, and high usage of chemicals. It results in the generation of huge quantities of waste and the shedding of microfibres into disposal [38]. It is estimated that 70% of the textile pollution is generated by the textile wet processing section. On categorization, pre-treatment of textiles is the most wastewater-producing section followed by textile dyeing, textile printing, and textile finishing industry. The effluent/waste produced in various processes is not up to the mark and far away from the global standard and can endanger human health and life as well. In some developing countries the effluents are directly discharged into the environment without any effluent treatment that influences the health of humans and animals. The development of sustainable production practices in all types of processes is equally important to avoid environmental and social challenges such as climate change, resource depletion, and inequality. These challenges have arisen because of different human activities. The environmental impacts of textile manufacturing can be reduced by taking different measures like the use of sustainable materials, innovative waterless processes, minimizing/managing waste, safe chemicals, and green approaches to energy. The circular economy/manufacturing is the concept that favors the sustainable manufacturing and processing of materials using the design of products such that the products can be reused or recycled.

3.2 Sustainable Approaches in Textile Processing

A lot of problems are associated with the wet processing of textiles that need to be minimized to protect the descendants and ensure their quality of life. Before approaching the final effects of wet processing, there is a need to prepare the textiles to make them absorbent, impurities-free, and homogeneous. To achieve these characteristics the textiles undergo pre-treatment processes that generate significant waste/effluent. The coloration processes also produce effluent rich with colors and auxiliaries. All these processes are performed at high temperatures using steam or dry heat. The heat energy is produced by burning fossil fuels emitting greenhouse gases in the environment. The sustainability in textile processing can be enhanced through the adoption of various practices and strategies such as use of biodegradable materials, use of green chemicals, natural colorants, enzyme processing, treatment of waste before disposal, reuse of materials and resources, recycling and upcycling of textile waste, energy efficient/renewable energy processes. Hasanbeigi and Price [39] expressed different technologies to reduce the generation of effluent and emissions.

3.3 Sustainable Approaches to Pre-treatment Processes

Textile materials possess various impurities either natural or added/acquired. Cotton, being a natural fiber, has many impurities presenting a hindrance to the absorption of various aqueous-based solutions. The consumption of cotton fibers, owing to their promising characteristics, is about 50 million tons every year and is expected to remain the same or above in the present century. Several challenges are faced in the coloration of textiles if processed without the removal of these impurities. Present challenges in pre-treatment processes [40]. Pre-treatment processes comprise desizing, scouring, bleaching, and mercerizing. These processes are carried out using highly alkaline chemicals, oxidizing agents, surfactants, chelating agents, and other auxiliaries. An intense washing of textiles is needed at the end of any pretreatment process to remove the degraded impurities and chemicals from textiles. Moreover, each step is carried out separately for efficient removal of impurities. It leads to the washing effluent with higher biological oxygen demand (BOD), chemical oxygen demand (COD), total dissolved solids (TDS), and various salts [41].

Currently, starch is mostly used as a sizing agent in the textile industry despite various disadvantages such as hydrophobicity, instability of size viscosity as a function of temperature in size cooking, rigidity of size film, and susceptibility to microorganism attack. Many attempts have been made and are still in the process of synthesizing biopolymer-based eco-friendly and water-soluble sizing material. It facilitates the easy removal of sizing agents reducing the consumption of energy and water [42]. PVA is a good alternative to starch having the characteristics of easy removal and reuse through PVA recovery plant. The drawback is that PVA is non-biodegradable and cost-intensive and still has impacts on the environment. The use of enzymes in the pretreatment of textiles has been widely investigated in the last decades.

In the scouring process, fabric is made free from natural impurities such as fat, oils, waxes, proteins, pectins, minerals, and other solid contaminants. The bleaching process is performed to eliminate the color of the fabric to make it white. The removal of these impurities enables the manufacturer to induce the required characteristics in fabric efficiently. The conventional scouring and bleaching processes involve harsh chemicals and are carried out at elevated temperatures [43]. Caustic soda and hydrogen peroxide are the main chemicals for scouring and bleaching processes, respectively. The recipe also comprises of wetting agent and sequestering agent. The eco-friendly alternative to these chemicals is enzymes. Enzymes are biocatalysts obtained from natural resources and are substrate-specific. These pose very little or no negative effect on the properties of the substrate and the environment. The enzyme processing is usually carried out at low temperatures reducing the consumption of fuel and cost of energy. Harsh chemicals are replaced by eco-friendly enzymes and cause a minimal load on the effluent. The lower contamination of the effluent can significantly reduce the cost of effluent treatment. Chawan et al. used the pro-biotic technology for scouring cotton fabrics and observed that the fabric treated through probiotic technology exhibited better results regarding weight loss, absorbency, and tear strength. The BOD, COD, and TDS of the effluent of probiotic-scoured fabrics were less than the conventionally treated fabric [44].

A good alternative to the conventional pretreatment processes can be the ozone processing of greige fabrics. Ozone treatment is carried out at a very low liquor ratio and thus can be an efficient process with low water consumption. Starch can be degraded through an oxidation process making it water soluble. The color molecules can also be degraded through oxidation to achieve a white fabric. Various intermediates such as peroxide, epoxide, per hydroxyl, and hydroxyl radicals are generated in the process that contributes to the oxidation of color in the bleaching of textiles [45]. Bahtiyari and Benli compared the conventional and ozone bleaching of cotton fabrics. They were of the conclusion that comparable whiteness could be achieved in both processes. However, the consumption of water in ozone bleaching was significantly lower than in conventional bleaching [46]. The effluent generated from washing will contain no or minimal chemical ingredients and will exhibit very low BOD and COD. The generation of ozone is achieved from oxygen so it's a quite sustainable process posing minimal impacts on the environment. The application of plasma processing to remove PVA from fabric can be found in the literature. The plasma treatment before bleaching and mercerizing also gives encouraging results. The plasma treatments owing to the dry processing, environment, and worker-friendly technique are a promising alternative [47]. However, the commercialization of plasma treatments in textiles is still a challenge that limits its scope in textile processing. One-step desizing, scouring, and bleaching processes are also helpful in reducing the effluent [41].

3.4 Sustainable Approaches in Textile Coloration

The coloration of textiles aims to produce an attractive look to the fabrics regardless of its impacts. The synthesis of synthetic dyes is a large-scale industry today producing new improved and bright hues having maximum yield. However, the impact of the colorants and their application auxiliaries on the environment has always been a concern [48]. Extensive research is being done to minimize the hazardous effects of coloration processes both in terms of chemical aspects and process modifications [49, 50].

The major damage to the environment induced by the textile industry is attributed to contaminated effluent disposal without treatment. The untreated effluent contributes about 80% of the emissions caused by the textile industry. Textile dyes are toxic and carcinogenic causing various diseases to living and harming to environment [51]. The annual production of synthetic dyes is about 7×10^7 tons globally. The share of dyes consumed by the textile industry is about 10^4 tons. The dyes poorly attached to the fibers are discharged into the water in the washing process [52].

3.5 Sustainable Textile Coloration Approaches

Sustainability in dyeing can be majorly achieved through innovations in dye structures, dyeing processes, and dyeing equipment. The dyeing processes consuming less amount of water, dyes, and auxiliaries will exhibit a lower impact on the environment. A good alternative to synthetic dyes is natural dyes extracted from renewable resources. A lot of research can be found in literature where a lot of natural dye resources are explored. Color pigments grown through microorganisms can be a good source for the coloring industry. These pigments have the potential to be applied on substrates without any chemicals. So, these can reduce the contamination of the effluents [53]. The limitations of natural dyes comprise the limited extraction yield, reproducibility of dye color, range of hues, and fastness characteristics. Pre and post-mordanting of fabrics are the techniques to be used to enhance the color fastness characteristics. The mordanting processes involve the chemical treatment that poses a question on the sustainable processing with natural dyes. The research work regarding the development of eco-friendly mordants for natural dyes is the focus of the scientists now. A strategy to modify the surface of the fiber is a good technique to achieve higher dyeability using no salt in exhaust dyeing [54].

A low liquor ratio dyeing was carried out using dioctyl sodium sulfosuccinate that makes micelle with dye molecules. The observations depict that the low liquor ratio dyeing saves about 50% water. The consumption of other auxiliaries was also less, however, the characteristics of dyed fabric were comparable with conventionally dyed fabric. The use of fewer chemicals reduces the effluent treatment cost too [55]. Lin et al. proposed a sustainable spray dyeing system to dye textile fabrics. The development can reduce the consumption of water and auxiliaries [56]. Cationization of cotton before dyeing is a wonderful technique to achieve darker shades for fabrics with a lower quantity of dye and chemicals. Dow has developed and patented a pretreatment process named ECOFAST Pure to produce a permanent positive charge on cotton surfaces that helps to reduce the amount of dye and water by 50%. Supercritical fluid dyeing of polyester fabrics is an environmentally friendly and cost-effective technique. The main attraction of this process corresponds to the water-less technology, low energy requirement, and recovery of the residual dyes. The amount of effluent is very low which opens its future in the textile industry [57, 58].

Electrochemical reduction of vat, sulfur, and indigo dyes presents the advantage of minimizing the reducing agents used to solubilize these dyes. This technique can also be used in discoloration of effluents [59]. Air dyeing techniques can reduce water and energy tremendously saving the environment. In this technique, a spray of color solution is imparted directly to textile fabrics. It will use less amount of chemicals and produce a darker shade saving the dye consumption and reducing the effluent. The reduction in emission of greenhouse gases and consumption of energy in air dyeing techniques are 84% and 87%, respectively [60].

Ultrasonic technology is also found to be a promising alternative in dyeing textiles. In the wet processing of textiles, ultrasound can resist the agglomeration of dyes through the mechanical oscillation of solid/liquid/gaseous media using the acoustic, thermal, and cavitation effects. The energy is transferred to the particles that exhibit enhanced migration characteristics. The air between the interstices of the fabric is removed by ultrasonic waves facilitating the adsorption and diffusion of dye molecules into the fibres. Using ultrasonic waves, a darker shade in dyeing is achieved which helps save the dyes, chemicals, and water [61, 62]. Microwaves have the potential to be utilized in the synthesis, processing, and treatment of a wide range of materials. The research work regarding the use of microwaves in the dyeing of different fibers (cellulose, wool, polyester) can be found in the literature. It can be used as part of the dyeing process or for the treatment of fibers before dyeing. One of the advantages of microwaves in textile processing is their ability of homogeneous and rapid heading of medium that can reduce the power consumption and time of process [63, 64].

Textile printing is a long process to produce a pattern on textile fabrics. It has a lot of steps involving toxic unsustainable chemicals. Digital printing is a good alternative textile printing technique that is suitable for both short and long-run fabrics. This technique eliminated the screen engraving process saving energy, water, and chemicals. Although digital printing has reduced the environmental impacts as compared to traditional printing, there is a still need to address the unsustainable impacts of digital printing. Synthetic inks used in printing are not sustainable and can be replaced with inks extracted from natural biomaterials [65].

3.6 Sustainable Textile Finishing Approaches

The inherent surface characteristics of textile fibers are limited and cannot fulfill the requirements of a particular scenario. To achieve the required characteristics, the surfaces of the fabrics are functionalized using different chemicals and techniques. Plasma treatment is one of the sustainable techniques applied on different surfaces to modify their characteristics [47]. Plasma processing is an environmentally friendly dry process involving no solvents or aqueous medium [66, 67]. It can be used for both natural and synthetic fibers to modify the various characteristics. The surface of polypropylene is quite inert owing to the absence of functional groups and low surface energy. In a study, polypropylene nonwoven sheets were treated with oxygen plasma to transform their surface from hydrophobic to hydrophilic. Upon printing, the treated fabric showed a higher color yield as compared to the untreated fabric. Wool can be treated by plasma for shrinkage resistance, absorbency, and dyeability [68]. The surface treatment of polyester for enhanced surface energy, adhesion of various molecules, superhydrophobicity, and other characteristics has been extensively reported in the literature [69–71].

The use of enzymes in functionalization of textile surfaces is also an environmentally friendly option. The enzymes are proteins in nature extracted from renewable resources (living organisms). The enzymes used in textiles are amylases, lipases, pectinases, proteases, cellulases, catalases, peroxidases, ligninases, collagenases, esterases, and nitrilases. Enzymes have numerous advantages working in mild conditions, being an alternative to polluting chemicals, being specific in action, being biodegradable, and being easy to control [72]. The use of enzymes for processing textiles is widely reported in the literature. These are used for the removal of protruding fibers, the size applied on warp yarn, natural impurities, natural color in fibers, modification of fibrous surfaces, decomposition of residual hydrogen peroxide, decomposition of hydrolyzed dye in water, and modification of textile surfaces [73]. The extensive use of enzymes can be found as a biopolishing agent for knits and denim garments where cellulases are used to hydrolyze the cellulosic fibers. The hydrolyzed cellulosic fibers under the effect of the mechanical action of pumice stones are removed from the surface of denim garments to induce an attractive fashionable look to the garments. Prajapati et al. used protease to create the controlled patterning of the dyed wool polyester blended fabric through the hydrolysis of wool fibers [74].

Textiles with bioactive characteristics have applications in end uses such as medical gowns, bandages, hospital bed sheets, wear for infants, socks, sportswear, etc. The presence of bacteria on textiles may cause cross-infection by harmful pathogens, unwanted odor, allergies, inflammation, and skin diseases. Synthetic antibacterial agents like triclosan, organometallics, metal ions, quaternary ammonium compounds, and phenols face the challenge of environmental concerns [75]. The use of natural antimicrobial compounds is a promising alternative to synthetic agents addressing the sustainability in antibacterial finishing processes [76, 77]. Plasma deposition and plasma sputtering is also a sustainable option to develop antibacterial textiles. Plasma-employed antiviral textiles have been developed for personnel protection [78].

3.7 Sustainable Approaches in Denim Processing

Denim garments are among the most famous wearables owing to their suitability for people of each age, various seasons, and occasions. The market for denim garments is expected to rise to 87.4 billion US$ in 2027. The real beauty/uniqueness of denim garments is linked to the fading/washing effects that are generated to give a vintage look [79]. Denim is dyed using indigo dyes that are considered a type of vat dye and need a reducing agent and alkali for solubilization and application on fabrics. The unfixed dye is also drained to the effluent causing turbidity and ecological damage. Moreover, different washing treatments are given to denim using various chemicals such as sodium hypochlorite, sodium permanganate, and large quantities of water. The use of unsustainable materials and plenty of water always put the denim industry under serious criticism [80, 81]. The use of electrochemical techniques, fruit extracts, glucose, and bacteria for the reduction of indigo dye are the sustainable alternatives [79]. Supercritical fluid dyeing of denim yarn can be a good alternative to conventional rope dyeing as it will have a minimal impact on the environment. The use of clean solvent and recovery at the end of the process addresses the sustainability aspect of the process [79]. The use of enzymes for wash-down effects in denim has already been established in the denim industry. The development of the laser fading and ozone bleaching processes has successfully addressed the sustainability issues in denim processing [82]. Laser fading is a dry, chemical-free process needing water only for rinsing or mild washing. Another advantage of laser fading is the minimal loss of the mechanical properties of fabric which can increase the life of the garments. Moreover, ozone-fading techniques use very small amounts of water without any chemicals to bleach the garments. A water jet fading system equipped with water recycling is another sustainable option for fading denim garments [83]. Plasma can be used in denim processing for oxidation of the denim fabrics to produce desired fading effects. Plasma technology, being operated in a dry environment involving no harmful chemicals, can be a good alternative to conventional fading processes for denim garments [84].

4 Sustainable Production Practices in Garment Manufacturing

Sustainable garment manufacturing is a step towards making the fashion industry one of the world's most significant contributors to reducing waste and global emissions. Visionist has implemented various elements in the assembly of a range to reduce waste and save money while producing excellent quality garments and working within transparent frameworks [85].

Promoting sustainability in the apparel industry involves several key strategies. First, the adoption of eco-friendly materials that require less water for cultivation and minimize harmful chemicals is crucial for reducing environmental impact. Secondly, embracing a circular economy approach, focusing on designing recyclable products, promoting reuse and repair, and recycling textile waste, is essential to minimize resource utilization and waste. Thirdly, companies should implement strategies to minimize water and energy consumption through optimized manufacturing processes, improved dyeing techniques, and investment in wastewater treatment facilities and energy-efficient technologies. Ensuring fair working conditions by enhancing wages, providing safe environments, and obtaining third-party certifications demonstrates a commitment to ethical labor practices. Lastly, establishing transparent supply chains by disclosing product origins, manufacturing processes, and labor conditions, along with conducting regular audits, is vital to meet consumer demands for ethical sourcing. These strategies collectively drive sustainability and responsible practices in the apparel industry [86, 87].

4.1 Eco Designed Garments

In the era of sustainability and responsible practices related to circular economy, it is important to start from design. Eco-designed products help in the later stages of recycling and reusing the material. Complex designs having multiple types of materials are difficult to segregate in recycling and hence circular economy is not achieved. Each type of material has various methodologies for recycling and when they are mixed, it makes them difficult to reuse. One other option is to segregate them before they are reused or recycled. However, it will create extra impact and it is not possible to separate them. For this reason, it is good to focus on eco-design and avoid using different types of materials in the manufacturing process [11].

Similarly, designs should be developed to reduce the production steps and time. Complex designs having a greater number of operations in the garment production lines will not only increase the production time but energy usage and labor cost will also be increased.

4.2 Water and Energy Consumption

Energy-efficient technologies are need of the day to save energy consumption. It is important to adopt energy-efficient production lines with a smaller number of machines. State-of-the-art modern machines reduce the energy consumption and also increase the production efficiency. Similarly, for denim production, it is important to optimize the process to save water. To obtain sustainable practices in the process, it is important to reduce water usage and adopt the techniques to save water. Denim washing is another major source of water usage in the garment industry. An optimized washing process should be adopted to reduce water waste. Waterless washing is another option that may be adopted. Chemical washing should be avoided to reduce the environmental impact produced due to hazardous chemicals discharged into the drain water.

4.3 Reduction of Waste

Garment production is the final stage of product development. Fabric is passed through the spinning, weaving, and textile process stages before it is finally stitched. Any waste at this stage is a total loss of resources consumed in all previous processes. Before the stitching process in the garment industry, spreading and cutting are performed according to the given design. CAD and other software are developed to reduce the cutting waste. State-of-the-art cutting practices can reduce waste and ultimately reduce the resources lost consumed in the previous stages of production. Similarly trained worker also contributes to waste reduction in the garment industry.

Proper training should be conducted to upgrade the workers' skills. It is important especially in the garment industry because any damage or loss here is non-transferable and impacts directly on sustainable practices.

4.4 Labor Performance and Social Impact

The garment industry is labor labor-oriented industry and production quality is directly related to labor behavior. Proper worker training and refresher courses are important to motivate them and reduce damaged production. Each damaged piece has an environmental impact on sustainable practice. A special focus should be on the worker's development and a friendly atmosphere in the industry. If proper social benefits are provided in the industry, it will be helpful to sustain the workers in their jobs instead of making an effort to find better job options. It will ultimately impact on their performance and the product quality [88]. Lastly, there is another significant social impact to be recognized, as the sustainable fashion market can empower individuals, including single mothers, by equipping them with the skills to transform used clothing into new items, thus providing them with a means to sustain their livelihoods [89].

These practices in the garment industry can promote sustainable development. However, there are some challenges that can hurdle these attempts. Firstly, sourcing the right materials and managing the supply chain. Secondly, users of sustainable fashion often express their satisfaction with clothing made from natural materials, as it has a soothing effect on the skin and can positively impact their overall well-being. This issue can create hurdles for using various materials in sustainable fashion products.

5 Conclusion

The adoption of sustainable production practices in textile manufacturing is no longer a choice but a necessity. As consumers become more environmentally and ethically conscious, the textile industry must respond by embracing sustainability throughout its operations. By selecting eco-friendly materials, improving energy efficiency, reducing waste, and fostering ethical practices, textile manufacturers can contribute to a more sustainable and responsible industry, meeting the demands of today's conscious consumers and securing a greener future for generations to come. Efforts to develop eco-design also enhance the chances to reuse and recycle the products and give them a second life. These practices will impact the overall green impact on the environment and motivate other industries to come forward for these initiatives to make the earth better for future living.

References

1. Desore A, Narula SA (2018) An overview on corporate response towards sustainability issues in textile industry. Environ Dev Sustain 20:1439–1459. https://doi.org/10.1007/s10668-017-9949-1
2. Luján-Ornelas C, Güereca LP, Franco-García M-L, Heldeweg M (2020) A life cycle thinking approach to analyse sustainability in the textile industry: a literature review. Sustainability 12:10193. https://doi.org/10.3390/su122310193
3. Sharma A, Narula SA (2020) What motivates and inhibits Indian textile firms to embrace sustainability? Asian J Sustain Soc Responsib 5:6. https://doi.org/10.1186/s41180-020-0032-8
4. Dafia CSN, Chen F, Sumo PD (2022) Guideline and strategies of textile industry on the sustainable development of Benin. Sustainability 14:12762. https://doi.org/10.3390/su141912762
5. Faysal GM, Azad TNS, Repon MdR et al (2022) Sustainable yarn production using left-over fabric from apparel industries. Heliyon 8:e11377. https://doi.org/10.1016/j.heliyon.2022.e11377
6. Focus T (2021) The sustainability in textile industry. Text. Focus. https://textilefocus.com/sustainability-textile-industry/. Accessed 24 Aug 2023
7. Sustainable practices in the textile industry | Wiley. Wiley.com. https://www.wiley.com/en-ie/Sustainable+Practices+in+the+Textile+Industry-p-9781119818915. Accessed 24 Aug 2023
8. Amutha K (2017) Sustainable practices in textile industry: standards and certificates. In: Muthu SS (ed) Sustainability in the textile industry. Springer, Singapore, pp 79–107
9. Cleaner production in the textile industry and its relationship to sustainable development goals. ScienceDirect. https://www.sciencedirect.com/science/article/abs/pii/S0959652619314313. Accessed 24 Aug 2023
10. Pandey R, Pandit P, Pandey S, Mishra S (2020) Solutions for sustainable fashion and textile industry. In: Recycling from waste in fashion and textiles. Wiley, pp 33–72
11. Harsanto B, Primiana I, Sarasi V, Satyakti Y (2023) Sustainability innovation in the textile industry: a systematic review. Sustainability 15:1549. https://doi.org/10.3390/su15021549
12. (2018) Sustainability solutions for spinning industry by operational excellence. Wazir. https://wazir.in/sustainability-solutions-for-spinning-industry-by-operational-excellence. Accessed 24 Aug 2023
13. Spinning a sustainable yarn: environmental sustainability and brand story in the Australian fashion industry. Intellect. https://intellectdiscover.com/content/journals/10.1386/infs.1.2.185_1?crawler=true. Accessed 24 Aug 2023
14. Ltd K (2020) 5 ways to include sustainable practices in the textile industry. Kingly Merch. https://wearekingly.com/en/5-ways-to-include-sustainable-practices-in-the-textile-industry/. Accessed 24 Aug 2023
15. Urooj SS| A (03:57:15+05:00) Textile sector: sustainable business practices. Brecorder. https://www.brecorder.com/news/40205980. Accessed 24 Aug 2023
16. Chain TV (2020) Sustainable strategies for cost effective spinning. In: Text. mag. text. news appar. news fash. news. https://textilevaluechain.in/in-depth-analysis/articles/textile-articles/sustainable-strategies-for-cost-effective-spinning/. Accessed 24 Aug 2023
17. Anand KT, Rajan AJ (2011) A study on environmental standards in spinning industries for sustainable development. In: International conference on green technology and environmental conservation (GTEC-2011). pp 61–67
18. Khairul Akter MMd, Haq UN, Islam MdM, Uddin MA (2022) Textile-apparel manufacturing and material waste management in the circular economy: a conceptual model to achieve sustainable development goal (SDG) 12 for Bangladesh. Clean Environ Syst 4:100070. https://doi.org/10.1016/j.cesys.2022.100070
19. Islam MM, Perry P, Gill S (2020) Mapping environmentally sustainable practices in textiles, apparel and fashion industries: a systematic literature review. J Fash Mark Manag Int J 25:331–353. https://doi.org/10.1108/JFMM-07-2020-0130

20. Costa C, Azoia N, Silva C, Marques E (2020) Textile industry in a changing world: challenges of sustainable development. UPorto J Eng 6:86–97. https://doi.org/10.24840/2183-6493_006.002_0008
21. www.fibre2fashion.com Sustainable packaging solutions for fashion industry. http://www.fibre2fashion.com/industry-article/9443/sustainable-packaging-solutions-for-fashion-industry. Accessed 31 July 2023
22. Fisher G (2020) Demand for sustainable packaging and materials is catching on. Int. Fiber J. https://www.fiberjournal.com/demand-for-sustainable-packaging-and-materials-is-catching-on/. Accessed 31 July 2023
23. Kiron MI (2013) Packaging textiles | Advantages and uses of textile packaging materials. Text. Learn. https://textilelearner.net/packaging-textiles/. Accessed 31 July 2023
24. Diabat A, Kannan D, Mathiyazhagan K (2014) Analysis of enablers for implementation of sustainable supply chain management—a textile case. J Clean Prod 83:391–403. https://doi.org/10.1016/j.jclepro.2014.06.081
25. Fogla H (2022) How do spinning cans and eco-sustainable textile production go hand in hand? Jumac Cans. https://www.jumaccans.com/blog/how-spinning-cans-and-eco-sustainable-textile-production-go-hand-in-hand/. Accessed 31 July 2023
26. Pattanayak AK (2020) Sustainability in fabric manufacturing. In: Sustainable technologies for fashion and textiles. Elsevier, pp 57–72
27. Alkaya E, Demirer GN (2014) Sustainable textile production: a case study from a woven fabric manufacturing mill in Turkey. J Clean Prod 65:595–603. https://doi.org/10.1016/j.jclepro.2013.07.008
28. Nguyen T-L, Shu M-H, Hsu B-M (2016) Extended FMEA for sustainable manufacturing: an empirical study in the non-woven fabrics industry. Sustainability 8:939. https://doi.org/10.3390/su8090939
29. Nayak R, Panwar T, Nguyen LVT (2020) Sustainability in fashion and textiles. In: Sustainable technologies for fashion and textiles. Elsevier, pp 3–30
30. Çay A (2018) Energy consumption and energy saving potential in clothing industry. Energy 159:74–85. https://doi.org/10.1016/j.energy.2018.06.128
31. Browne M, Allen J, Rizet C (2006) Assessing transport energy consumption in two product supply chains. Int J Logist Res Appl 9:237–252. https://doi.org/10.1080/13675560600859243
32. Branchetti S, Petrovich C, Ciaccio G et al (2019) Energy consumption characterization based on a self-analysis tool: a case study in yarn manufacturing. In: Proceedings of the 8th international conference on smart cities and green ICT systems. SCITEPRESS - Science and Technology Publications, pp 40–50
33. Palamutcu S (2015) Energy footprints in the textile industry. In: Handbook of life cycle assessment (LCA) of textiles and clothing. Elsevier, pp 31–61
34. Ozer B, Guven B (2019) Energy efficiency implementation in a fabric finishing factory in turkey. In: 2019 4th international conference on smart and sustainable technologies (SpliTech). IEEE, pp 1–5
35. Tayyab M, Jemai J, Lim H, Sarkar B (2020) A sustainable development framework for a cleaner multi-item multi-stage textile production system with a process improvement initiative. J Clean Prod 246:119055. https://doi.org/10.1016/j.jclepro.2019.119055
36. Wanniarachchi T, Dissanayake K, Downs C (2020) Improving sustainability and encouraging innovation in traditional craft sectors: the case of the Sri Lankan handloom industry. Res J Text Appar 24:111–130. https://doi.org/10.1108/RJTA-09-2019-0041
37. Lee KE (2017) Environmental sustainability in the textile industry. In: Sustainability in the textile industry. pp 17–55
38. The environmental price of fast fashion. Nature Reviews Earth & Environment. https://www.nature.com/articles/s43017-020-0039-9. Accessed 30 July 2023
39. Hasanbeigi A, Price L (2015) A technical review of emerging technologies for energy and water efficiency and pollution reduction in the textile industry. J Clean Prod 95:30–44. https://doi.org/10.1016/j.jclepro.2015.02.079

40. Asaduzzaman, Miah MR, Hossain F, Li X et al (2016) A study on the effects of pre-treatment in dyeing properties of cotton fabric and impact on the environment. J Text Sci Eng 06. https://doi.org/10.4172/2165-8064.1000274
41. Imran MA, Hussain T, Memon MH, Abdul Rehman MM (2015) Sustainable and economical one-step desizing, scouring and bleaching method for industrial scale pretreatment of woven fabrics. J Clean Prod 108:494–502. https://doi.org/10.1016/j.jclepro.2015.08.073
42. Polymers | Free full-text | Synthetized potato starch—a new eco sizing agent for cotton yarns. https://www.mdpi.com/2073-4360/11/5/908. Accessed 30 July 2023
43. del Valle LJ, Oños M, Garriga P et al (2006) Bioscouring of cotton fiber with polygalacturonase induced in Sclerotium rolfsii using cellulose and glucose-pectin. Text Res J 76:400–405. https://doi.org/10.1177/0040517506063386
44. Chavan SV, Tyagi K, Sharma A, Sadannavar MK, Girmal SB (2020) Probiotic technology—a novel approach in scouring of textiles. J Text Sci Eng 10. https://doi.org/10.37421/jtese.2020.10.416
45. Panda SKBC, Sen K, Mukhopadhyay S (2021) Sustainable pretreatments in textile wet processing. J Clean Prod 329:129725. https://doi.org/10.1016/j.jclepro.2021.129725
46. Bahtiyari Mİ, Benli H (2020) Comparison of ozone-based cold bleaching processes with conventional pretreatment of cotton. Ozone Sci Eng 42:450–460. https://doi.org/10.1080/01919512.2019.1704218
47. Zille A, Oliveira FR, Souto AP (2015) Plasma treatment in textile industry. Plasma Process Polym 12:98–131. https://doi.org/10.1002/ppap.201400052
48. Textile dyeing industry: environmental impacts and remediation. SpringerLink. https://link.springer.com/article/10.1007/s11356-019-07137-z. Accessed 30 July 2023
49. Eco-friendly and community sustainable textile fabric dyeing methods from Thai buffalo manure: from pasture to fashion designer—Suthasini Bureekhampun, Chanida Maneepun, 2021. https://journals.sagepub.com/doi/full/10.1177/21582440211058201. Accessed 30 July 2023
50. Lara L, Cabral I, Cunha J (2022) Ecological approaches to textile dyeing: a review. Sustainability 14:8353. https://doi.org/10.3390/su14148353
51. Lellis B, Fávaro-Polonio CZ, Pamphile JA, Polonio JC (2019) Effects of textile dyes on health and the environment and bioremediation potential of living organisms. Biotechnol Res Innov 3:275–290. https://doi.org/10.1016/j.biori.2019.09.001
52. Al-Tohamy R, Ali SS, Li F et al (2022) A critical review on the treatment of dye-containing wastewater: ecotoxicological and health concerns of textile dyes and possible remediation approaches for environmental safety. Ecotoxicol Environ Saf 231:113160. https://doi.org/10.1016/j.ecoenv.2021.113160
53. Tuli HS, Chaudhary P, Beniwal V, Sharma AK (2015) Microbial pigments as natural color sources: current trends and future perspectives. J Food Sci Technol 52:4669–4678. https://doi.org/10.1007/s13197-014-1601-6
54. Niu T, Wang X, Wu C et al (2020) Chemical modification of cotton fabrics by a bifunctional cationic polymer for salt-free reactive dyeing. ACS Omega 5:15409–15416. https://doi.org/10.1021/acsomega.0c01530
55. Mamun Kabir SMd, Sk S, Koh J (2021) Sustainable low liquor ratio dyeing of cotton with C.I. Reactive Blue 21 using dioctyl sodium sulfosuccinate. Text Res J 91:1083–1093. https://doi.org/10.1177/0040517520971363
56. Lin L, Zhu W, Zhang C et al (2021) Combination of wet fixation and drying treatments to improve dye fixation onto spray-dyed cotton fabric. Sci Rep 11:15403. https://doi.org/10.1038/s41598-021-94885-z
57. Abate MT, Seipel S, Yu J et al (2020) Supercritical CO_2 dyeing of polyester fabric with photochromic dyes to fabricate UV sensing smart textiles. Dyes Pigments 183:108671. https://doi.org/10.1016/j.dyepig.2020.108671
58. Goñi ML, Gañán NA, Martini RE (2021) Supercritical CO_2-assisted dyeing and functionalization of polymeric materials: a review of recent advances (2015–2020). J CO2 Util 54:101760. https://doi.org/10.1016/j.jcou.2021.101760

59. Bechtold T, Turcanu A (2009) Electrochemical reduction in vat dyeing: greener chemistry replaces traditional processes. J Clean Prod 17:1669–1679. https://doi.org/10.1016/j.jclepro.2009.08.004
60. Kant R (2011) Textile dyeing industry an environmental hazard. Nat Sci 4:22–26. https://doi.org/10.4236/ns.2012.41004
61. Tissera ND, Wijesena RN, de Silva KMN (2016) Ultrasound energy to accelerate dye uptake and dye–fiber interaction of reactive dye on knitted cotton fabric at low temperatures. Ultrason Sonochem 29:270–278. https://doi.org/10.1016/j.ultsonch.2015.10.002
62. Ma X, Wei Y, Wang S et al (2020) Sustainable ultrasound-assisted ultralow liquor ratio dyeing of cotton fabric with natural turmeric dye. Text Res J 90:685–694. https://doi.org/10.1177/0040517519878793
63. New trend in textile coloration using microwave irradiation. https://jtcps.journals.ekb.eg/article_30593.html. Accessed 20 July 2023
64. Popescu V, Astanei D-G, Burlica R et al (2019) Sustainable and cleaner microwave-assisted dyeing process for obtaining eco-friendly and fluorescent acrylic knitted fabrics. J Clean Prod 232:451–461. https://doi.org/10.1016/j.jclepro.2019.05.281
65. Thakker AM, Sun D (2022) Engineering sustainable inks from natural biomaterials for digital printing. Sustain Clim Change 15:272–278. https://doi.org/10.1089/scc.2022.0010
66. Wen L, Jeong D-C, Javid A et al (2015) Conductive polythiophene-like thin film synthesized using controlled plasma processes. Thin Solid Films 587:66–70. https://doi.org/10.1016/j.tsf.2014.12.030
67. Jeong D-C, Javid A, Wen L et al (2016) Low-temperature plasma polymerization of dicyclopentadiene for anti-corrosion properties. Polymer 92:133–139. https://doi.org/10.1016/j.polymer.2016.03.094
68. Fatkullina RR, Yarullin RM, Abutalipova LN, Khadysov BA (2022) Effect of low-temperature plasma treatment on the properties of a complex wool-containing material. J Phys Conf Ser 2379:012002. https://doi.org/10.1088/1742-6596/2379/1/012002
69. Polymers | Free full-text | Surface modification by the DBD plasma to improve the flame-retardant treatment for dyed polyester fabric. https://www.mdpi.com/2073-4360/13/17/3011. Accessed 19 July 2023
70. Morshed MN, Behary N, Bouazizi N et al (2019) Surface modification of polyester fabric using plasma-dendrimer for robust immobilization of glucose oxidase enzyme. Sci Rep 9:15730. https://doi.org/10.1038/s41598-019-52087-8
71. Sun Y, Ouyang B, Rawat RS, Chen Z (2020) Rapid and stable plasma transformation of polyester fabrics for highly efficient oil-water separation. Glob Chall 4:1900095. https://doi.org/10.1002/gch2.201900095
72. A green approach for modification and functionalization of wool fabric using bio- and nano-technologies. SpringerLink. https://link.springer.com/article/10.1007/s10098-022-02385-z. Accessed 19 July 2023
73. Besegatto SV, Costa FN, Damas MSP et al (2018) Enzyme treatment at different stages of textile processing: a review. Ind Biotechnol 14:298–307. https://doi.org/10.1089/ind.2018.0018
74. Prajapati CD, Smith E, Kane F, Shen J (2019) Selective enzymatic modification of wool/polyester blended fabrics for surface patterning. J Clean Prod 211:909–921. https://doi.org/10.1016/j.jclepro.2018.11.079
75. Javid A, Raza ZA, Hussain T, Rehman A (2014) Chitosan microencapsulation of various essential oils to enhance the functional properties of cotton fabric. J Microencapsul 31:461–468. https://doi.org/10.3109/02652048.2013.879927
76. Emam HE (2019) Antimicrobial cellulosic textiles based on organic compounds. 3 Biotech 9:29. https://doi.org/10.1007/s13205-018-1562-y
77. Wolela AD (2020) Antibacterial finishing of cotton textiles with extract of citrus fruit peels. Curr Trends Fash Technol Text Eng 6:1–7
78. Tanasa F, Teaca C-A, Nechifor M et al (2023) Highly specialized textiles with antimicrobial functionality—advances and challenges. Textiles 3:219–245. https://doi.org/10.3390/textiles3020015

79. Critical review on sustainability in denim: a step toward sustainable production and consumption of denim. ACS Omega. https://pubs.acs.org/doi/10.1021/acsomega.2c06374. Accessed 17 July 2023
80. Employing a biochemical protecting group for a sustainable indigo dyeing strategy. Nature Chemical Biology. https://www.nature.com/articles/nchembio.2552. Accessed 17 Jul 2023
81. Superiority of sustainable ozone wash over conventional denim washing technique. Inpressco. https://inpressco.com/superiority-of-sustainable-ozone-wash-over-conventional-denim-washing-technique/. Accessed 17 July 2023
82. Investigation of color fading and fabric-touch test for jeans through ozonation. Ozone: Sci Eng 43(3). https://www.tandfonline.com/doi/abs/10.1080/01919512.2020.1796584?journalCode=bose20. Accessed 17 July 2023
83. Samanta KK, Basak S, Chattopadhyay SK (2017) 12—Environmentally friendly denim processing using water-free technologies. In: Muthu SS (ed) Sustainability in denim. Woodhead Publishing, pp 319–348
84. Radetić M, Jovančić P, Puač N et al (2009) Plasma-induced decolorization of indigo-dyed denim fabrics related to mechanical properties and fiber surface morphology. Text Res J 79:558–565. https://doi.org/10.1177/0040517508095612
85. Sustainable garment manufacturing I Visionise I Ethical
86. Rosen MA, Kishawy HA (2012) Sustainable manufacturing and design: concepts, practices and needs. Sustainability 4:154–174. https://doi.org/10.3390/su4020154
87. Islam MdT, Jahan R, Jahan M et al (2022) Sustainable textile industry: an overview. Non-Met Mater Sci 4. https://doi.org/10.30564/nmms.v4i2.4707
88. Wijethilake C, Upadhaya B, Lama T (2023) The role of organisational culture in organisational change towards sustainability: evidence from the garment manufacturing industry. Prod Plan Control 34:275–294. https://doi.org/10.1080/09537287.2021.1913524
89. Shafie S, Kamis A, Ramli MF et al (2021) Fashion sustainability: benefits of using sustainable practices in producing sustainable fashion designs. Int Bus Educ J 14:103–111. https://doi.org/10.37134/ibej.vol14.1.9.2021

Life Cycle Assessment of Textile Products

Raja Muhammad Waseem Ullah Khan and Khubab Shaker

Abstract Life Cycle Assessment is an approach to help organizations, and policymakers to make informed decisions for modifications of their process and practices, keeping in view their environmental impact. There are different approaches of LCA evaluating the effect of either a particular process or whole supply chain. The main stages of LCA include defining scope, inventory analysis, impact assessment, and interpretation. Textile companies have widely used LCA, as a tool to evaluate the environmental impact of denim production. These studies demonstrate that denim production has a considerable environmental impact, however, sustainable measures can assist to lessen this impact. Adopting sustainable practices, implementing water and energy-efficient technology, and promoting circular economy principles will assist to lessen the environmental effect of denim production and promote a more sustainable business.

1 Introduction

Environmental issues such as global warming, waste disposal greenhouse effect, and the change in climate, etc. have fostered the development of different programs for environmental protection. Various tools and indicators have been developed including Life Cycle Assessment (LCA), e.g., material flow analysis environmental impact assessment environmental risk assessment ecological footprints to assess the standard environmental impacts. LCA assessed the utilization of resources and environmental impacts throughout the life cycle of a product for example extraction of raw material, its processing, useful life and waste management (disposal and recycling) [1], as shown in Fig. 1. It is an effective tool for consumers, businesses, and policymakers to make educated decisions about the environmental impact of their actions [2, 3].

R. M. W. U. Khan
Department of Textile Engineering, National Textile University, Sheikhupura Road, Faisalabad, Pakistan

K. Shaker (✉)
Department of Materials, National Textile University, Sheikhupura Road, Faisalabad, Pakistan
e-mail: khubab@ntu.edu.pk

S. R. Batool et al. (eds.), *Circularity in Textiles*, Textile Science and Clothing Technology, https://doi.org/10.1007/978-3-031-49479-6_6

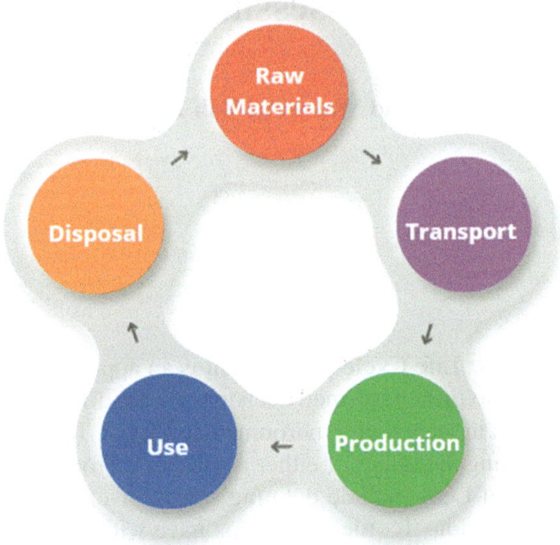

Fig. 1 Schematic of the entire life cycle of a product

It can further be used to facilitate product design, manufacture, and use, as well as to identify areas for improvement to reduce overall environmental impact [4, 5]. In this chapter, an overview of the life cycle assessment ends its benefits in the textile industry are discussed.

1.1 Life Cycle Assessment

There are four different stages in the LCA process including scope and goal identification, inventory analysis, impact analysis, and interpretation. In the goal and scope identification step, the aims of the study are defined, system boundaries are defined, and functional units are established [6]. At the stage of inventory analysis, data about all inputs and all outputs of the system is collected. In the impact assessment step the environmental impacts of the product development are evaluated using different environmental indicators like greenhouse gases emission land use and water consumption [7]. The interpretation stage entails analyzing the impact assessment results and developing conclusions regarding the long-term viability of the system. LCA is an iterative process that refines the results as more data is obtained, and each analysis provides more useful information to improve the subsequent LCAs.

Policymakers and companies can use LCA to reduce the environmental effect of their products and operations. LCA has applications in a wide range of industries, including manufacturing, construction, transport, and agriculture [8, 9]. LCA can be used in the manufacturing sector to identify opportunities to reduce a product's environmental effects throughout its life cycle. LCA can be used by a manufacturing unit for alternative solutions to reduce a product's carbon footprint by using more efficient manufacturing techniques or using more sustainable materials [10, 11].

LCA can be used in transportation to assess the environmental impact of various means of transportation such as vehicles, trains, and airplanes. This can assist policymakers and consumers in making informed decisions about selecting the mode of transportation depending on environmental effects. In agriculture, LCA can be used to examine the environmental impact of different agricultural practices, such as conventional versus organic farming. This can help farmers make informed decisions about which strategies to employ based on environmental impact [12]. LCA can be used in construction to assess the environmental impact of various building materials and construction methods. This can assist architects and builders in selection of materials and procedures to utilize in terms of environmental effects [13].

1.2 Importance of LCA in Textiles

Textiles and clothing represent one of the largest manufacturing sectors with a global export of 548.8 billion USD in 2021. Denim is one of the most common textile fabrics used to produce jeans, skirts, and jackets, and has a market size of 70.71 billion USD [14, 15]. It is recommended, as with other sectors and international norms, that the LCA of denim fabric be performed to identify areas of substantial impact.

This chapter overviews the LCA approach and its significance in assessing the sustainability of denim fabric production. The main processes involved in denim fabric production include cotton cultivation, yarn spinning, weaving, dyeing, and finishing, and each of these steps has a substantial environmental impact. For example, the cultivation of cotton has negative impacts on the soil and water quality in terms of the use of a large number of pesticides and water requirements [16, 17]. The spinning of yarn and weaving of fabric consumes a lot of energy while dyeing and finishing involves the use of huge amount of water and chemicals that are hazardous to the environment [18, 19].

2 Life Cycle Assessment Methodology

2.1 Different Approaches of LCA

LCA has six different approaches, as discussed below.

2.1.1 Cradle-to-Grave

It involves the complete life cycle assessment of a product, that starts from resource extraction (termed as cradle) and extends to its use phase and disposal at the end of life (termed as grave).

2.1.2 Cradle-to-Gate

This type of LCA involves the assessment of partial stages of the product life cycle, ranging from resource extraction (called cradle) to the factory gate (i.e., before the transportation of the product to the consumer) [7].

2.1.3 Cradle-to-Cradle

It is a sort of life cycle evaluation in which the end of the life of a product is the recycling process. Recycling is done to lessen the environmental impact of a product by indicating sustainable production, usage, and disposal procedures that strive to include social responsibility in product development [20].

2.1.4 Gate-to-Gate

This type of life cycle assessment examines only one value-added process of the entire production, i.e., from the factory's entry gate to its exit gate.

2.1.5 Wheel-to-Wheel

The impact of fuel used for cars and transportation is calculated in the wheel-to-wheel LCA type. The first stage investigates feedstock or fuel production and processing, as well as fuel delivery or energy transfer. The distribution or transmission of energy is referred to as the upstream stage, while the stage dealing with vehicle operations is referred to as the downstream stage.

2.1.6 Ecology-Based

The ecology-based LCA quantifies the services provided by economic commodities throughout their life cycle. This kind was designed to provide a guideline for competent human activity management by analysing the direct and indirect influence on ecological-based resources and surrounding ecosystems.

2.2 *Stages of Life Cycle Assessment*

The LCA follows the methodology explained in ISO 14040-14,043, which is the most comprehensive way to analyze the environmental impacts of textile products [6]. It is crucial for promoting sustainable development and analysing its effect on reduce negative environmental impacts. It does not, however, take into account the social and economic dimensions of sustainability [21]. LCA of a product/process is conducted in four key stages, as shown in Fig. 2.

2.2.1 Goal and Scope Definition

The aims of the study are specified, and the scope is established during this phase. The aims may include establishing environmental implications of a product/service, comparing the environmental impacts of various products/services, or identifying areas for improvement. The scope includes the system boundary, its functional units, and data requirements.

2.2.2 Inventory Analysis

Inventory analysis includes compiling a complete list of all inputs and outputs associated with the product or service. This includes the raw materials, energy inputs, and emissions produced by the product or service during its entire life cycle. Surveys, literature reviews, and other methods are commonly used to collect the data needed for this stage.

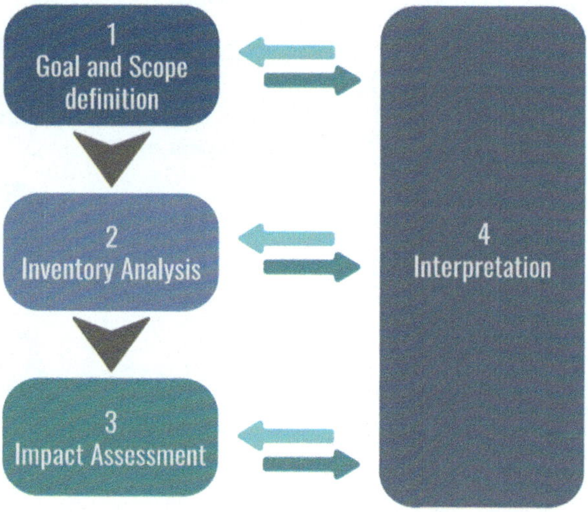

Fig. 2 Stages of life cycle assessment

2.2.3 Impact Assessment

This stage comprises evaluating the potential environmental impacts of the inventory-identified inputs and outputs. This is frequently performed using pre-defined impact categories such as greenhouse gas emissions, acidification, eutrophication, and human toxicity.

2.2.4 Interpretation

At the interpretation phase, the impact analysis and impact assessment results are concluded into the most relevant and actionable report. This step may also include sensitivity analysis to assess the robustness of the results and identify areas where further data is required.

2.3 Inputs and Outputs of LCA

The inputs and outputs of LCA vary depending on the product or service being evaluated, but in general, the following inputs and outputs are considered in the LCA, as shown in Fig. 3.

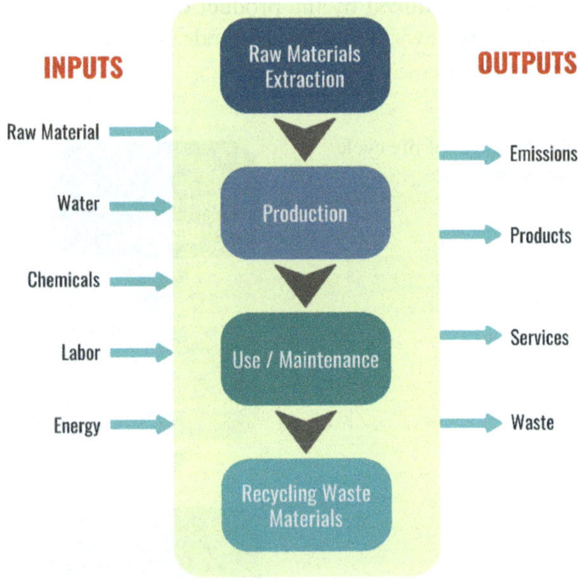

Fig. 3 Inputs and outputs of LCA

2.3.1 Inputs

- Raw materials: Natural resources used to create a product/service, e.g., metals, minerals, fossil fuels, etc.
- Energy: Expended in the extraction, manufacture, transportation, usage, and disposal of a product/service.
- Water: Used in the production of a particular product/service.
- Chemicals: Used in the production process of a product/service, e.g., solvents, adhesives, coatings, etc.
- Labor: Workforce required to produce the product/service [6, 7].

2.3.2 Outputs

- Emissions to air: Gases and particles released into the atmosphere during the production, use/maintenance, or disposal of the product/service, e.g., CO_2, NO_x, particulate matter, etc.
- Emissions to water: Pollutants released into the water during the production, use/maintenance, or disposal of a product/service, e.g., heavy metals, nutrients, organic compounds, etc.
- Waste: Generated during the production, use/maintenance, or disposal of a product/service, e.g., solid waste, wastewater, hazardous waste, etc.
- Products: Actual products produced during the production process.
- Services: Provided during the product or service use phase, e.g., maintenance, repair, and disposal services [6, 7].

LCA covers these inputs and outputs throughout the life cycle of a product from the generation of raw materials two the final disposal or recycling of developed products or use products including services. It gives a complete understanding of the overall performance of a product or service by quantifying the performance associated with the environment and these inputs and outputs.

3 Environmental Impacts of Textile Products

A few variables play an important role in the life cycle assessment of a textile product. Figure 4 identifies four main variables for LCA of a textile product including sustainability, fabric production, processability, and hazards.

- Sustainability: Sustainability is a major variable in this research because it has a direct impact on the environment. It is assumed as water-saving and less use of chemicals.
- Production: It has a vital role in the whole LCA of a textile product, hence it has a major impact on the environment.

Fig. 4 Variables considered for LCA of textile product

Table 1 Electricity emission factors for Pakistan [10]

Consumption category	Kg CO_2/kWh	Kg CH_4/kWh	Kg N_2O/kWh
Electricity (generated)	4.73×10^{-1}	1.38×10^{-5}	2.43×10^{-5}
Electricity (consumed)	6.16×10^{-1}	1.80×10^{-5}	3.16×10^{-5}

- Processability: It is studied from fiber to end product, how much the process is sustainable and reproducible.
- Hazards and drawbacks: These are also key factors, which should be in consideration while product LCA [6, 14, 22, 23].

All the above-mentioned impacts are analyzed based on energy consumption and greenhouse gas (GHG) emissions, abiotic depletion, eutrophication depletion/water pollution, waste generation, disposal, and land pollution [24].

3.1 Energy Consumption and GHG Emissions

Many aspects play an important role in climate change through the emission of different chemicals, gasses, and constraints. The term which is used to calculate the effect of these factors on the environment is called "Carbon footprints". Table 1 shows how much CO_2 is produced by 1 kWh of electricity [10].

3.2 Abiotic Depletion

Abiotic depletion is a category of environmental impact in LCA, that refers to the depletion of non-renewable resources from the earth's crust, such as minerals, ores, and fossil fuels, due to human activities. This impact category is also referred to as "depletion of non-renewable resources" or "mineral resource depletion." Abiotic depletion is frequently analyzed in LCA by determining the total amount of a certain resource utilized throughout the life cycle of a product/activity, including extraction, processing, transportation, and disposal. The impact is then expressed as an equivalent amount of a reference mineral, which is usually a metal or mineral used in industrial applications [6, 25].

Abiotic Depletion Potential (ADP) quantifies the impact of abiotic depletion and is measured in kilograms Sb-eq (antimony equivalent) or megajoules (MJ) and represents the amount of energy or antimony necessary to restore the depleted resource. The impact of abiotic depletion in LCA can be reduced by using more sustainable and efficient production methods, minimizing the use of non-renewable resources, and increasing the use of recycled or renewable resources [16].

3.3 Eutrophication Depletion/Water Pollution

The eutrophication phenomenon is induced by a higher nutrient level in the water body resulting from industrial processes. This heavy nutrient level supports the growth of aquatic plants, for example, algae, on the surface of water or in slow-moving aquatic habitats. The excessive growth of algae results in a layer that prevents sunlight from penetrating the water, leading to reduced oxygen levels. The algae perform photosynthesis, which further reduces the water's oxygen content, thus impacting aquatic life. Additionally, the microbial activity on dead algae at the bottom of the lake also contributes to the deoxygenation level, exacerbating eutrophication. These factors resist sunlight and oxygen from penetrating the water, which ultimately results in a reduced diversity of aquatic life, including fish and plants [26]. In the LCA assessment of environmental and ecosystem damage, eutrophication is expressed at the final stage. Although phosphorus is a basic nutrient for the growth of plants, excessive levels of phosphorus can cause plants to uptake the external Trojan before all the phosphorus is depleted, leading to the increase in the growth of algae in freshwater bodies [14, 27].

Plants require a range of nutrients, including nitrogen and phosphorus, in addition to water, sunshine, and CO_2, to grow. Typically, plants absorb nutrients from the soil via their roots. However, if the soil is poor or if there has been significant broaching or leaching, farmers may add fertilizers to increase plant growth. Since plants do not efficiently utilize all the nutrients in the soil, fertilizers are enriched with additional nutrients to boost productivity. It can be challenging to determine how much fertilizer a crop requires, given the variability of soil quality over small distances. As a precaution, farmers often add more fertilizer than needed. Unfortunately, excess nutrients are typically washed away by rain or irrigation and end up in ponds, lakes, and reservoirs. When these nutrients combine with water, they promote the rapid growth of algal plants, like how crops thrive in fields. Algal blooms can quickly cover an entire lake with layers of plant growth. While it may seem beneficial to have thriving plants in the environment, an overgrowth of algae creates a floating layer that prevents sunlight from penetrating the water, which negatively impacts the underwater plants' ability to participate in photosynthesis. Consequently, these plants must rely on internal energy reserves to survive. When the water's capacity to support a certain amount of life is exceeded, the excessive plants, phytoplankton, and algae die off and settle at the bottom of the water [1, 6, 9, 28].

In this scenario, the corpses are consumed by bacteria and other decomposers in a chemical process that uses up oxygen. Normally, the amount of dead organic matter in the environment remains relatively constant, resulting in stable oxygen levels. However, when a large amount of organic matter is introduced, such as through a plume of pollution, the excess material depletes almost all of the oxygen in the water during decomposition, leaving none for the aquatic organisms that rely on it. Consequently, fish and other creatures that require dissolved oxygen can suffocate and die. This can lead to an increase in decomposition and the need for more oxygen, exacerbating the situation. Recovery of the affected body of water can be a lengthy process and can depend on factors such as the amount and type of nutrients introduced, the size of the body of water, and the types of organisms present [29]. In lakes, the suppression of native species can provide opportunities for invasive species to take over. In the ocean, a lack of oxygen can cause corals to bleach and even die. Such events have negative impacts on various ecosystems around the world and contribute to a decrease in biodiversity. Clear-cutting practices that release nutrients to the soil can also result in nutrient-rich runoff, adding to the potential for such harmful plumes to occur [16, 30, 31].

3.4 *Waste Generation, Disposal, and Land Pollution*

When consumer's favorite clothing items no longer meet their needs, they face choices such as donating, reselling, or throwing away the item. Disposal is an important part of the adult life cycle, and the behavior of the consumer at this stage is the same important as purchasing behavior, according to Fletcher [32]. Young consumers feel that fast-growing fashion promotes a throw-away culture, where products lose their worth, prompting consumers to replace and discard products prematurely [33]. Even if manufacturers and designers continuously improve sustainable fashion, the effort will be null and void if consumers are unaware of how to make a positive environmental impact. Manufacturers and designers can utilize life cycle assessment (LCA) data to identify environmental issues and advise consumers about how to close the loop.

According to the CTR (2015) and EPA (2009), the United States generates 25 billion lbs. of textiles per year, which translates to around 82 lbs /resident, however, only 15% of that is donated or recycled, with the remaining 85 percent ending up in landfills (2009). To reduce waste going to landfills and positively impact the environment, recycling and donating at the disposal stage are encouraged. Reduce, reuse, and recycle (3 R's) are the three most popular trash management techniques. To reduce, one must purchase fewer things, and the EPA (2015) advises choosing goods with less packaging. Utilizing a product for the same or a different purpose, mending or refurbishing clothing, or giving away old items are all examples of reuse. Reducing and reusing are advantageous in many ways, such as eliminating pollution, saving money and energy, lowering greenhouse gas emissions, and extending the useful life of things [32].

Finally, recycling entails gathering and processing waste materials to create new products. Recycling conserves resources and consumes less energy than manufacturing new items. The mechanical method of extracting fibers from fabrics is used to manufacture new clothing.

Recycling programs, such as Cotton Inc.'s "Cotton from Blue to Green" and Patagonia's use of recycled materials, help reduce waste and lessen the environmental footprint. Patagonia uses recycled polyester, nylon, and wool in its product lines to reduce its reliance on oil and the amount of land required for sheep grazing [34]. Recycling garments protects landfills from being filled with non-decomposable synthetic textiles and woolen textiles. Recycling has endless possibilities and numerous environmental and economic benefits [32].

4 Case Study: LCA of Textile Products

There has been an increasing interest in performing LCA of textile products, particularly in recent years owing to enhanced awareness of their environmental impacts. However, the number of textile industries that have undertaken a life cycle assessment (LCA) varies by location, country, and industry sector. The Sustainable Garment Coalition (SAC), a consortium of prominent apparel and footwear firms, has created a tool named Higg Index, which incorporates an LCA module. Individual corporations, such as Nike and Levi Strauss & Co., have also conducted LCAs to assess the environmental impact of their products [35].

In addition to corporations, academic studies (LCA) have been done on textile products. These studies have looked at various aspects of the product, such as the environmental impact of cotton production, the impact of textile dyeing and finishing processes, and the usage of recycled materials. Overall, while the number of textile industries that have conducted LCAs may not be known, there has been a growing interest and effort to evaluate the environmental impact of the textile industry through LCA studies [22, 23]. Several companies have conducted LCA of denim fabric, e.g., Levi Strauss & Co. H&M Group, Gap Inc, Kontoor Brands (Wrangler and Lee), Nudie Jeans, G-Star RAW, Patagonia, AG Jeans, and MUD Jeans. These companies have undertaken LCA studies to identify the environmental impact of denim production, including raw material extraction, textile manufacturing, transportation, utilisation, and disposal. The outcomes of these analyses assist businesses in identifying areas where they may reduce their environmental effect and establish more sustainable operations [22, 34, 36–38].

4.1 LCA of Denim by Levi Strauss & Co.

LCA of denim products (jeans and jackets) was performed by Levi Strauss & Co. to study the environmental impacts, ranging from raw material extraction to end-of-life disposal. It was discovered that cotton production, indigo dyeing, and finishing operations had the greatest environmental impact. Cotton cultivation was shown to have the greatest impact in terms of water use, energy use, and greenhouse gas (GHG) emissions, according to the study. To address these issues, Levi Strauss & Co. adopted several sustainability practices, including the use of more sustainable cotton farming approaches, the reduction of water usage in the dyeing and finishing processes, and the investment in renewable energy sources. Through initiatives such as its "Buy Better, Wear Longer" campaign, the corporation also focused on extending the life of its products. Overall, the LCA analysis emphasized the industry to adopt sustainable production practices [5].

5 Processes Involved in the Scope of LCA

LCA is based on the whole cycle of products ranging from the extraction of raw material to the disposal at the end of service life. The intermediate steps including textile production and processing, distribution & transportation, and use & maintenance are also included in it.

5.1 Raw Material Extraction and Production

A widespread misconception about textile fibers is that synthetic fibers are terrible for the environment as compared to natural fibers. This is not exactly correct, because each type of fiber has its unique set of sustainability issues during production. Cotton, silk, flax, and wool are the four most often used natural fibers in the apparel industry and cultivated rather than being manufactured.

Manufactured fibers, on the other hand, are produced from a variety of raw resources, including natural and synthetic polymers. There are substantial concerns about sustainability during the production process. Cotton, for example, necessitates enormous amounts of water and pesticides, whereas emissions to the air and water are prevalent in the manufacture of synthetic and regenerated fibers. Natural fiber production can have a negative influence on water, while the production of synthetic fibers necessitates a substantial quantity of energy and nonrenewable resources [5].

5.2 Textile Production and Processing

The manufacturing step requires the transformation of fibers into fabrics and garments, according to the final product design. The suitable fibers are processed into yarn, which is subsequently utilized to make the fabric. The fabric is manufactured by weaving or knitting the yarns into a planar structure. It is then subjected to numerous finishing steps, including desizing, scouring, bleaching, dyeing/printing, etc. to get the desired appearance [39]. Cut-Make-Trim (CMT) is the last stage of production, where the fabric is converted into the garment. CMT, in contrast to the preceding steps, requires more manual labor to complete the garment.

Nonetheless, all stages of production significantly impact the environment, with fiber production and processing being a significant source of environmental impacts. The washing, dying, finishing, and many other wet processes involved in fiber production and fabric finishing also have impacts on the quality of water, air, and land. Therefore, to promote sustainability, manufacturers and designers need to evaluate the process used in fiber and yarn production critically and all the intervening steps required to develop fabric from fibers [5].

5.3 Distribution and Transportation

The environmental impact of distribution and transportation is significant due to the large amounts of fuel consumed during these processes. With rising fuel prices, the cost of freight charges has also gone up. However, transportation does not only occur after the manufacturing of a product but is a crucial factor that occurs at every step of its life cycle.

This is especially true for the fashion sector, where raw ingredients are frequently cultivated in one country, fabric production is done in another, and garment assembly takes place in still another. As a result, the garment frequently travels across international borders before it reaches the store floor. Reduced packaging is crucial throughout the distribution phase since it not only raises the price of the product but also generates more waste that eventually ends up in landfills. Although it is difficult to quantify the influence of transportation on the LCA, it is an important factor that must be considered in the quest for sustainability [5].

5.4 Use and Maintenance

This aspect of LCA is aimed at the end user and is beyond the control of the manufacturer. This stage emphasizes the need for consumers to purchase only what is necessary and take care of their garments to extend their longevity. Many companies, such as Levi Strauss & Co. with their "Care tag for our planet" program, actively encourage customers to take good care of their clothes by washing in cold water and washing less frequently. The reason for this is that, in some textile products, the consumer use phase has a greater environmental impact as compared to the production phase, e.g., clothing, workwear, and household textiles, etc. Nevertheless, there is little attention given to the effects of laundering and garment care by designers, and designing for the reduction of consumer use stage impact is not mentioned [5, 32].

Studies reveal that home laundering techniques contribute significantly to the environmental impact of the consumer use stage.

For example, Franklin Associates performed LCA of a polyester knit blouse in 1993 and discovered that washing and drying account for 82% of energy consumption, 66% of solid waste, more than half of air emissions (83% CO_2), and significant volumes of waterborne effluents [32]. Significant environmental benefits can be realized by improving and changing laundering habits, especially in the case of furnishings and carpets. Energy, water, and detergent consumption during washing, as well as energy usage during drying and ironing, are the main laundering concerns [5, 32]. Therefore, when discussing sustainability, it is crucial for manufacturers to consider the garment's use throughout its life cycle.

5.5 End of Life

Individuals who are genuinely concerned about protecting the environment may have strong opinions on how to handle what they perceive as "waste." The LCA process now includes the end user and is no longer the sole responsibility of the manufacturer. Consumers can extend the lifespan of their clothing by passing it on to others or sending it for reselling or recycling. This approach promotes responsible consumption and encourages people to only buy what they need while taking care of their possessions. However, despite efforts from companies like Levi Strauss & Co. to promote responsible washing and disposal, denim clothing is still washed frequently and often ends up in landfills after just one year.

As mentioned above, to reduce the environmental impact of waste the three R's-reduce reuse and recycle -are mainly used strategies. Reusing an item includes using it for the same purpose or using the same product in a new way. Donating unnecessary or unwanted items to different charity groups or friends and family is a common way to reuse used products.

Reduce and reuse practices can help to minimize pollution generated by the production of new materials and it ultimately contributes to energy savings, and reduced greenhouse gas emissions. Recycling needs collecting and processing waste materials in order to turn them into new useable products. This approach conserves resources and consumes less energy as compared to the manufacturing of new products (Table 2).

Table 2 Findings of life cycle impact analysis for LEVI'S 501® jeans [5]

		Fiber	Fabric assembly	Cut, sew, finish	Sundries & packaging	Transport logistics retail	Consumer care	End of life	Total
Climate change (kg CO_2-e)	Value	2.9	9.0	2.6	1.7	3.8	12.5	0.9	33.4
	Share	9%	27%	8%	5%	11%	37%	3%	100%
Water consumption (liters)	Value	2,565	236	34	77	10	860	0	3,781
	Share	68%	6%	1%	2%	0%	23%	0%	100%
Eutrophication (PO_4-e g)	Value	18.0	5.5	2.9	7.9	3.1	7.9	3.5	48.9
	Share	37%	11%	6%	16%	6%	16%	7%	100%
Land occupation (m^2/year)	Value	9.3	0.2	0.0	0.5	0.3	1.7	0.0	12.0
	Share	78%	1%	0%	4%	2%	14%	0%	100%
Abiotic depletion (mg Sb-e)	Value	19.9	7.2	1.9	118.5	4.4	17.9	0.1	29.1
	Share	12%	4%	1%	70%	3%	11%	0%	100%

5.6 *Key Findings of the Case Study*

According to the LCA studies, the most significant contributors to the environmental impact of denim manufacture are cotton cultivation, its dyeing, and finishing. The Indigo dye has been identified as a major contributor to water pollution and other environmental impacts. Furthermore, denim fabric transportation to retail sites and end-of-life disposal have been identified as substantial contributors to the environmental effect of denim manufacture [22, 40].

Several environmentally friendly practices have been implemented to lessen the environmental impact of denim fabric production. Organic cotton, natural dyes, the utilization of water and energy-efficient technology, and denim recycling/upcycling are just a few examples. Furthermore, by minimizing waste and encouraging reuse and recycling, the adoption of circular economy concepts can assist to reduce the environmental effect of denim fabric production [5, 41].

5.7 *Limitations of LCA*

Although LCA is a useful tool for analyzing the environmental effects of products, it has some significant limitations. One limitation is that the data used for analysis limits its precision. If there are errors in the data collection, the LCA results will be unreliable. Another concern is that LCA does not take social or economic aspects into account. For example, a product with a lower environmental impact may be more expensive to produce, which may be prohibitively expensive for some consumers or businesses [3, 40, 41].

6 Conclusion and Perspectives

LCA help organizations, consumers, and policymakers to make informed decisions about modifications of their process and practices, keeping in view their impact. Textile companies have widely used LCA, as a tool to evaluate the environmental impact of denim production. These studies demonstrate that denim production has a considerable environmental impact, however, sustainable measures can assist to lessen this impact. Adopting sustainable practices, implementing water and energy-efficient technology, and promoting circular economy principles will assist to lessen the environmental effect of denim production and promote a more sustainable business. While LCA has some limitations, it is nonetheless an effective tool for promoting sustainability and lowering the environmental impact of products and services.

In recent times, numerous communications have been published on LCA studies, each claiming impartiality. To communicate effectively, it is essential to strictly adhere to the reference standards set by the SR EN ISO 14040 series, along with the required procedures and records. Implementation can be aided by a consultant who can provide evidence-based solutions. A vision is necessary for the overall management, linking the action system, product assembly, sub-system components, and parts.

References

1. Rebitzer G, Ekvall T, Frischknecht R et al (2004) Life cycle assessment. Environ Int 30:701–720. https://doi.org/10.1016/j.envint.2003.11.005
2. Henry BK, Russell SJ, Ledgard SF, et al (2015) LCA of wool textiles and clothing. In: Handbook of life cycle assessment (LCA) of textiles and clothing. Elsevier, pp 217–254
3. Wenzel H (1998) Application dependency of LCA methodology: key variables and their mode of influencing the method. Int J Life Cycle Assess 3:281–288. https://doi.org/10.1007/BF02979837
4. Koszewska M (2015) Life cycle assessment and the environmental and social labels in the textile and clothing industry. In: Handbook of life cycle assessment (LCA) of textiles and clothing. Elsevier, pp 325–344
5. Levi Strauss & Co. (2015) the life cycle of a jean. Understanding the environmental impact of a pair of Levi's 501 jeans. Levi Strauss & Co., pp 16–50
6. ISO 14040 (1997) Environmental assessment–life cycle assessment - Principles and framework. International Standard Organisation pp 1–20
7. Hackett T (2015) A comparative life cycle assessment of denim jeans and a Cotton T-shirt : the production of fast fashion essential items from cradle to gate
8. Kalliala E, Talvenmaa P (2000) Environmental profile of textile wet processing in Finland. J Clean Prod 8:143–154. https://doi.org/10.1016/S0959-6526(99)00313-3

9. Zhang Y, Liu X, Xiao R, Yuan Z (2015) Life cycle assessment of cotton T-shirts in China. Int J Life Cycle Assess 20:994–1004. https://doi.org/10.1007/s11367-015-0889-4
10. Khan WM, Siddiqui S (2017) Estimation of greenhouse gas emissions by household energy consumption: a case study of Lahore, Pakistan. Pakistan J Meteorol 14:65–83
11. Muthu SS (2015) Environmental impacts of the use phase of the clothing life cycle. In: Handbook of life cycle assessment (LCA) of textiles and clothing. Elsevier, pp 93–102
12. Muralikrishna IV, Manickam V (2017) Environmental management life cycle assessment. Environ Manage 57–75. https://doi.org/10.1016/B978-0-12-811989-1.00005-1
13. Asim M, Uddin GM, Jamshaid H et al (2020) Comparative experimental investigation of natural fibers reinforced light weight concrete as thermally efficient building materials. J Build Eng 31:101411. https://doi.org/10.1016/j.jobe.2020.101411
14. Schrott W, Paul R (2015) Environmental impacts of denim manufacture. In: Denim. Elsevier, pp 563–580
15. Sandin G, Peters GM (2018) Environmental impact of textile reuse and recycling—a review. J Clean Prod 184:353–365. https://doi.org/10.1016/j.jclepro.2018.02.266
16. La Rosa AD, Grammatikos SA (2019) Comparative life cycle assessment of cotton and other natural fibers for textile applications. Fibers 7:101. https://doi.org/10.3390/fib7120101
17. Duque Schumacher AG, Pequito S, Pazour J (2020) Industrial hemp fiber: a sustainable and economical alternative to cotton. J Clean Prod 268:1–13. https://doi.org/10.1016/j.jclepro.2020.122180
18. Pardo Martínez CI (2010) Energy use and energy efficiency development in the German and Colombian textile industries. Energy Sustain Dev 14:94–103. https://doi.org/10.1016/j.esd.2010.02.001
19. Jena B, Das BP, Khandual A et al (2015) Ecofriendly processing of textiles. Mater Today Proc 2:1776–1791. https://doi.org/10.1016/j.matpr.2015.07.020
20. Nawab Y, Sapuan SM, Shaker K (2021) Composite solutions for ballistics, 1st edn. Cambridge, USA
21. Jolliet O, Margni M, Charles R et al (2003) IMPACT 2002+: a new life cycle impact assessment methodology. Int J Life Cycle Assess 8:324. https://doi.org/10.5479/si.00963801.32-1531.411
22. García Sánchez M, Padilla-Rivera A, Güereca LP (2023) Social life cycle assessment of Mexico city's water cycle. Adv Sustain Syst. https://doi.org/10.1002/adsu.202300024
23. Risner D, Kim Y, Nguyen C, Siegel JB, Spang E (2023) Environmental impacts of cultured meat: a cradle-to-gate life cycle assessment Derrick. https://doi.org/10.1101/2023.04.21.537778
24. Roy P, Nei D, Orikasa T et al (2009) A review of life cycle assessment (LCA) on some food products. J Food Eng 90:1–10. https://doi.org/10.1016/j.jfoodeng.2008.06.016
25. Esteve-Turrillas FA, de la Guardia M (2017) Environmental impact of recover cotton in textile industry. Resour Conserv Recycl 116:107–115. https://doi.org/10.1016/j.resconrec.2016.09.034
26. Heijungs R, Kleijn R (2001) Numerical approaches towards life cycle interpretation five examples. Int J Life Cycle Assess 6:141
27. Periyasamy AP, Wiener J, Militky J (2017) Life-cycle assessment of denim. Elsevier Ltd
28. van der Werf HMG, Turunen L (2008) The environmental impacts of the production of hemp and flax textile yarn. Ind Crops Prod 27:1–10. https://doi.org/10.1016/j.indcrop.2007.05.003
29. Ecocompass (2016) Content ▶ Background ▶ LCA methodology ▶ Selected results of literature survey ▶ Conclusions & Outlook
30. Ozturk E, Koseoglu H, Karaboyaci M et al (2016) Sustainable textile production: cleaner production assessment/eco-efficiency analysis study in a textile mill. J Clean Prod 138:248–263. https://doi.org/10.1016/j.jclepro.2016.02.071
31. Eryuruk SH (2015) Life cycle assessment method for environmental impact evaluation and certification systems for textiles and clothing. In: Handbook of life cycle assessment (LCA) of textiles and clothing. Elsevier, pp 125–148
32. Fletcher K (2014) Sustainable fashion and textiles
33. Doyle SA, Moore CM, Morgan L (2006) Supplier management in fast moving fashion retailing. J Fash Mark Manag Int J 10:272–281. https://doi.org/10.1108/13612020610679268

34. Patagonia (2015) Recycled Nylon. https://www.patagonia.com/our-footprint/recycled-nylon.html
35. Bianco I, Gerboni R, Picerno G, Blengini GA (2022) Life cycle assessment (LCA) of MWool® recycled wool fibers. Resources 11:1–16. https://doi.org/10.3390/resources11050041
36. Buckner CA, Lafrenie RM, Dénommée JA et al (2018) Complementary and alternative medicine use in patients before and after a cancer diagnosis. Curr Oncol 25:275–281. https://doi.org/10.3747/co.25.3884
37. Wu M, Sadhukhan J, Murphy R et al (2023) A novel life cycle assessment and life cycle costing framework for carbon fibre-reinforced composite materials in the aviation industry. Int J Life Cycle Assess 28:566–589. https://doi.org/10.1007/s11367-023-02164-y
38. Gheewala SH (2023) Life cycle assessment for sustainability assessment of biofuels and bioproducts. Biofuel Res J 10:1810–1815. https://doi.org/10.18331/BRJ2023.10.1.5
39. Rukhaya S, Yadav S, Rose NM, et al (2021) Sustainable approach to counter the environmental impact of fast fashion. ~517. Pharma Innov J 10:517–523
40. Zimdars C, Haas A, Pfister S (2018) Enhancing comprehensive measurement of social impacts in S-LCA by including environmental and economic aspects. Int J Life Cycle Assess 23:133–146. https://doi.org/10.1007/s11367-017-1305-z
41. Giang NP, Dao MA, Ha N (2022) Life cycle assessment approach. Fact Affect Appl Lca Enterprises 8:318–329. https://doi.org/10.5281/zenodo.6402371

Recycling in Textiles

Muzzamal Hussain, Munir Ashraf, Hafiz Muhammad Kaleem Ullah, and Saba Akram

Abstract This chapter briefly presents the need of recycling of textiles and why there is so much emphasis globally to recycle the textiles. The recycling process include different steps from waste collection to sorting, recycling, yarn and fabric manufacturing. The sorting of textile waste is very crucial step as the quality of final recycled fibers depends on the quality of input feed stock. In this chapter the sorting of textiles wastes and available commercial solutions are discussed in details. This chapter also discuss in details the available recycling technologies including mechanical, chemical, biological and thermo-mechanical recycling including their advantages and limitations. This chapter also have details regarding recycling strategies of different feed stock based on the type of material and material compositions e.g., cotton, cotton/polyester and thermoplastic fibers.

1 Introduction

It has been reported that the textile production has doubled since 2000 due to higher demand of textile products. This increased demand in reality is the reflection of increased purchasing power of people worldwide. For instance, the purchasing power of Chinese has increased massively during the last two decades. Resultantly, the utilization of garments has decreased significantly as they have more number of garments at a particular time. This leads to disposal of large number of underutilized textiles which end up in landfills or get incinerated. On one side the requirements for new materials are increasing exponentially and on the other side, they are massively

M. Hussain (✉)
National Center for Composite Materials, National Textile University, Faisalabad 37610, Pakistan
e-mail: muzzamal313@gmail.com

M. Ashraf · S. Akram
Functional Textiles Research Group, School of Engineering and Technology, National Textile University, Faisalabad 37610, Pakistan

H. M. Kaleem Ullah
Centre Européen des Textiles Innovants (CETI), 59200 Tourcoing, France

S. R. Batool et al. (eds.), *Circularity in Textiles*, Textile Science and Clothing Technology, https://doi.org/10.1007/978-3-031-49479-6_7

underutilized. At the moment 87% of the total fibers used to make clothing are either landfilled or incinerated, this causes a lost opportunity [1]. Only 13% are recycled in a way that 12% is open loop recycling (cascaded applications) and less than 1% is being recycled back to textiles. Therefore, the present business model of textiles is called linear as the textile products are disposed of despite having significant value.

The present linear business model of textiles has a huge environmental impact. Clothes are massively underutilized which causes a substantial loss of resources [1]. Such a lower rate of recycling and massive underutilization are increasing the burden on available resource streams. The fiber consumption has doubled as compared to 2000 and will increase three-fold by 2050 [2]. If we continue to the linear model of consumption, Green House Gas (GHG) Emissions will increase from 2–6 to 26% by 2050 and we will add 22 million tons of microplastic to the environment. These impacts can only be reduced by shifting to a circular economy, which focuses on the utilization of material to the full extent in the first place, and at the end of use, these will not be disposed of but will be recycled back to new garments. Recycling is one of the most important components of a circular economy [3].

There are different types of textile waste which are available at different stages of the textile value chain as shown in Fig. 1. A huge material resource can be saved by recycling this waste. If dynamics of textile recycling are considered the majority of post-industrial waste is recycled in some way, while distribution and post-consumer waste are recycled in low quantities, which poses a very serious challenge.

If the textile waste is further characterized based on its value it can be subdivided into three main categories as shown in Fig. 2. Some textile waste is not usable at all as it contains different types of oil marks and stains, some are recyclable, and some are reusable. It should be taken into account that textile products should be used as long as they can be before being sent for recycling. The large portion of textiles is

Pre-Consumers	Post-Industrial	Fibre, filament, yarn, woven, knitted, braided, non-woven etc.
		Source: Fibre & Filament production, spinning, weaving, knitting, non-woven, finishing and garments
	Distribution	Garments, home textiles, technical textiles and non-woven
		Source: Brands, retailers, distributers
Post-Consumer	Garments, Home textiles, technical textiles, non-woven	
	Source: Final Consumer	

Fig. 1 Types of textile waste available at different stages of textile value chain

Fig. 2 Categories of textile waste based on its value [1]

actually reusable. It is very important to know that when a textiles should be recycled and in which conditions.

Textile recycling is very complex as some textiles are easy to recycle while others are difficult. There are different textile waste streams available based on the type of textiles e.g., apparel, home textiles, and technical textiles as shown in Fig. 3. Some are easily recyclable like apparel and home textiles and technical or industrial textiles are very difficult to recycle due to very complicated fabrication. Further, some textiles are used for short periods like face masks, and non-woven gowns, and some are used for longer periods e.g., curtains, and firefighter uniforms. There will be different standards and methods to recycle different type of textiles. Industrial or technical textiles are by far the least recycled textiles among all.

In apparel, everything can be recycled, which shows a huge opportunity to decrease the burden on virgin materials but on the contrary, there are a lot of factors that are limiting the recycling of textiles. Figure 4 shows the reasons identified for the low recycling in the textiles and apparel sector.

Fig. 3 Type of textile waste streams available for recycling

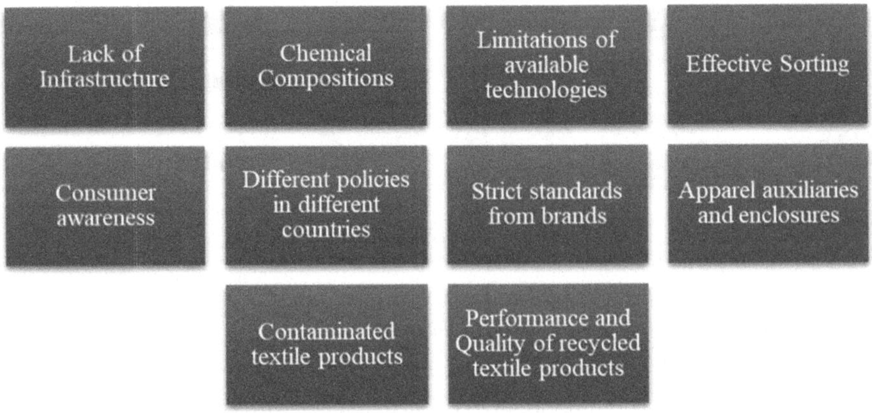

Fig. 4 Reasons for low recycling in the textile sector [4]

2 Challenges in Recycling

2.1 Limited Textile Recycling Infrastructure

The textile collection and recycling supply chain is a multi-step process. Several collection points, a huge infrastructure of recycling, and intensive operations are required. All this is impossible without the input of sufficient capital. This is one of the biggest constraints in the recycling of Textiles [5]. Further, It is treated as municipal waste in many parts of the world and that's why ends up in landfills. The recycling of textiles is not integrated into the main industry due to the linear model of the economy. Recycling is just complementing the main textile industry at the moment.

2.2 Chemical Composition

A major portion of Textile products are composed of more than one type of material. Materials from different origins are usually combined to achieve various functional and performance objectives. These blends are difficult to separate and may require several distinct processes which increase recycling costs and complexities [6].

2.3 Effective Sorting

Efficient sorting of textiles is a very important step for further recycling processes. Sorting is one of the bottlenecks of textile recycling and advancement is required in this sector, fully operational, practical, automated, advanced, and efficient technologies are still required [6].

2.4 Apparel Auxiliaries and Enclosures

Garments mostly contain secondary materials in the form of leather, zippers buttons, etc. These materials cannot be recycled with textiles using the same processes, they add more complexities to textile recycling and must be removed before the material reclamation process initiates [7].

2.5 Contaminated Textile Products

Textile products are inherently contaminated with dyes and finishes when taking them in the context of recycling perspective, the story of contamination does not end here, during the service life of textiles they are further contaminated with oils, stains, and chemicals. These contaminants degrade the quality of recycled materials [8].

2.6 Performance and Quality of Recycled Textile Products

Recycled and reclaimed materials from textile recycling processes may not be of the same quality as their virgin counterparts, they may offer reduced strength, fast aging, and poor aesthetics. This significantly impacts the usability and applications of recycled materials in high-quality products [9].

2.7 Consumer Awareness

Textiles are the most widely used and most important products after food, Everyone uses textile products and the majority of the consumers may not know the environmental impacts the textile products may cause. Educating consumers about recycling and persuading them to recycle their discarded textiles can be challenging. The majority of textile consumers still trash their textile waste rather than recycle it [10]. The consumers also perceive the recycled products as inferior quality, which makes it more difficult to accelerate the recycling of textiles.

2.8 Limitations of Available Technologies

The available technologies for recycling downgrade the properties of fibers as compared to virgin raw material which was used to make the product at first instance. This thing shakes the confidence of both brands and consumers in textile products made with recycled products. Therefore, there is a need for strong investment in the new technologies to recycle the textiles with the highest properties retention.

2.9 Different Policies in Different Countries

In some parts of the world, textile waste is considered trash, and in some parts of the world, there is no such policy to deal with textile waste. Therefore, there is a need for strong legislation to cater to this big problem of resource waste. Further, the policies should be uniform around the globe to get the best out of it.

2.10 Strict Standards from Brands

One of the reasons the recycling industry has not flourished at the moment is that brands have strict standards for products which are making it very difficult to increase the portion of recycled content in the product.

There are different technologies available for textile waste recycling including mechanical, chemical, and thermomechanical recycling. These technologies have their pros and cons, Mechanical recycling is the most widely used technology but the issue with it is that it deteriorates the properties of textile fibers, then chemical and thermomechanical recycling are available but those are still evolving. Despite the significant available potential of materials recovery by recycling, it is still not done at the level of requirement.

In the following section, the available techniques used for textile recycling are presented. But before discussion about the recycling technologies, the most important step comes about the sorting of textiles. The feed stock sorted effectively will result in better quality fibers.

3 Textile Sorting for Recycling

Textile wastes collected separately cover on average 30% of the textiles placed in the market each year [11]. Mostly, textiles are being collected through banks with indoor collection systems such as at first or second-hand retail stores. With some exceptions, the majority of textiles collected separately are reused in global second-hand markets. On the other hand, at collection points, consumers also dispose of textiles that are not reusable or re-wearable because of their intensive use, cleanliness, damage, or poor quality. These are referred to as post-consumer textile wastes. The volume of collected textiles is rapidly increasing due to the growing fashion industry. To reuse or to prepare for recycling, these textiles will need to be sorted properly [12]. Currently, textile sorting is mainly done manually and is a main financial driver for textile sorting activities [11, 13]. However, because of the increasing volume of post-consumer textile waste, manual sorting is not an optimum solution for textile recycling. Moreover, textile recycling techniques such as mechanical/thermomechanical and chemical textile recycling processes require the identification of feedstock and are free of pollution for efficient processes and high-quality recycling outputs [14]. To feed such recycling systems, the manual sorting must be replaced by a semi or fully-automatic sorting system that separates the textiles based on their material type and color. Different projects have been initiated to automate the textile sorting systems such as SIPTex, Fibersort, REISKAtex, and Resyntex projects [15–17]. The technologies being developed during these projects are based on optical and near-infrared (NIR) spectroscopic techniques which sort the textiles depending on their fiber composition, and color, and also detect metal pieces. In NIR spectroscopy, the textile is illuminated with a broad infrared spectrum of near-infrared light which can be absorbed, transmitted, reflected, or scattered by the textile sample. As a result, it will produce a characteristic spectrum which is then compared to the predefined database and helps to detect the composition of the textile material. However, NIR has some advantages and disadvantages [11]. Table 1 shows some examples of these technologies.

Table 1 Example of textiles sorting with their used technologies [18]

Industry	Sorting technology
Valvan [19, 20] (Belgium, Netherlands, UK)	Industrial scale sorting using near-infrared (NIR) spectroscopy (i.e. FIBERSORT project)
IVL Swedish environmental research institute [21] (Sweden)	Automated sorting method similar to packaging sorting based on optical sensors that can detect different fiber materials
Telaketju [22] (Finland)	Small-scale identification and sorting technique using near-infrared (NIR) spectroscopy (i.e. REISKAtex® project)
HKRITA [23] (Hong Kong)	Automated sorting system (residual metal detection, color detection algorithm) with robotics AGV and intelligent conveyor control
CETIA [24] (France)	Automated sorting system (residual metal detection, color detection algorithm) with robotics AGV and intelligent conveyor control

3.1 Advantages

- Higher efficiency and accuracy than manual sorting.
- Additional tagging of the garment is not required.
- Inexpensive compared to other spectroscopic technologies.

3.2 Disadvantages

- Difficult to identify fiber composition covered by darker colors and finishes (i.e. dyes or detergents).
- Unable to recognize the low content of fiber in blends, especially elastane.
- Weak infrared light which only penetrates the outermost layer of the garment.

4 Mechanical Recycling

Mechanical recycling of textiles mostly involves post-industrial textile waste and sometimes post-consumer waste broken down into individual fibers by using aggressive mechanical processes. Mechanical recycling of textiles is commonly used technique worldwide despite its limitations. One of the reasons for its popular usage is unavailability of alternative methods at that scale where they can cater to huge demand. Mechanical recycling has existed for a long time and is widely used in the cottage industry to recycle many types of fibers from natural to synthetic. Further, the process of mechanical recycling is very simple.

4.1 Mechanical Recycling Process

- Textile Waste is collected and sorted based on quality, color, and fiber type.
- The sorted textile waste is cut into smaller pieces, and attachments and enclosures are removed from textile waste if any. The shredding process converts the large pieces of clothing into tiny pieces that are then suitable for further processing. Shredding is done at different scales e.g., commercial shredding, milling, and low scale shredding [25]. The fabric shredder works on the shear and cut principle to complete the shredding process of textile waste. The motor of the shredding machine drives the knife, and the blade of opposite cutters cuts the waste material. The crushed material discharged from the bottom of the machine. The waste fabrics are shredded twice by the shredding machine for the preparation of fiber webs.

Fig. 5 Schematics of the typical mechanical recycling process of textiles

- Tiny pieces of fabrics are then opened in a carding machine, this machine opens the material into fiber-to-fiber level individually. It develop a homogeneous web. These fiber webs are further utilized solely or in combination to develop the yarn.
- The reclaimed fibers are then mixed with virgin material if required and then fed to the conventional spinning process, this process is utilized for yarn manufacturing. Mostly the yarn prepared from recycled fibers is coarse and sometimes mixing enables the manufacturing of finer yarn qualities too.
- The recycled yarn is converted into fabric either by weaving or knitting process. The yarn once converted into fabric can be used to manufacture any textile product, either clothing or home textiles, etc.

Mechanical Recycling of Textiles mostly follows the following route as shown in Fig. 5.

4.2 Mechanical Recycling Techniques

There are different techniques used to mechanically recycle textiles, (1) Fibers to needle punching non-woven (2) Fibers to spun yarn.

Mechanical recycling uses two waste streams (a) Soft waste which is spinning process waste, e.g., card dropping, comber noil or ring dropping (b) Hard waste which can be pre-consumer or post-consumer, e.g., cutting waste, or used garments etc. To

Fig. 6 Typical process routes of mechanical recycling of textile waste

recycle both hard and soft waste different processes are adopted. The soft waste is taken from the spinning process, after processing and passing through different rollers and removal of short fibers and trash, the fibers are sent back to the spinning process again. The hard waste however is recycled through a rigorous process as shown in Fig. 6.

The hard waste as discussed earlier comes from different streams, if cutting waste of 100% cotton is recycled, it will be processed through a route as shown in Fig. 6. If the post-consumer waste is processed, it is collected, sorted, and then sent to the recycling industry which will again follow the process as shown in Fig. 6. Sometimes, the seam portion is removed from waste to avoid fiber mixing, especially for 100% cotton fiber [26].

The textile waste needs some pretreatment before shedding including sorting, washing, and cutting off [25, 27].

4.3 Factors Effecting the Mechanical Recycling

There are also some other parameters and materials that are important for the quality of output of the shredding machine [25]. These include material properties for example moisture and strength etc., conditioning of textile waste, size distribution, shredding box size, blade direction, blade length, power of motor, shredding time in terms of repetitions and card of shredding machine etc. This process has also some limitations, due to its aggressiveness, it is hard to maintain the fiber quality. Material heterogeneity is also a big concern for attaining a single type of output from the mechanical recycling process. This process is only used efficiently for one type of textile waste material [27, 28].

4.4 Advantages of Mechanical Recycling

The following are the advantages of mechanical recycling of textiles [29, 30].

- Mechanical Recycling is a straightforward approach to textile recycling. It does not require specialized knowledge-based processes and a trained workforce. Already present processes and workforce may be utilized to recycle textiles mechanically.
- It helps in the conservation of resources by reducing the requirement of virgin materials needed to manufacture textile products.
- It is the most widely used textile recycling process, it directly helps to recycle a lot of textile waste that may otherwise be landfilled.
- Mechanical recycling requires less energy compared to virgin materials.

4.5 Disadvantages of Mechanical Recycling

Despite being widely used, mechanical recycling has various limitations and challenges associated with it as well [29, 31].

- Sorting is the first step of mechanical recycling, sometimes mere sorting may not be enough to completely separate different types of materials from each other, For example, textiles are mostly blends of different materials, a common example is a cotton-polyester blend. The first and most widely faced implication of recycling is combating the impact of various materials and blend ratios in different articles.
- The process of collection, grading, sorting, processing, and logistics involved in the supply chain of recycling may impact the economic viability of the recycled products, sometimes the profit margins may not be competitive compared to virgin materials.
- The quality of recycled raw materials may be lower than virgin raw materials.
- The raw material reclaimed from mechanical recycling are mostly contaminated by dyes and finishes that may affect the appearance and processing of recycled products.

5 Chemical Recycling

Chemical recycling is a process that converts polymers back into their monomers. A series of chemical procedures are adopted for the chemical recycling of textiles to dissolve/depolymerize the fibers into monomer/solution to make either an innovative fiber compound or extract one compound from a mixture. Most often, the output product is of similar quality, just like its virgin counterparts, without significant loss in physical properties by chemical recycling. It is far more developed than mechanical recycling when it comes to technology because it uses chemicals, enzymes, a regulated atmosphere, etc. for its procedure, thus possessing additional advantage of reduced restriction in the fabric form like catering, knits, and woven to a broad range of products such as home decors, auto-parts, and jackets, etc. [32].

Chemical treatment is a potential process for the recycling of textiles. In this method, chemicals are used for the degradation of complex polymers of textiles to produce small molecules. Moreover, an oligomer, crude feedstock, or monomer in the form of liquid or gas are all the products of chemical textile recycling. [33, 34].

Tertiary recycling or chemical recycling enables the retrieval of more worthy products from polymeric waste than incineration and overpowers a few issues that restrict mechanical recycling. A more restrictive definition may define chemical recycling as "the production of valuable chemical products from polymeric waste materials through economically feasible procedures". Chemical recycling products are reintroduced easily into the production cycle without any market saturation problems; another advantage is that the crude products resulting from a chemical breakdown can be utilized without further purification [32].

Easiest to depolymerize are resins with condensation types such as polyurethanes (PU), polycarbonate (PC), polyamide (PA), polyester (PET), etc. In these materials, the molecule's bonding is such that if suitable pressure and heat are applied in the occurrence of a chemical reactive agent, they disrupt into shorter chains relatively in a manageable way. Breakdown technologies of these polymers, mainly aminolysis, methanolysis, hydrolysis, and glycolysis have already been proven to be cost-effective comparatively. The recovered products from such chemical procedures can be either monomers with a purity degree preferable for re-polymerization or oligomers mixture with reactive end moieties. The overview of chemical recycling technique is represented in Fig. 7.

Fig. 7 Overview of chemical recycling techniques for textile waste

Depolymerization of addition type polymers (acrylates, styrenics, etc.) is of great interest for the recovery of monomer through unzipping the bonds precisely. This needs an appropriate selection of catalysts and conditions for a reaction. When unzipping of monomers occurs from polymer chain at a lower rate than active site transfer reactions, a complex product mixture is formed from low to medium molecular weight. Generally, it is not economically viable to derive well-defined chemical products from this mixture directly, which can be utilized instead either as a fuel or restored inside petroleum processing. The difference between mechanical and chemical recycling is that chemical recycling uses operations to reuse waste polymers by degrading them to its constituents. Feedstock recycling and depolymerization-based chemical recycling procedures break down the long hydrocarbon chains of a material into shorter hydrocarbon fractions or monomers utilizing catalytic, thermal, or chemical procedures. In contrast, purification deals with utilization of solvents for eliminating additives from the polymers [3]. Thus, chemical recycling technique has been emerged as outstanding technique to recycle the textile waste.

5.1 Types of Chemical Recycling Techniques

5.1.1 Pyrolysis

Pyrolysis can be defined as the thermal decomposition of feedstock in the absence of oxygen causing the gas, liquid, and solid fractions [35]. Therefore, textile pyrolysis may be an option for waste materials disposal while enabling useful product recovery. Textiles are a potent feedstock source in municipal waste streams at a low weight percentage. Pyrolysis works as a promising technology for recycling textiles that can be utilized for the degradation of solid waste carbon-polymers to become three products of pyrolysis in the gas, solid and liquid states [36–38]. Diverse textile materials which has not been sorted before could be used in pyrolysis [39]. Pyrolysis without oxygen reallocated O/H/C elements from the waste of organic compounds into pyrolysates exhibiting three phases. Syngas (CO and H_2 combination) produced through thermochemical procedures were used as a raw material or direct fuel to generate alcohols and other hydrocarbons [40, 41]. Pre-treatment is not required in the pyrolysis process and is a potent process for contaminated waste treatment. Such features make pyrolysis more effective procedure than other procedures such as bio-chemical, which need various chemicals and repeatedly produce lot of waste which should be re-landfilled and require more time to operate at large scale [42, 43]. Table 2 summarizes the diversified yields of the gas, liquid, and solid products generated through the pyrolysis of textile waste. Generally, liquid products generated from the textile waste are oxygen compounds and mono-polyaromatic, consisting of hydrocarbons like ketones, carboxylic acids, aldehydes, and alcohols [44]. The tar formation and oil production from the pyrolysis procedures is based on the operation conditions and raw materials [45]. Pyrolysis has been categorized into different types as follows:

Table 2 Yields of the gas, liquid, and solid fractions produced through the pyrolysis of textile waste

Textile material	Catalyst	Temperature	Products	Yield			References
				Solid	Liquid	Gas	
Flax, 100%	$ZnCl_2$	450	Activated carbon	44.80	NA	NA	[46]
Egyptian banknote ELCBs cotton, 100%	NA	600	1,3-Dioxolane-2-propanal, 2-methyl, furfural, 2-furanmethanol, 4,4-ethylenedioxy-1-pentylamine	17.26	30.28	52.46	[42]
Cotton, 100%	Na_2CO_3	600	Furanes, ketones	16.25	29.49	54.26	[47]
Cotton, 70% Polyester, 30%	Dye-originating Heavy Metals	500–700	Activated carbon	17.79	37.59	44.62	[42]
Flax, 100%	H_3PO_4	450	Activated carbon	39.20	NA	NA	[46]
Egyptian banknote ELCBs cotton, 100%	NA	500	Toluene, furfural, 2-furanmethanol, 2-furancarboxaldehyde, 5-methyl	20.74	39.75	39.51	[42]
Hemp, 100%	$ZnCl_2$	450	Activated carbon	41.60	NA	NA	[46]
Cotton, 100%	NA	700	Double bond carboxyl and carbonyl liquid	12.50	74	13.50	[39]
Low grade biomass fiber (Flax, 100%)	NA	900	Activated carbon	20	55	25	[48]
Biomass fiber waste Jute, 100%	NA	800	Activated carbon	24.60	59.60	15.90	[49]
Coir, 100%	NA	800	Activated carbon	37.40	47.40	18.20	[49]
Abaca, 100%	NA	800	Activated carbon	28.60	48.10	23.60	[49]

- **Catalytic pyrolysis:** Catalytic pyrolysis typically involves using solid catalysts like metal oxides, zeolites, or activated carbon. Such catalysts can assist in enhancing the pyrolysis procedure yield through the promotion of cracking and reactions of reforming which generate smaller molecules. It also improves process selectivity via preferring the production of a specific product like olefins or aromatics. Kwon et al. used Co-based catalyst for the catalytic pyrolysis of textile waste to produce CH_4 and syngas in the presence of CO_2 environment. This process has converted 80 wt% of textile waste into CH_4 and syngas [50]. Wang et al. investigated catalytic pyrolysis of textile waste over zeolites which produces furanic substituents, toluene, xylene and benzene by Diels-Alder reaction. The co-pyrolysis of textile waste has 80% selectivity to toluene, xylene and benzene [51].
- **Flash pyrolysis:** It is another type of pyrolysis that can be utilized to recycle the textiles chemically. In this procedure, the waste from the textile is heated rapidly to the highest temperatures (approx. 500–700 °C) and then rapidly quenched to generate a liquid product. High yield liquid products produced through flash pyrolysis have high heating value, making it a capable technology for making biofuels from waste textiles. In a recent study, flash pyrolysis has been conducted on textile residue of wool, short fiber and card waste separately. Results revealed that the card and short fiber waste has produced large amounts of mass within the smaller temperature range as compared to wool. The results of gas chromatography showed that different gaseous species have been produced in flash pyrolysis. Card waste and short fiber waste emitted CO and CO_2 while wool pyrolysis generated O_2. Furthermore, polycyclic aromatic hydrocarbons and non-aromatic organic substances have also been produced during the flash pyrolysis process of textile waste [52].
- **Slow pyrolysis:** This pyrolysis is of the conventional type, which involves a long time of residing and a slow heating rate. Biomass is pyrolyzed in slow pyrolysis for almost 400–500 °C temperature having a rate of heating about 0.1 to 1 °C/s. The estimated time for slow pyrolysis ranges from 5 to 30 min. Solid carbonaceous char is produced via slow pyrolysis. Gaseous and liquid products are generated in small quantities. In a study, slow pyrolysis of textile substances such as cotton and silk were performed at 573–1173 K temperature under nitrogen atmosphere. The effect of pyrolysis temperature was examined on the yield and properties of pyro-oil, gas and char. The results revealed that pyrolysis was deactivated at 673 K temperature and different products were obtained from the slow pyrolysis of the textile substances [53].

5.1.2 Hydrolysis

Hydrolysis is a chemical breakdown procedure which occurs because of a reaction with water. In most cases, the hydrolysis procedure must be pretreated using acids, alkalis, and ionic liquids because pretreatment is essential in hydrolysis that changes the yield percentage. Different types of hydrolysis are discussed as following:

- **Acid hydrolysis:** Acid hydrolysis is a process in which acids are used to disintegrate the target material into smaller parts. In this chemical reaction, an acid gives a proton to the target material for it to react with it and break the bonds of molecules and break it down into smaller parts. It is possible to perform acid hydrolysis using concentrated acids, such as sulphuric, nitric, or phosphoric, as well as diluted acids at high pressures. Researchers have investigated acid pretreatments by sulfuric and phosphoric acid even in the presence of cellulose solvents [54, 55]. Acid pretreatments can increase the availability of cellulose by hydrolyzing polymeric structures within hemicellulose into monomers, thus increasing biodegradability by breaking down polymeric structures into monomers [56]. Several challenges exist, including production of bi-products, high cost process, and the need of anti-corrosion instruments [57].
- **Alkaline hydrolysis:** Alkaline hydrolysis is the pretreatment of textile waste with different hydroxides, such as ammonium, sodium, calcium, and potassium, to enhance the glucose content in the final product. Alkaline hydrolysis improves the lignin solubility and reduces the crystallinity of the cellulose in textile waste, increasing the overall digestibility of the cellulose. Alkaline hydrolysis also increases glucose output and decreases the production of inhibitors of fermentation [52, 53]. Kumagai et al. have done the recycling of polyvinyl coated PET fabric by alkaline hydrolysis. Firstly, fabrics were treated with sodium hydroxide for hydrolysis which converts PET fibers into sodium terephthalate while PVC can be separated by filtration. The results reveal that the PET fibers hydrolyzed at 180 °C and yield terephthalic acid [54]. Table 3 represents the different conditions and products by the hydrolysis of textile materials.
- **Enzymatic hydrolysis:** Enzymatic hydrolysis is a process of the breakdown of different compounds in the presence of suitable enzyme following its chemical reaction with the water. Enzymatic hydrolysis has been extensively used to recycle the textile waste. Two types of enzymes are used for polyester hydrolyzation: lipases and cutinases. Among them, cutinases enzymes are present naturally that hydrolyze the cutin's ester bindings, a substance present in plants having similar properties to polyester [67]. Lipids like waxes and fats are hydrolyzed via lipases. For instance, they are used for hydrolysis of triglycerdes and diglycerides inside the human body. It has been shown that commercially available products of these enzymes can hydrolyze polyester. Currently, the yields are small, but there are ways to increase the hydrolysis of polyester, for instance, cutinase enzymes fusing to binding molecules. A main issue for all such processes of polyester hydrolysis is that the huge molecules of protein cannot enter inside polyester material; thereby, the hydrolysis happens mostly on the material's surface, restricting the limiting speed significantly of a reaction. In addition, the hydrolysis process duration, as observed within previous examples, is within months scale, thus, making it difficult to develop the economical processes. Polyester recycling, therefore, with enzymatic hydrolysis, is not feasible currently. Also, some enzymes could hydrolyze potentially polyacrylonitriles and polyamides, but such hydrolysis procedures have the same limitation as enzymatic hydrolysis of polyester. The two biggest obstacles are the restricted catalytic activity due to restricted

Table 3 Hydrolysis of different textile materials

Textile materials	Type of hydrolysis	Catalyst	Temperature (°C)	Products	References
Polyester (PET)	Alkaline	NaOH	90	Terephthalic acid	[58]
Cellulose-based waste textile	Acid	H_2SO_4	30	Glucose	[59]
Pre and post-consumer textile waste	Acid	H_2SO_4	40	Cellulose acetate	[60]
65% PET and 35% cotton	Enzymatic	Humicola insolens cutinase	55	Terephthalic acid and glucose	[61]
Textile waste	Alkaline	NaOH	–	Cellulosic fibers/PET	[62]
Cotton and PET	Enzymatic	Cellulose	55	–	[63]
Cotton and PET	Alkaline	Benzyltributylammonium chloride	90	Terephthalic acid and Ethylene glycol/ cellulosic fibers	[64]
Cotton and PET	Enzymatic	Cellulase and β–glucosidase	50	Glucose	[65]
PET	Chemical/ enzymatic	Humicola insolens cutinase	250	Terephthalic acid and ethylene glycol	[66]

attack by enzymes is restricted to the substrate surface [68]. Vecchiato et al. hydrolyzed viscose fibers consisting of flame-retardant pigment. The decomposition of viscose fibers happened almost completely at 50 °C for 24 h. They have also revealed that there was no effect of enzymatic treatment on the flame retardant pigment and pigment could be reused [69]. In addition to the cellulases enzymes, which hydrolyze cellulose, it is also reported that a few enzymes are able to hydrolyze the synthetic fibers such as polyester and biodegradable polymers like polylactic acid [70].

5.1.3 Hydrothermal Method

A hydrothermal decomposition process occurs at high temperatures and pressures via chemical crystallization engineering techniques. Water is the primary ingredient of the reaction. It is one of the most promising alternatives to the traditional method of converting carbon polymer waste into liquid, solid, and gas phases. An autoclave reactor is used to catalyze organic acids [71]. During this process only water is used and there is no need for any specific pretreatment. The hydrothermal process is further sub divided into five categories based on temperature and pressure ranges. First one is called hot water extraction, while the second type is the hot water extraction in the presence of pressure, third type is called hot liquid water treatment, while the fourth type is called hydrothermal carbonization and the last type is called hydrothermal liquefaction. Powder formation and particle size are controlled by hydrothermal conditions, which require temperatures below 280 °C [72]. Furthermore, hydrothermal process has some limitations as well such as it requires high temperature and pressure, additional the recycling through hydrothermal method required more time [73]. With an organic acid catalyst, cotton-polyester textile wastes can be recycled hydrothermally in following steps: (i) separating the polycotton fiber waste into fragments, (ii) making dispersion of mixture in aqueous phase with an organic acid catalyst. Temperature for the process is set up to 140 °C at high pressure. There is a 99% recycling rate for the polyester fiber aggregates and an 81% recycling rate for the cotton fiber fragments [74]. It is easy for cotton to dissolve in citric acid and degrade in strong acids. However, cotton was not affected by strong alkaline solutions [75]. Qi et al. achieved highly hydrophobic hydrochar by using $FeCl_3$ as a catalyst by lowering the temperature of hydrothermal carbonization of cotton textile waste. As a consequence of $FeCl_3$ and hydrothermal carbonization of cotton waste occurring simultaneously, more furfural compounds were produced as derivatives, leading to a side reaction pathway [76].

5.1.4 Glycolysis

Glycolysis is a degrading procedure that can be utilized to break down a big molecule of textiles into smaller ones. Recycling of textile components like PET fiber and polyurethane is commonly done using this method. PET fiber is the most often used textile fabric due to its excellent mechanical properties and low cost which lead to the production of PET waste in large amount [77, 78]. Glycolysis process has various advantages such as less time of reaction and it uses less amount of energy. PET fibers in glycolysis gradually processed without any catalyst [79, 80]. Glycolysis of PET waste remained challenge to degrade without any catalyst. Researchers have been focused to develop efficient and sustainable catalyst for the degradation of textile fibers macromolecules to the monomers of bis(2-hydroxyethyl) terephthalate (BHET) and other compounds [81]. Guo et al. uses Ag–Al double oxide catalyst for the recycling of polyester. Their results showed that the obtained yield of BHET using Mg–Al double oxides as catalyst was about 82 mol%. In addition, regenerated PET fibers via polymerization have similar mechanical properties and spinnability to the pure PET fibers [82]. Hommez et al. has studied the depolymerization of nylon 6 by glycolysis that was performed at 250 °C with phosphoric acid. The main products were N-(2-hydroxyethyl)-caprolactam, caprolactam, and linear oligomers [83]. Conventionally, boiled ethylene glycol was used for the glycolysis of polyester fibers under atmospheric pressure in the presence of different metal catalyst such as zinc acetate at 96 °C. The yield of monomer and conversion of PET fiber was about

80 and 100% respectively [84]. Furthermore, the yield of BHET by glycolysis of PET fibers with sodium sulfate, potassium sulfate, acetic acid and lithium hydroxide was about 65.72, 64.42, 64.42 and 63.50 respectively [85].

5.2 *Advantages and Limitations*

The advantages and limitations of the different processes of chemical recycling are presented in Table 4. Research has been done on chemical polymer recycling for cellulose-based synthetics and blends of textile fibres. Mechanical methods are used to destroy textile fibers like shredding, and their breakdown followed by a process namely chemical dissolution. Often, harmful solvents are being utilized, and polymer is retained while fibers are spun and regenerated. Additionally, the dyes must be detached from the textile waste. Recently, ionic liquids have been utilized to restore dyed post-consumer textile waste inside new life cycle and to lower ecological impact [86]. Moreover, conversion of cotton to viscose is the well-known example of textile chemical recycling. Industry leaders like Birla Cellulose and Lenzing adapted the cotton conversion to viscose technology, starting with the depolymerization of 100% cotton textile materials. However, the synthesized polymer chain yield is reduced compared to the pure process. Therefore, it is essential to blend the regenerated fibers with virgin viscose fibers in order to achieve a high yield [87].

Table 4 Advantages and disadvantages of different chemical recycling techniques

Chemical recycling technique	Advantages	Disadvantages	References
Pyrolysis	• Simple process • High efficiency • Small footprint • Different raw materials can be used for pyrolysis	• Production of hazardous chemicals • More energy required • High temperature require	[88, 89]
Hydrolysis	• Environmental protective • Mild conditions required	• Time taking • Low selectivity • Pretreatment required	[90]
Glycolysis	• Less energy require • Cost-effective • High yield • Scalability	• Slow process without catalyst • Less selectivity	[82, 91]
Hydrothermal	• Less oxygen content • Reduced ash • Low temperature • More heating values	• Time taking • Non-uniform and impure	[76, 92, 93]
Gasification	• High rate of reaction • Production of clean syngas	• More energy required • High temperature required	[94, 95]

Natural fibers are dispersed in a mixture containing Schweizer's reagent and sodium hypochlorite (NaClO) as solvents for cellulosic fibers. Geotextiles are installed in situations with vegetated and sloped ground-this process reduces pedestalling and soil movement [96, 97]. Another application of textile chemical recycling is the utilization of textile waste as packaging. The nanofiber cellulose textile waste can also be used to improve the strength performance of biopolymer. The combination of Cellulose Nano-Fibrillated Fiber (CNF) and polylactic acid for packaging applications has revealed outstanding reinforcement ability, elongation, and tensile modulus [98, 99]. Textile chemical recycling usually necessitates integrated energy and material flow inside the chemical industry [47]. Yousef et al. has conducted a study for cotton recovery from textile wastes and its regeneration for forming a new substance. The prime procedures for cotton recovery were liquefying dimethyl sulfoxide, leaching nitric acid, and diluting HCl for bleaching. The carbon footprint, recovery rate, and this technology economic act were found to be 1466 USD/ton, 1534 CO_2 eq/ton, and 93%, respectively, if textile waste is utilized as a renewable resource for thermal energy production [28].

5.3 *Chemical Recycling Techniques Used in Industry*

Different industries across the world use chemical techniques to recycle textiles and convert them into valuable products. The list of different companies which use chemical recycling techniques are listed in Table 5. A company named as Worn Again Technologies adapted unique chemical technique to recycle the poly/cotton into PET and cellulosic material as shown in Fig. 8. Briefly, in this procedure polyester is separated from the polycotton at high temperature which leaves behind the solid cotton and the solution of polyester. After that, polyester is recovered from the solvent and spun back into the polyester yarn by correcting its molecular weight. The solid cotton is dissolved in ionic liquid, which converts into the pulp to make cellulosic fibers again [100].

6 Biological Recycling Techniques

Biological recycling is the chemical decomposition of complicated and large compounds into simpler and smaller ones induced through the biochemical action of microorganisms for reuse. In biodegradation, waste products are decomposed into smaller pieces and finally transformed into compounds that are either utilized again in the biological cycles or may accumulate on the earth through a cascade of chemical and physical processes [101]. The structural complexity of textile products, particularly garment items, is very high. Semi-crystalline substructures are formed when polymers mix to make fibers, which are then tightly twisted into yarns.

Table 5 Chemical recycling techniques used at industrial level [32]

Type of chemical recycling	Company	Type of material	Country
Enzymatic treatment	Ambercycle	Textile	Los Angeles
Polymer dissolution	Ioncell	Polyester and cotton	Finland
Separation technology	BlockTexx	Polyester and cotton	Australia
Dissolution polymer solvent separation	Worn Again Technologies	Polyester and cotton	UK
Hydrolysis	FENC	Textile and PET	Taiwan
Chemical recovery	Lenzing	Pre-consumer cotton and post-consumer garments	Austria
Subcritical water treatment	Tyton BioSciences	Cotton pulp, polyester poly-cotton blends	Danville

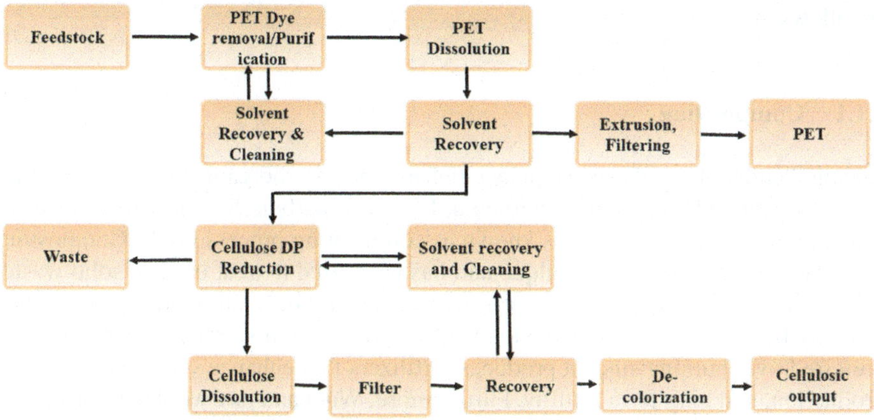

Fig. 8 Chemical recycling technique of poly/cotton adapted by Worn Again Technologies [100]

These fabrics are then sewn to make garments, and throughout the process, chemicals, dyes, and other finishes are used to achieve different performance effects. Although, certain products may be entirely bio-recycled moderately bio-recycled, or not bio-recycled at all under a specific set of conditions. Therefore, addressing the bio-recycling of products as an attribute depends on the treatment circumstances and how much product separated into its components is necessary before or after disposal. However, it is necessary that the disassembled, disintegrated, or degraded components are completely absorbed through biological or other mechanisms, or a combination of these mechanisms, as well as total recovery and recycling of disassembled, disintegrated, or degraded components [102, 103].

The biological cycle relies on resources that have the ability to break down and create nutrients, ultimately converting them into fresh renewable resources. In the biological cycle, the concept of circularity centers around utilizing "biological nutrients" like food waste to revitalize the soil, employing methods like anaerobic digestion and compositing. However, when it comes to textiles, composting is an inefficient utilization of resources and not be integrated into the circular economy system that aims to retain the worth of textiles in the system. According to the Biomimicry Institute, a suggestion is made to combine the technical and biological cycles within the biological cycle, provided that only biomaterials are generated and they can be composted at the end of their lifespan. In this particular scenario, microbial processes serve as a transitional technology, facilitating the dispersal of materials into the natural environment [103].

6.1 Types of Biological Recycling Techniques

Biological recycling has been categorized into different types, which are discussed as follows:

6.1.1 Compositing

A natural and biological procedure in which aerobic microorganisms convert organic textile constituents into water, ammonia, heat, and carbon dioxide when nitrogen, water, and oxygen are present is known as compositing process [103]. Composting is a relatively inexpensive technique and has been used to treat organic solid waste, animal, and human wastes, but not textile waste in larger amount. Waste textiles can only be composted if they are made of natural or semi-synthetic fibers, and it is preferred over burning since it produces fertilizers that enhance soil organic matter, production, and long-term fertility. Furthermore, Wool and raw wool have also been utilized to improve the growth of plants through compositing technique as it produces fertilizer [104]. In a study, composting has limitations when it comes to processing natural and biodegradable fibers, researchers prefer it because it produces zero waste and lower processing costs. It is possible to recycle biodegradable constituents of textile solid waste into nappies, sheets, agro mulching and car interiors that are recycled to compost. Textiles made from biodegradable polymers, like polylactic acid, can be composted. The compost that is obtained is helpful for fertilizing soil, although the process produces greenhouse gases like carbon dioxide and is very slow [105]. Compositing is a simple and environmentally friendly technique that can reduce the volume of organic waste by approximately 50% during the active composting phase. Cotton waste presents a notable challenge in waste management today, and composting has been considered as an alternative to avoid directly disposing of cotton trash in landfills [106]. Composting and vermicomposting of cotton waste have the potential to serve as a valuable and sustainable nutrient source over an extended period [102]. The Vermicomposting process utilized the earthworms for the conversion of waste into compost, resulting in a soil that is more fertile than that produced by traditional composting methods. Compost and vermicompost samples showed comparable bacterial diversity when cotton waste was used as a substrate. In contrast, the vermicompost samples produced superior humus due to their higher concentration of bacterial isolates compared to the compost samples [107].

A study was conducted to explore the vermicomposting of cotton textile waste in the form of willow waste obtained from ginning factories. The harvested willow waste was blended with a mixture consisting of cow dung slurry, cellulase, amylase enzymes (derived from cow dung), and a solution containing effective microorganisms to initiate the vermicomposting procedure. The agitation of the mixture starts and water was sprayed over it. After 20 days, the waste has been undergone complete decomposition, and earthworms were introduced into the system. The vermicomposting process continued until the color of the waste mixture converted into light brown that usually occurred within 14 days. The resulting vermicompost was subsequently employed for enhancing plant growth in potted plants. It was observed that plants cultivated using vermicompost derived from willow waste displayed a significant increase in leaf area index, root and shoot length surpassing the growth observed in the control pot [108].

6.1.2 Anaerobic Digestion

A process in which the degradation of macromolecules such as proteins, carbohydrates and lipids produce small-chain acids and eventually methane and CO_2 when methanogenic bacteria are present is known as anaerobic digestion (AD) [109]. Typically, anaerobic digestion of agricultural and food waste produces methane by combining it with water treatment plant inoculum or manure. Additionally, it has proven to be effective in converting resultant products into energy that can be utilized to power the same process from which the resultant products originated. Anaerobic digestion has been documented in the utilization of cellulose-rich feedstock like wheat straw and textile waste [86]. Table 6 represents the biorecycling of different textile materials.

Anaerobic digestion is frequently employ to manage the biodegradable portion of organic waste in order to produce biogas. More than 50% of the cellulose in cotton made it a suitable substrate for biological conversion. Studies on Anaerobic digestion producing methane-rich biogas from cotton waste have been done throughout the last ten years. Anaerobic digestion of cotton waste including cottonseed hull, cotton stalks and cotton oil cake lead to the production of biogas. Cotton waste from spinning mills has the potential to be employed as a substance for the anaerobic digestion. In a study, where medical cotton industry waste was subjected to AD under thermophilic conditions with cattle manure as the inoculum which yield biogas 92% approximately. In anaerobic digestion systems, the pre-treatment methods have ability to increase the breakdown of complex organic matter, leading to an enhance in the quality and production of biogas. The digestion process can be enhanced using different pre-treatment process, such as chemical, mechanical, biological, and thermal ones. Waste jeans consists of 60% cotton and 40% polyester and virgin

Table 6 Biorecyling of textile materials

Type of bio-recycling method	Textile material	Pretreatment	Product	References
Compost	Textile waste with green or paper waste	–	CO_2	[103]
Anaerobic digestion	Cellulosic	Na_2CO_3	Methane	[110]
Anaerobic digestion	Wool	Liquid nitrogen	Methane	[111]
Anaerobic digestion	Wool	Enzymatic thermochemical	Methane	[112]
Enzymatic depolymerization	Blended cellulosic	NaOH/urea	Glucose/PET	[113]
Fermentation	Wool	–	Amino acid	[114]
Fermentation	Blend cellulosic	NaOH/urea	Cellulase PET fibers	[65]

cotton waste products pre-treated with 0.5 M Na_2CO_3 at 150°C for 120 min yielded methane production of 328.9 mL CH_4/g VS and 361.1 mL CH_4/g VS, respectively [55]. Viscose//polyester and cotton/polyester blended textiles produced maximum methane production about 80% with a 20 g/L cellulose loading in both single-stage and two-stage batch reactor digestions [115].

6.1.3 Enzymatic Depolymerization

Enzymatic depolymerization is a promising method for the biorecycling of textiles, offering an environmentally friendly approach to tackle textile waste. This process utilizes specific enzymes to break down the polymers present in textiles, such as cellulose, polyester, and nylon, into their constituent monomers that can be reused to produce new fibers or other products. Several enzymatic processes that make use of cellulosic biomass require a chemical pretreatment before cellulose is converted into sugars enzymatically. Different methods have already been established for lignocellulosic biomass pretreatment, including size reduction, acid or alkaline solutions, ammonia, high-pressure treatment, microwave treatment, and ionic liquids [116]. The pretreatment of cellulosic textiles is not advanced but gaining much attention. They mainly involve size reduction and the use of concentrated caustic alkaline solutions, resembling the process of mercerization [117]. In comparison to typical cellulosic feedstocks, pretreating cellulose from textile waste with alkaline solutions is usually more effective at lower temperatures, around −20 °C, and in the presence of substances like urea or thiourea. Researchers have been utilized alkaline pretreatment at low temperatures for blended textile waste to produce cellulose. Glucose recovery from textile waste has been achieved by pretreating the waste in alkaline environment at −20 °C. The cooling step during pretreatment was reported as the most energy-intensive and detrimental stage in the life cycle analysis [118]. Previous studies have explored pretreatments with phosphoric acid and sodium carbonate, which have shown a reduced impact on the environment and a decrease in cotton crystallinity [55]. Organic solvents such as ionic liquids or N-methylmorpholine-N-oxide are newer and greener options that can be utilized to dissolve the cellulose. The ionic liquid [AMIM]Cl has been effective to pretreat the cotton waste for enzymatic hydrolysis, however its high cost poses a challenge for commercialization. Viable use of ionic liquids has been demonstrated with biomass, and a similar approach could be applied to textile feedstocks [119]. Further investigation is necessary in order to mitigate the cost and environmental consequences associated with pretreating textile waste. This will facilitate the advancement of economically feasible biochemical recycling method for discarded textiles. Navone et al. reported the utilization of enzymes for recycling wool textile waste in combination with other textile waste materials. They employed a commercially available enzyme specifically designed to solubilize the wool constituents from the textile waste [113].

6.1.4 Fermentation

Fermentation is a biochemical method that induces alterations in organic substances by means of enzyme activity. In practical terms, fermentation involves purposefully harnessing microorganisms like bacteria, fungi, and eukaryotic cells to create beneficial products for human use. Fermentation process has been categorized into following types:

- **Solid state fermentation:** Solid-state Fermentation (SSF) is a bioprocess method that conducted in the absence of water, while maintaining required moisture levels to facilitate the metabolic functions and growth of microorganisms. Following situations mimic the native environment from which these microorganisms are derived [120]. SSF has been employed in industrial sector to yield enzymes, particularly cellulases, for various biochemical processes. Additionally, SSF has proven useful in generating other enzymes associated in breaking down lignocellulosic substrates, utilizing cellulosic-rich residues as the substrate [121]. Hu et al. conducted a study in which textile waste was employed as a substrate material for SSF using Trichoderma reesei. The objective was to yield cellulases capable of efficiently breaking down pretreated blended textile waste into individual sugar units, facilitating the regeneration of insoluble polyester component. Additionally, the researchers enhanced their findings by optimizing the pretreatment and hydrolysis processes, allowing for the separation of textile cellulose from the insoluble polyester [65].
- **Submerged fermentation:** Submerged fermentation is the more commonly utilized method, wherein a liquid serves as a medium for the reaction, and desired compounds are released into the fraction of the liquid. This approach is employed to generate various fuel molecules and chemicals from hydrolyzed

cellulose-rich waste, including citric acid, succinic acid, ethanol, and lactic acid [122]. Koutinas et al. have discussed the utilization of waste materials rather than virgin resources as feedstock for fermentation. The literature also contains references to submerged fermentation using textiles [123]. In a recent study, researcher employed submerged fungi fermentation to yield cellulases from the textile waste, while yeast fermentation was documented to produce the bioethanol and succinic acid using cellulosic textiles [124].

6.2 Advantages and Limitations

Bio-based processes offer several advantages, including low energy requirements, utilization of safe chemicals and different solvents, and reliance on renewable carbon instead of fossil fuels. However, it is pertinent to mention that while these procedures are operated under mild conditions, this alone does not guarantee overall sustainability. Both economic and environmental aspects need to be considered. Before processing, feedstocks usually undergo pretreatment involving the usage of chemicals, specific equipment, and energy. The utilization of commercial enzymes contributes substantially to both financial expenses and environmental consequences. These factors have impeded the progress of economically viable biofuel production from lignocellulosic feedstock, and similar challenges can be anticipated while using textile material as feedstock. Another problem affecting the scaling-up of biorefineries for bioproduct production is concentration. As compared to other procedures, feedstocks are typically impure, and enzymatic reactions could be effective but result in low sugar concentrations that are not economically viable for further fermentation. One main advantage of end-of-life textiles is that cotton consists of 100% cellulose unlike lignocellulosic feedstock such as straw [86].

Biorecycling methods are often limited to specific types of textiles, particularly natural fibers like cotton or wool. Synthetic fibers, such as polyester or nylon, are more challenging to biodegrade or recycle through biological processes. As a result, the potential for biorecycling is limited to a subset of textile materials, which restricts its overall effectiveness in addressing the textile waste problem. Additionally biorecycling aims to reduce the environmental impact, it still requires energy and resources to carry out the process effectively. The collection, sorting, and treatment of textile waste demand energy inputs, including transportation, machinery, and infrastructure. Biorecycling textile waste on a large scale can be challenging due to logistical and operational issues. Collecting and processing textile waste from various sources, such as households, industries, or commercial establishments, requires well-established collection systems and infrastructure. Additionally, achieving high-efficiency rates in biorecycling processes, such as achieving complete biodegradation or maintaining consistent product quality, can be difficult and may require continuous research and development.

6.3 *Biological Recycling Techniques Used in Industry*

Biological recycling is not much common for textile industry however, some industries has establish this process to recycle the textile waste. A company named as "Carbios" uses an enzyme to recycle the polyethylene terephthalate (PET). Carbios claims that their enzyme can effectively break down polyester materials, allowing for the recovery of almost all polyester present in textile waste, including blended fabrics. This enables the creation of high-quality material with a virgin-like state, capable of being recycled repeatedly. In theory, this process facilitates a truly circular approach to recycling [125]. Moreover, the French startup Carbios has recently entered into an agreement with On, Patagonia, Puma, and Salomon. The objective of this partnership is to speed up the commercialization of Carbios' bio-recycling technology specifically designed for textiles [126].

7 Thermo-mechanical Recycling

Thermomechanical recycling is also referred to as polymer mechanical recycling. It is a process where synthetic polymeric material is melted through high pressure and temperature and converted into granules [32, 127]. In textile thermomechanical recycling, synthetic post-industrial or post-consumer textile waste is melted and transferred into new fibers or reshaped into other forms [32, 127].

However, there are several steps involved to achieve the complete textile-to-textile solution using the thermomechanical process as shown in Fig. 9.

- Collection of post-industrial or post-consumer textile wastes.
- Sorting of textile wastes based on their type i.e. material, color, etc., and removal of nontarget materials such as metals.
- Reduction of size through crushing, grinding, shredding, or pulling.
- Cleaning.
- Densification to prepare the feedstock for thermomechanical recycling.
- Degranulation of textile waste through the thermomechanical process.
- Yarn production through the melt extrusion process.
- Yarn post texturization.
- Recycled fabric production through knitting or weaving.

Thermomechanical recycling of textiles is done following the polymer extrusion process [18]. The process contains an extruder that applies heat and shear to melt the textile waste and convert it into pellets as shown in Fig. 8. It is important to note that, the post-industrial or post-consumer textile wastes being recycled through this process must be a thermoplastic material and free from metals. In addition, For example, pure synthetic textile materials such as polyamide 6, polyethylene terephthalate (PET), thermoplastic polyurethane, and elastane can be recycled through

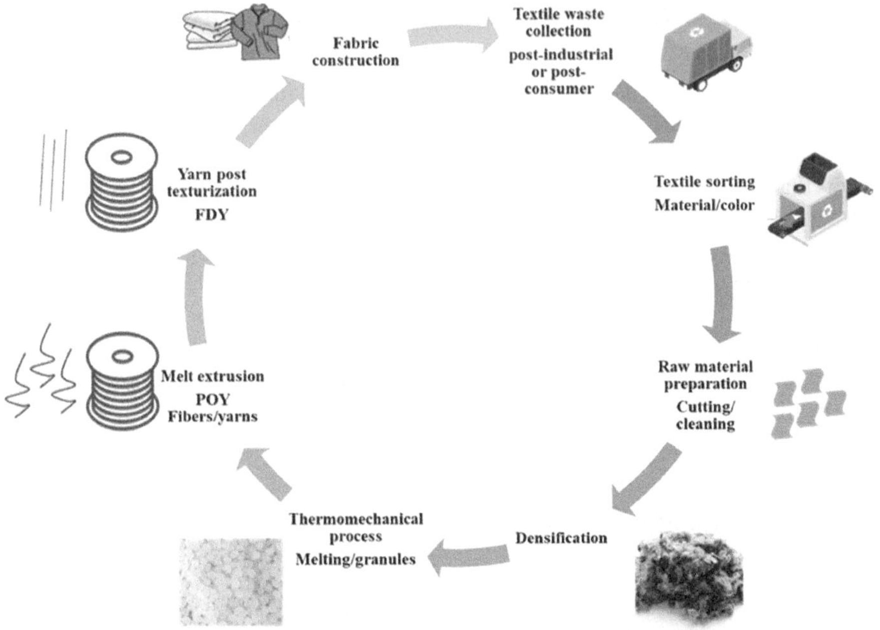

Fig. 9 Demonstration of different steps involved in textile-to-textile solution through a thermomechanical process

the thermomechanical process [127, 128]. However, recycling thermoplastic blend materials can be very challenging to process if polymer materials exhibit different melting points or flow properties [129]. The extrusion process includes the following basic elements as shown in Fig. 10.

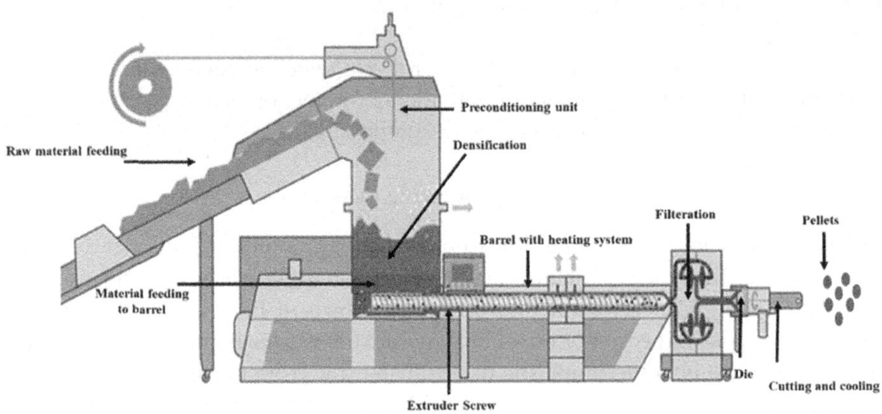

Fig. 10 Industrial thermomechanical recycling process through melt-extrusion

a. Feeder or hopper.
b. A screw that helps to melt and transport the melt.
c. Barrel heating system to melt the textile material.
d. A Die to convert the melted stream to pellet form.

The melting occurs in the first part of the extrusion process and then melted thermoplastic material is transported through the screw. At the end of the extruder, the melted material passes through the die hole which defines the shape (mostly circular) of this melted string coming out of the extruder. The resulting recycled thermoplastic string is then cooled down by air or water and chopped into the desired pellet size (a few millimeters). These pellets are called re-granulate and can be used for further processing such as to produce mono or multiple filament yarns or fibers.

Thermomechanical recycling provides a promising opportunity for a complete textile-to-textile solution. However, there are different pros and cons in this process as follows:

Advantages

- Cost-effective, efficient, and well-known process.
- Easy to implement.
- Fiber or filaments can be produced for different textile applications.

Disadvantages

- Degradation of the material.
- Challenging to process the blend materials.
- Sensitive to contained contaminants.
- High sorting cost.

7.1 Challenges in Thermo-Mechanical Recycling

Thermoplastic textile waste being recycled through the thermomechanical recycling process experiences polymer degradation caused by high processing and high shear during melt processing [130, 131]. The degradation of polymers leads to variations in properties such as mechanical (e.g., reduction of elongation at break), rheological (for example, a decrease in viscosity occurs with each reprocessing cycle), thermal (melting temperature, crystallization, etc.) physical (surface properties, color, etc.) [132]. For example, recycled PET recovered from this process encounters a loss of physical properties due to degradation, and contamination during use cycles and processing and, therefore, often used in under-value applications [133]. Since 2015, the market value of apparel made out of recycled PET yarn produced from plastic bottles has increased by 58% [14]. PET post-consumer bottles exhibit high intrinsic viscosity (IV) which is decreased during the thermomechanical recycling process due to polymer hydrolytic and thermal decomposition that happens during the processing stage [134]. On the other hand, the melt-spinning process requires a comparatively lower IV of PET to spun into yarns. Hence, if the IV of recycled PET is decreased to a certain level and provides a suitable IV for melt-spinning, recycled PET from bottles can be spun into fibers. This presents a successful example of open-loop recycling. Therefore, thermomechanical recycling of PET bottles into textile yarns is feasible and often practiced. However, it is much more challenging to maintain the quality of respun PET in thermomechanical recycling due to loss of physical properties [133].

Moreover, recycling post-industrial or post-consumer textile wastes which contain a blend of thermoplastic polymer is quite difficult to perform due to differences in their melt flows/viscosities, melting point, thermal properties, and recommended processing temperatures. For example, post-industrial textile wastes from the socks industry are usually made up of polyamide (PA) and thermoplastic polyurethane (TPU) threads. These two thermoplastic polymer materials are far apart in their melting points and melt flow properties [135]. While processing this textile waste through the thermomechanical process would require a relatively higher processing temperature as PA melts at a higher temperature as compared to TPU. Consequently, TPU will suffer from high thermal degradation and the resultant polymer blend, later, will exhibit poor performance properties during the melt-spinning process and at the yarn scale [135].

As mentioned above, textile waste must be properly sorted and free from metallic materials. The collection and sorting of post-consumer textile waste are comparatively expensive, time consuming, and difficult as compared to post-industrial textile waste [13]. This is due to the fact that post-consumer textile waste may contain varied material compositions or contamination that would be more difficult to thermomechanically recycle back into textile fibers [127].

8 Conclusion and Way Forward

The production and disposal of textile products impact the environment in multiple ways, whether it's pollution, GHG emissions, landfills, or resource depletion, Textiles are the major contributor to all these. Textiles are also massively underutilized and often discarded when they have sufficient value. Restructuring the supply chain of textiles into a circular fashion via recycling is one of the most promising strategies for mitigating the impact of textiles on the environment. Recycling of the textiles directly diverts the textiles from landfills and incinerators that are harmful to the environment otherwise. On the one hand, disposing of textiles in conventional ways imposes a burden on the environment and ultimately on living beings whereas on the other hand, recycling can drive the creation of more industries, more job opportunities, economic uplift, sustainability, and more room for research. There are different ways to recycle the textiles products including mechanical, chemical, thermo-mechanical and biological. Some process are very mature and used more frequently while others are evolving. The chemical and biological recycling are evolving very rapidly as there is more and more demand to use the recycled content rather using virgin feed stock.

Finally, it would not be wrong to conclude that recycling cannot be successful without consumer awareness, as the consumer is always the starting point of this cycle.

References

1. Foundation EM (2017) A new textiles economy: Redesigning fashion's future
2. Wagaw T, Babu KM (2023) Textile waste recycling: a need for a stringent paradigm Shift.24723444231188342
3. Liu X, Wang S, Li Y, Ali S, Khan YA, Salem S, et al (2022) Recycling in textile sector: a new circular economy approach towards ecology and environmental sustainability. 10:929710
4. Wang Y (2006) Recycling in textiles. Woodhead publishing
5. Majumdar A, Ali SM, Agrawal R, Srivastava S (2022) A triple helix framework for strategy development in circular textile and clothing supply chain: an Indian perspective. 367:132954
6. Stefan DS, Bosomoiu M, Stefan M (2022) Methods for natural and synthetic polymers recovery from textile waste. 14(19):3939
7. Weber S, Weber O, Habib K, Dias GM (2023) Textile waste in Ontario, Canada: opportunities for reuse and recycling. 190:106835
8. Cao H, Cobb K, Yatvitskiy M, Wolfe M, Shen H (2022) Textile and product development from end-of-use cotton apparel: a study to reclaim value from waste. 14:8553
9. Akter MMK, Haq UN, Islam MM, Uddin MA (2022) Textile-apparel manufacturing and material waste management in the circular economy: a conceptual model to achieve sustainable development goal (SDG) 12 for Bangladesh. 4:100070
10. Sinha P, Dissanayke D, Abeysooriya RP, Bulathgama B (2022) Addressing post-consumer textile waste in developing economies. 113(9):1887–1907
11. Circle Economy, EigenDreads, and Fashion for Good (2022) Sorting for circularity Europe
12. Étude de caractérisation des flux entrants et sortants de centres de tri (2023)
13. Köhler A, Watson D, Trzepacz S, Löw C, Liu R, Danneck J, et al (2021) Circular economy perspectives in the EU textile sector. Publications Office of the European Union Luxembourg
14. Mulvihill M, Horotan A (2018) Safer chemistry innovation in the textile and apparel industry
15. Technology|FibersortTM (2023) https://www.fibersort.com/en/technology
16. Cura K, Rintala N, Kamppuri T, Saarimäki E, Heikkilä P (2021) Textile recognition and sorting for recycling at an automated line using near infrared spectroscopy. 6(1):11
17. Technical monitoring on optical sorting and textile recognition technologies at a European level Summary (2020)
18. Le K (2018) Textile recycling technologies, colouring and finishing methods. 23–50
19. Sorting|Valvan. Accessed Access
20. Valvan Baling Systems FIBERSORT leaflet. Accessed Access.
21. Sustainable research and consultancy–IVL.se
22. About Telaketju|Telaketju
23. Homepage|HKRITA
24. CETIA|From goods to materials
25. Muthu SS (2018) Sustainable innovations in recycled textiles. Springer
26. Eppinger E (2022) Recycling technologies for enabling sustainability transitions of the fashion industry: status quo and avenues for increasing post-consumer waste recycling. 18(1):114–128
27. Beltrán FR, Arrieta MP, Moreno E, Gaspar G, Muneta LM, Carrasco-Gallego R, et al (2021) Evaluation of the technical viability of distributed mechanical recycling of PLA 3D printing wastes. 13(8):1247
28. Yousef S, Tatariants M, Tichonovas M, Sarwar Z, Jonuškienė I, Kliucininkas L (2019) A new strategy for using textile waste as a sustainable source of recovered cotton. 145:359–369

29. Juanga-Labayen JP, Labayen IV, Yuan Q (2022) A review on textile recycling practices and challenges. 2(1)174–188
30. Utebay B, Celik P, Cay A (2023) Valorization of fabric wastes through production of recycled cotton yarns by compact ring and open-end rotor spinning. 409:137135
31. Lu L, Fan W, Meng X, Xue L, Ge S, Wang C, et al (2023) Current recycling strategies and high-value utilization of waste cotton. 856:158798
32. Damayanti D, Wulandari LA, Bagaskoro A, Rianjanu A, Wu H-S (2021) Possibility routes for textile recycling technology. 13(21):3834
33. Valerio O, Muthuraj R, Codou A (2020) Strategies for polymer to polymer recycling from waste: current trends and opportunities for improving the circular economy of polymers in South America. 25:100381
34. Scheirs J (2006) Overview of commercial pyrolysis processes for waste plastics. 381–433
35. Pawelczyk E, Wysocka I, Gębicki J (2022) Pyrolysis Combined with the dry reforming of waste plastics as a potential method for resource recovery—a review of process parameters and catalysts. 12(4):362
36. Kwon EE, Lee T, Ok YS, Tsang DC, Park C, Lee J (2018) Effects of calcium carbonate on pyrolysis of sewage sludge. 153:726–731
37. Lee T, Nam I-H, Jung S, Park Y-K, Kwon EE (2020) Synthesis of nickel/biochar composite from pyrolysis of Microcystis aeruginosa and its practical use for syngas production. 300:122712
38. Wu HS (2018) Pyrolysis kinetic of alkaline and dealkaline lignin using catalyst. 25(1):7
39. Miranda R, Sosa_Blanco C, Bustos-Martinez D, Vasile C (2007) Pyrolysis of textile wastes: I. Kinetics and yields.80(2):489–495
40. Lee J, Kim K-H, Kwon EE (2017) Biochar as a catalyst. 77:70–79
41. You S, Ok YS, Chen SS, Tsang DC, Kwon EE, Lee J, et al (2017) A critical review on sustainable biochar system through gasification: energy and environmental applications. 246:242–253
42. Yousef S, Eimontas J, Striūgas N, Trofimov E, Hamdy M, Abdelnaby MA (2020) Conversion of end-of-life cotton banknotes into liquid fuel using mini-pyrolysis plant. 267:121612
43. Damayanti D, Wulandari YR, Wu H-S (2020) Product distribution of chemical product using catalytic depolymerization of Lignin. 15(2):432–453
44. Molino A, Chianese S, Musmarra D (2016) Biomass gasification technology: the state of the art overview. 25(1):10–25
45. Devi L, Ptasinski KJ, Janssen FJ (2003) A review of the primary measures for tar elimination in biomass gasification processes. 24(2):125–140
46. Williams PT, Reed AR (2004) High grade activated carbon matting derived from the chemical activation and pyrolysis of natural fibre textile waste. 71(2):971–986
47. Barışçı S, Öncel MS (2014) The disposal of combed cotton wastes by pyrolysis. 11(3):255–266
48. Williams PT, Reed AR (2003) Pre-formed activated carbon matting derived from the pyrolysis of biomass natural fibre textile waste. 70(2):563–577
49. Reed AR, Williams PT (2004) Thermal processing of biomass natural fibre wastes by pyrolysis. 28(2):131–145
50. Kwon D, Yi S, Jung S, Kwon EE (2021) Valorization of synthetic textile waste using CO2 as a raw material in the catalytic pyrolysis process. 268:115916
51. Wang J, Jiang J, Ding J, Wang X, Sun Y, Ruan R, et al (2021) Promoting Diels-Alder reactions to produce bio-BTX: Co-aromatization of textile waste and plastic waste over USY zeolite. 314:127966
52. Ruiz B, Fuente E, Pérez A, Taboada-Ruiz L, Sanz JM, Calvo LF, et al (2023) Employment of conventional and flash pyrolysis for biomass wastes from the textile industry with sustainable prospects. 105864
53. Chowdhury R, Sarkar A (2012) Reaction kinetics and product distribution of slow pyrolysis of Indian textile wastes. 10:1
54. Sasaki C, Nakagawa T, Asada C, Nakamura Y (2020) Microwave-assisted hydrolysis of cotton waste to glucose in combination with the concentrated sulfuric acid impregnation method. 11:4279–4287

55. Hasanzadeh E, Mirmohamadsadeghi S, Karimi K (2018) Enhancing energy production from waste textile by hydrolysis of synthetic parts. 218:41–48
56. Keskin T, Abubackar H, Arslan K, Azbar N (2019) Biohydrogen production from solid wastes. Biohydrogen. Elsevier, Amsterdam, The Netherlands
57. Dimos K, Paschos T, Louloudi A, Kalogiannis KG, Lappas AA, Papayannakos N, et al (2019) Effect of various pretreatment methods on bioethanol production from cotton stalks. 5(1):5
58. Bengtsson J, Peterson A, Idström A, de la Motte H, Jedvert K (2022) Chemical recycling of a textile blend from polyester and viscose, part II: mechanism and reactivity during alkaline hydrolysis of textile polyester. 14(11):6911
59. Sanchis-Sebastiá M, Ruuth E, Stigsson L, Galbe M, Wallberg O (2021) Novel sustainable alternatives for the fashion industry: A method of chemically recycling waste textiles via acid hydrolysis. 121:248–254
60. Costa C, Viana A, Silva C, Marques EF, Azoia NG (2022) Recycling of textile wastes, by acid hydrolysis, into new cellulosic raw materials. 153:99–109
61. Kaabel S, Arciszewski J, Borchers TH, Therien JD, Friščić T, Auclair K (2023) Solid-state enzymatic hydrolysis of mixed PET/cotton textiles. 16(1):e202201613
62. Peters GM, Sandin G, Spak BR (2019) Environmental prospects for mixed textile recycling in Sweden. 7(13):11682–11690
63. Boondaeng A, Keabpimai J, Srichola P, Vaithanomsat P, Trakunjae C, Niyomvong N (2023) Optimization of textile waste blends of cotton and PET by enzymatic hydrolysis with reusable chemical pretreatment. 15(8):1964
64. Palme A, Peterson A, de la Motte H, Theliander H, Brelid H (2017) Development of an efficient route for combined recycling of PET and cotton from mixed fabrics. 3(1):1–9
65. Hu Y, Du C, Leu S-Y, Jing H, Li X, Lin CSK (2018) Valorisation of textile waste by fungal solid state fermentation: An example of circular waste-based biorefinery. 129:27–35
66. Quartinello F, Vajnhandl S, Volmajer Valh J, Farmer TJ, Vončina B, Lobnik A, et al (2017) Synergistic chemo-enzymatic hydrolysis of poly (ethylene terephthalate) from textile waste. 10:6:1376–1383
67. Longhi S, Cambillau C (1999) Structure-activity of cutinase, a small lipolytic enzyme. 1441(2–3):185–196
68. Piribauer B, Bartl A (2019) Textile recycling processes, state of the art and current developments: a mini review. 37(2):112–119
69. Vecchiato S, Skopek L, Jankova S, Pellis A, Ipsmiller W, Aldrian A, et al (2018) Enzymatic recycling of high-value phosphor flame-retardant pigment and glucose from rayon fibers. 6(2):2386–2394
70. Banerjee A, Chatterjee K, Madras G (2014) Enzymatic degradation of polymers: a brief review. 30(5):567–573
71. Yildirir E (2015) Chemical recycling of waste plastics via hydrothermal processing. University of Leeds
72. Wikberg H, Grönberg V, Jermakka J, Kemppainen K, Kleen M, Laine C, et al (2015) Hydrothermal refining of biomass: an overview and future perspectives. 14(3):195–207
73. Tang K, Zhang Y, Lin D, Han Y, Chen C-TA, Wang D, et al (2018) Cultivation-independent and cultivation-dependent analysis of microbes in the shallow-sea hydrothermal system off Kueishantao Island, Taiwan: unmasking heterotrophic bacterial diversity and functional capacity. 9:279
74. Keh EYM, Yao L, Liao X, Liu Y, Cheuk K, Chan A (2022) Method for separating and recycling a waste polyester-cotton textile by means of a hydrothermal reaction catalyzed by an organic acid. Google Patents
75. Kawamura K, Sako K, Ogata T, Tanabe K (2020) Environmentally friendly, hydrothermal treatment of mixed fabric wastes containing polyester, cotton, and wool fibers: application for HMF production. 11:100478
76. Qi R, Xu Z, Zhou Y, Zhang D, Sun Z, Chen W, et al (2021) Clean solid fuel produced from cotton textiles waste through hydrothermal carbonization with FeCl3: upgrading the fuel quality and combustion characteristics. 214:118926

77. Dadash Ziyaei M, Barikani M, Honarkar H (2020) Recycling of polyethylene terephthalate (PET) via glycolysis method for synthesis waterborne polyurethane. In: Eco-friendly and smart polymer systems, vol 13. Springer, pp 520–523
78. Guo Z, Adolfsson E, Tam PL (2021) Nanostructured micro particles as a low-cost and sustainable catalyst in the recycling of PET fiber waste by the glycolysis method. 126:559–566
79. Al-Sabagh A, Yehia F, Eshaq G, Rabie A, ElMetwally A (2016) Greener routes for recycling of polyethylene terephthalate. 25(1):53–64
80. Heiran R, Ghaderian A, Reghunadhan A, Sedaghati F, Thomas S, Haghighi AH (2021) Glycolysis: an efficient route for recycling of end of life polyurethane foams. 28:1–19
81. Al-Sabagh AM, Yehia FZ, Eissa A-MM, Moustafa ME, Eshaq G, Rabie A-RM, et al (2014) Glycolysis of poly (ethylene terephthalate) catalyzed by the Lewis base ionic liquid [Bmim][OAc]. 53(48):18443–18451
82. Guo Z, Eriksson M, de la Motte H, Adolfsson E (2021) Circular recycling of polyester textile waste using a sustainable catalyst. 283:124579
83. Hommez B, Goethals EJ (1998) Degradation of nylon-6 by glycolysis. Part 1: identification of degradation products. 35(9):1489–1505
84. Hu Y, Wang Y, Zhang X, Qian J, Xing X, Wang X (2020) Synthesis of poly (ethylene terephthalate) based on glycolysis of waste PET fiber. 57(6):430–438
85. Shukla S, Harad AM (2005) Glycolysis of polyethylene terephthalate waste fibers. 97(2):513–517
86. Ribul M, Lanot A, Pisapia CT, Purnell P, McQueen-Mason SJ, Baurley S (2021) Mechanical, chemical, biological: Moving towards closed-loop bio-based recycling in a circular economy of sustainable textiles. 326:129325
87. Saha S Textile recycling: the chemical recycling process of textiles
88. Zhang H, Gao Z, Liu Y, Ran C, Mao X, Kang Q, et al (2018) Microwave-assisted pyrolysis of textile dyeing sludge, and migration and distribution of heavy metals. 355:128–135
89. Athanasopoulos P, Zabaniotou A (2022) Post-consumer textile thermochemical recycling to fuels and biocarbon: a critical review. 834:155387
90. Jagadeesh P, Mavinkere Rangappa S, Siengchin S, Puttegowda M, Thiagamani SMK, Hemath Kumar M, et al (2022) Sustainable recycling technologies for thermoplastic polymers and their composites: a review of the state of the art. 43(9):5831–5862
91. Sert E, Yılmaz E, Atalay FS (2019) Chemical recycling of polyethlylene terephthalate by glycolysis using deep eutectic solvents. 27:2956–2962
92. Xu Z, Qi R, Xiong M, Zhang D, Gu H, Chen W (2021) Conversion of cotton textile waste to clean solid fuel via surfactant-assisted hydrothermal carbonization: mechanisms and combustion behaviors. 321:124450
93. Duman G (2021) Preparation of novel porous carbon from hydrothermal pretreated textile wastes: effects of textile type and activation agent on structural and adsorptive properties. 43:102286
94. Yasin S, Massimo C, Rovero G, Behary N, Perwuelz A, Giraud S, et al (2017) An alternative for the end-of-life phase of flame retardant textile products: degradation of flame retardant and preliminary settings of energy valorization by gasification. 12(3):5196–5211
95. Santos SM, Assis AC, Gomes L, Nobre C, Brito P (2022) Waste gasification technologies: a brief overview. Waste MDPI 140–165
96. Rickson R (2006) Controlling sediment at source: an evaluation of erosion control geotextiles. 31(5):550–560
97. Broda J, Przybyło S, Gawłowski A, Grzybowska-Pietras J, Sarna E, Rom M, et al (2019) Utilisation of textile wastes for the production of geotextiles designed for erosion protection. 110(3):435–444
98. Yang Z, Li X, Si J, Cui Z, Peng K (2019) Morphological, mechanical and thermal properties of poly (lactic acid)(PLA)/cellulose nanofibrils (CNF) composites nanofiber for tissue engineering. 34:207–215
99. Rizal S, Olaiya FG, Saharudin N, Abdullah C, NG O, Mohamad Haafiz M, et al (2021) Isolation of textile waste cellulose nanofibrillated fibre reinforced in polylactic acid-chitin biodegradable composite for green packaging application. 13(3):325

100. Saha S (2020) Textile recycling: the chemical recycling process of textiles. Accessed Access
101. Joutey NT, Bahafid W, Sayel H, El Ghachtouli N (2013) Biodegradation: involved microorganisms and genetically engineered microorganisms. 1:289–320
102. Mahitha U, Devi GD, Sabeena MA, Shankar C, Kirubakaran V (2016) Fast biodegradation of waste cotton fibres from yarn industryusing microbes. 35:925–929
103. Biyada S, Merzouki M, Dėmčėnko T, Vasiliauskienė D, Urbonavičius J, Marčiulaitienė E, et al (2020) Evolution of microbial composition and enzymatic activities during the composting of textile waste. 10(11):3758
104. Nyika J, Dinka M (2022) Sustainable management of textile solid waste materials: the progress and prospects. 62:3320–3324
105. Islam MS, Islam JM (2021) Sources and fates of textile solid wastes and their sustainable management. In: Handbook of solid waste management: sustainability through circular economy. Springer, pp 1–21
106. Hamawand I, Sandell G, Pittaway P, Chakrabarty S, Yusaf T, Chen G, et al (2016) Bioenergy from cotton industry wastes: a review and potential. 66:435–448
107. Selvi C, Koilraj A (2015) Bacterial diversity in compost and vermicompost of cotton waste at Courtallam, Nellai District in Tamilnadu, India. 4(9):582–585
108. Aishwariya S, Amsamani S (2012) Evaluating the efficacy of compost evolved from biomanaging cotton textile waste. 6(4):941–952
109. Oreggioni GD, Luberti M, Reilly M, Kirby ME, Toop T, Theodorou M, et al (2017) Technoeconomic analysis of bio-methane production from agriculture and food industry waste. 123:81–88
110. Juanga-Labayen J, Yanac K, Yuan Q (2021) Effect of substrate-to-inoculum ratio on anaerobic digestion of treated and untreated cotton textile waste. 18:287–296
111. Kuzmanova E, Zhelev N, Akunna JC (2018) Effect of liquid nitrogen pre-treatment on various types of wool waste fibres for biogas production. 4(5):e00619
112. Kabir MM, Forgács G, Horváth IS (2013) Enhanced methane production from wool textile residues by thermal and enzymatic pretreatment. 48(4):575–580
113. Navone L, Moffitt K, Hansen K-A, Blinco J, Payne A, Speight R (2020) Closing the textile loop: enzymatic fibre separation and recycling of wool/polyester fabric blends. 102:149–160
114. Fakhfakh N, Ktari N, Siala R, Nasri M (2013) Wool-waste valorization: production of protein hydrolysate with high antioxidative potential by fermentation with a new keratinolytic bacterium, Bacillus pumilus A1. 115(2):424–433
115. Jeihanipour A, Aslanzadeh S, Rajendran K, Balasubramanian G, Taherzadeh MJ (2013) High-rate biogas production from waste textiles using a two-stage process. 52:128–135
116. Beig B, Riaz M, Naqvi SR, Hassan M, Zheng Z, Karimi K, et al (2021) Current challenges and innovative developments in pretreatment of lignocellulosic residues for biofuel production: a review. 287:119670
117. Shuhua W, Xiaoying Y, Xiaogang C, Wensheng H, Mei N (2020) Recycling of cotton fibers separated from the waste blend fabric. 17(4):520–531
118. Li X, Hu Y, Du C, Lin CSK (2019) Recovery of glucose and polyester from textile waste by enzymatic hydrolysis. 10:3763–3772
119. Brandt-Talbot A, Gschwend FJ, Fennell PS, Lammens TM, Tan B, Weale J, et al (2017) An economically viable ionic liquid for the fractionation of lignocellulosic biomass. 19(13):3078–3102
120. Thomas L, Larroche C, Pandey A (2013) Current developments in solid-state fermentation. 81:146–161
121. Verma N, Kumar V, Bansal M (2021) Valorization of waste biomass in fermentative production of cellulases: a review. 12:613–640
122. Sheldon RA (2014) Green and sustainable manufacture of chemicals from biomass: state of the art. 16(3):950–963
123. Koutinas AA, Vlysidis A, Pleissner D, Kopsahelis N, Garcia IL, Kookos IK, et al (2014) Valorization of industrial waste and by-product streams via fermentation for the production of chemicals and biopolymers. 43(8):2587–2627

124. Wang H, Kaur G, Pensupa N, Uisan K, Du C, Yang X, et al (2018) Textile waste valorization using submerged filamentous fungal fermentation. 118:143–151
125. Carbios Enzymatic recycling: removing the constraints of current processes. https://www.carbios.com/en/enzymatic-recycling. Accessed Access
126. Clouse C (2022) Bio-recycling gets fashionable with enzymes that will eat your shoes. https://www.greenbiz.com/article/bio-recycling-gets-fashionable-enzymes-will-eat-your-shoes. Accessed Access
127. Worrell E, Reuter MA (2014) Handbook of recycling: state-of-the-art for practitioners, analysts, and scientists. Newnes
128. Mondragon G, Kortaberria G, Mendiburu E, González N, Arbelaiz A, Peña-Rodriguez C (2020) Thermomechanical recycling of polyamide 6 from fishing nets waste. 137(10):48442
129. Bahlouli N, Pessey D, Raveyre C, Guillet J, Ahzi S, Dahoun A, et al (2012) Recycling effects on the rheological and thermomechanical properties of polypropylene-based composites. 33:451–458
130. Felgel-Farnholz A (2021) Thermo-mechanical recycling: study on the performance of unstablized and stabilized polyolefins during extrusion
131. Colin X, Tcharkhtchi A (2013) Thermal degradation of polymers during their mechanical recycling
132. Dorigato A (2021) Recycling of polymer blends. 4(2):53–69
133. Le K (2018) Textile recycling technologies, colouring and finishing methods
134. Dal Lago E, Boaretti C, Piovesan F, Roso M, Lorenzetti A, Modesti M (2018) The effect of different compatibilizers on the properties of a post-industrial PC/PET blend. 12(1):49
135. Kunchimon SZ, Tausif M, Goswami P, Cheung V (2019) Polyamide 6 and thermoplastic polyurethane recycled hybrid Fibres via twin-screw melt extrusion. 26:1–14

Digitalization in the Textile Sector for Circularity

Ayesha Kanwal, Muhammad Anwaar Nazeer, and Shahid Rasul

Abstract The ever-growing textile industry is one of the biggest contributors to environmental pollution and resource depletion. The current linear model is highly unsustainable which results in underutilization of resources and waste generation. To mitigate the effects of the current system, a circular system needs to be in place that provides products with an extended life cycle without value loss and are made of recycled or renewable materials. Industry 4.0 a crucial role in this transition by providing real time data and analysis, which subsequently allows for data driven decision making. This article investigates the implementation of a circular economy through digital transformation in the textile industry, concentrating on areas such as repair, reverse logistics, product-as-a-service models, recycling, as well as Designing for circularity, which uses additive manufacturing and 3D technologies. Also discussed is the role of IoT and Big data, including traceability, blockchain, AI, and AR/VR technologies. The combination of digitalization and circular economy principles has the potential to transform the textile industry, making it more sustainable, efficient, and prosperous.

1 Introduction

1.1 Textile Industry's Environmental Impact and Circular Economy

Textile falls in the category of human survival necessity. The unfortunate advent of fast fashion has turned it into a cheap luxury rather than an essential, doubling the production in the last 15 years. Unarguably textile has been a great part of the

A. Kanwal · M. A. Nazeer (✉)
School of Engineering and Technology, National Textile University, Faisalabad 37610, Pakistan
e-mail: Mnazeer13@ku.edu.tr

S. Rasul (✉)
Faculty of Engineering and Environment, Northumbria University, Tyne NE1 8ST, UK
e-mail: shahid.rasul@northumbria.ac.uk

S. R. Batool et al. (eds.), *Circularity in Textiles*, Textile Science and Clothing Technology, https://doi.org/10.1007/978-3-031-49479-6_8

economy and social relations, however, USD 500 billion or more in value is lost to underutilization annually due to the fabricate-make-and-dispose model of the textile sector. This linear system of production is detrimental to the environment with 98 million tonnes of non-renewable resources used per year. The textile sector alone utilizes 93 billion cubic meters of water annually and contributes 20% to industrial water contamination worldwide. Along with greenhouse gas (GHG) emissions of 1.2 billion tonnes, half a million tonnes of plastic microfibres or more are released into the ocean every year due to the washing of plastic-based textiles, which disrupts all aspects of the environment [1]. If a circular economy is not adapted promptly, the textile sector alone would have utilized more than 26% of the carbon budget of the 2 °C goal of the Paris climate agreement by 2050 [2].

To mitigate the effects of the linear production system, a new system must be implemented. A circular economy can be defined as an economy of closed cycles that is restorative or regenerative by design, employs systems thinking at its foundation, and shifts towards the use of renewable energy. Industry 4.0, on the other hand, is known as the "era of cyber-physical systems" and integrates computing, digitization, networking, and physical processes into traditional manufacturing and industrial platforms [1]. Industry 4.0 and the circular economy both aim to enhance products and processes while optimizing resource consumption and costs. Figure 8.1 illustrates that while the circular economy can drive the transition of manufacturing industries towards systemic sustainability, Industry 4.0 can drive innovation and accelerate the digital transformation toward intelligent and resilient manufacturing enterprises. In simplest terms, circular economy (CE) for the textile sector means the products have a prolonged life cycle, can be recycled into a new product without value loss, and are made from either recycled or renewable materials. It is focused on eradicating substances of concern from the loop, minimizing immoderate inventory, annihilating the concept of pre-consumer waste, and recycling post-consumer waste without value loss [3]. With less than 1% of recycling and USD 500 billion lost in value in the linear economy, textile materials retain their value and never end up as waste in the circular economy, bringing environmental, societal, and economic prosperity [1]. According to the world economic forum, we must quicken the shift to a circular economy in order to accomplish our global climate objectives by 2050 and this can only be achieved through digitalization [4].

1.2 Circular Economy and Digitalization

With the 4th industrial revolution and the desire for sustainability, digitalization for circularity has become an important research topic revolving around digital networks and technologies that provide real-time data, analysis, and decision-making. The real-time capabilities and decentralized interoperability of digitalization can lead the textile industry toward circularity [5]. For instance, knowing the environmental impact associated with each stage of a product's life cycle collected by the Internet of Things (IoT) can assist businesses in making defensible choices regarding how

Fig. 8.1 Digitalization in the textile sector for circularity

to lower their carbon footprint and move towards circularity. Digitalization can also help the textile sector to reduce waste and optimize production processes, resulting in more sustainable practices [6].

2 Digital Transformation for Circular Economy

Implementation of a circular economy requires change on an organizational level where all stakeholders participate to maximize the benefits of this opportunity. Increased awareness of the evils of the linear production system has made all stakeholders yearn for change but the attention has gone toward the reduction of the negative impact of the present-day linear system [1]. To reduce the negative impact of the current system, new circular and digitally enabled business models need to be designed, as seen in Table 8.1. This calls for businesses to be connected, creating

Table 8.1 Sustainable operations for circular economy and digital enablers

Operation	Digital enabler	Effect on circular economy
Repair	3D printing	Extended product life, conserving resources, and minimizing waste
Reverse logistics	Traceability devices	Recovered material for reuse, repair, or recycling aiding value creation in the textile industry
Product as a service (PaaS)	Traceability devices	Efficient use of resources and discouraged product ownership resulting in a low carbon footprint
Recycling and sorting	Near-infrared technology	Accurate sorting of waste for material recovery and value creation. Increased material flow and decreased use of virgin materials

interdependencies that enable optimal material use [7]. If business models such as resale, rental, repair, and remake market share reach 23% by the year 2030, the fashion industry will produce up to 16% less carbon dioxide than it currently does [8].

2.1 *Circular Repair Solutions*

Recycling, although the most popular "R", is not the most sustainable option. To promote reuse, multiple brands like Houdini offer a warranty for repair or replacement which is one of the most efficient policies. This policy not only benefits the environment but also benefits the company in the form of customer loyalty and data [1]. Recent technological developments, such as 3D printing, have facilitated and encouraged the repair of broken items rather than their disposal. As some 3D-printed garments are constructed using links, it is very simple to replace or repair them. Similarly, 3D-printed textile patches can be used for repairs, as these patches can mimic the properties of the garment that needs to be fixed in order to create a seamless repair [9, 10].

2.2 *Reverse Logistics and Circularity*

Supply chain management programs such as reverse logistics can be combined with take- back business model to encourage circularity. Reverse logistics take products from their site of consumption to their site of origin [11]. For instance, Marks and Spencer incentivized take-back by rewarding customers with shopping vouchers, resulting in the creation of an entire collection of coats from the returned items [12]. This example also goes to show that customers play an important role in the value chain [5]. When it comes to reverse logistics, having a knowledgeable and involved customer base is essential. In order for reverse logistics to be successful, it is necessary for customers to be aware of the environmental and economic value of the textile item. Digital tools such as traceability tags can greatly assist reverse logistics in the tracking of these items. Traceability tags contain information regarding the fabric's material and chemical composition, as well as any damages or alterations made during its life cycle, making it easier to determine a sustainable end-of-life procedure.

2.3 *Product as a Service for Sustainable Consumption*

The product-as-a-service (PaaS) model encourages the extended use of garments, thereby preventing resource underutilization. Offering clothing as a service encourages organizations to prioritize its durability and repairability, as they would want it to last as long as possible. The European Union's push for subscription-based business models has made it an integral part of the circular economy concept [6]. Product as a service is beneficial not only for the environment as it promotes sharing and reuse of resources but for the consumer as well, allowing them to experience more and own less. The owner is responsible for the care of the garment liberating its customer from cleaning and repairing responsibilities [13]. Subscription-based companies leverage IoT to monitor the life cycle of their products to make informed decisions about reassigning, recycling, or disposal. For instance, companies like Wozabal use Radio-frequency identification (RFID) tags to monitor repair cycles and losses [14]. Similarly, Garmex, a rental service uses traceability tags at every stage of the rent cycle to monitor how many times a textile has been used [15].

2.4 *Recycling and Near Infrared Sorting for Resource Recovery*

A lot of value is lost in recycling but it is a necessary evil as not everything can be repaired and reused. The processes that are used to recycle textiles can be categorized as chemical, mechanical, and thermal. The chemical process includes the depolymerization of polymeric fibres and the dissolution of natural fibres. On the other hand, mechanical processes include pre-treatments while thermal includes (the conversion of PET pellets, chips, or flakes into fibres by melt extrusion [16]. Regardless of the route you choose, it is essential to sort textiles for efficient processing. Sorting is mostly done manually, making it a costly and laborious activity. Automating and digitizing this process can encourage recycling, resulting in a reduced environmental impact. In order to differentiate between different fibres, a huge amount of data needs to be collected and analysed. Initially, spectral data and digital photos were studied for sorting but failed to provide a 100% accurate result [17, 18]. However, Riba suggests analysing textile samples using an Attenuated total reflection- Fourier transform infrared (ATR-FTIR) spectrometer without the need for any pre-treatment, to get 100% accuracy. Once analysed, the resulting spectrum is digital data representing the interaction between infrared radiation and the sample in the form of abundant data points which call for mathematical algorithms such as principal component analysis (PCA), Canonical variate analysis (CVA), and k nearest neighbour (k-NN) to be processed [16]. However, REISKAtex sorting algorithm involves building a library with the spectral data of textile samples. It suggests comparing the spectral data of an unidentified material to the spectral data available in the library. Fourier transform infrared (FT-IR) analyser, which is a spectral library can be used to test the accuracy of the match. REISKAtex confirms 100% accurate recognition of pure cotton, polyester, wool, viscose, and a few blends of CO/PES [19]. FIBERSORT by Valvan uses a similar concept as REISKAtex. Although these technologies are highly successful, extensive research is required to sort multi-coloured and complex blend garments [1]. Recycling helps businesses save money on virgin material, making it a win–win situation for the business and the environment.

3 Designing for Circular Economy

Designing for circularity is a direct synonym for designing to be used more. The "R" imperative goes hand in hand with designing for circularity thus design greatly influences the lifecycle of the product. For instance, 64% of the clothing produced worldwide includes essentials like outerwear, undergarments, jeans, and hosiery which are not subject to the latest trends. It only makes sense to design these items to be functionally durable. On the other hand, to increase emotional durability, the product needs to have a timeless design, and customer involvement in the production process is a key enabler that has been made easy by 3D technologies [1]. For Example, Digital fashion, a collaborative project with the likes of Houdini, sells locally produced, customized garments based on the consumer's avatar allowing customers to create a long-lasting emotional connection with the product [20]. It is safe to say that functional and emotional durability prevents clothing from being discarded in landfills. Similarly, designing for repair and multiple uses is equally important just as Melissa Ortuno designs with removable parts, making it simple to repair. The current system of production does not take into consideration, what happens when the first life cycle of a textile product is over therefore designing for disassembly is crucial. Clothing needs to be designed in a way that its components of aesthetic and functional use can be disassembled for different recycling processes. Rothenborg's design is a great example of how a design for circularity should be. He designed a coat with 2 layers of mono-materials that can be disassembled for recycling into mono streams. The piece is made with removable sleeves and a hood for use throughout the year [21].

3.1 *Wear2 Technology for Disassembly*

Disassembly of garments for recycling is unheeded, as the process is labour intensive causing clothing to end up in the landfill, due to the inhomogeneous nature of the materials used. Designing Mono Material garments like Puma's athletic wear which is made of polyester entirely with all accessories and thread being polyester can be limiting. However, Wear2 technology enables functional and aesthetic accessories to be removed, for all components to end up in their respective recycling cycles. This works by using a thread that is extremely sensitive to electromagnetic radiation. Once exposed to a microwave, it loses its mechanical properties and allows for easy disassembly without being constraining [12, 22]. Wear2 partners with digital platforms that encourage sustainable design by providing an expert opinion. Their circular return systems allow you to trace your garments through a digital passport for disassembly and circularity by registering for the Global Sustainable Enterprise System (GSES) along with a Quick response (QR) label attached to your clothing. They also facilitate transport between you, your supplier, and sustainable processors creating a full circle of sustainability [23, 24].

3.2 *Additive Manufacturing and On-demand Production*

Due to its potential for repair, remanufacturing, reduced resource consumption, mass customization, and localized production, additive manufacturing can fundamentally align us with the circular economy [25]. Additive manufacturing also known as 3D printing is a layered manufacturing process with numerous advantages over subtractive methods such as annihilating design constraints with reduced cost and footprint, making it an excellent sustainable solution. It further Facilitates in-house and on-demand manufacturing which also enables reduced carbon emissions. Additive manufacturing allows for low-volume production, which is considered a limitation in the textile industry but not in the case of circularity. Even though additive textile manufacturing is a relatively new and complex term, it has a very simple production process. A person can simply start by scanning the body and check the fit on an avatar followed by localized additive manufacturing, all while creating a minimal footprint. Additive manufacturing is categorized into two types: manufacturing onto textiles and manufacturing textiles. Additive manufacturing onto textiles can allow customers to 3d print functional or aesthetic elements on the surface of a fabric which results in added personalization, increasing the emotional or functional durability of the garment. This however comes with constraints, such as adhesion issues and effects on the drape of the fabric. On the other hand, materials such as polylactic acid (PLA) and Filaflex have enabled the 3D printing of textile-like materials [26]. It is possible to convert traditional textiles into digital code that 3D printers can then print. Fused deposition modelling (FDM) allows for the creation of pliable and soft textiles and the most common FDM materials for fashion products are polylactic acid (PLA), polyamide (nylon), and thermoplastic polyurethane (TPU), which offers a vast array of colour options [27]. The material versatility of additive manufacturing enables the optimal use of waste material as feedstock. Therefore extrusion-based additive manufacturing is compatible with a variety of recycled materials [28]. This technology also enables design for disassembly, repair, and recycling, such as the Modeclix project in which NylonPA12 textiles were disassembled and reassembled using simple links [29]. Similarly, Adidas' use of additive manufacturing in projects such as FutureCraft has been exceptional. Collecting data on athletes, Adidas collaborated with an American 3D printing company to develop a midsole for precise and individualized performance which takes us a step closer to mass customization [13].

Additive manufacturing facilitates on-demand production, resulting in low waste, reduced logistic emission, and efficient use of material. Inaccurate demand forecasting can leave a business with unsold inventory, therefore Centralized database systems like ERP systems allow businesses to access real-time data which facilitates a pull system of production by making demand forecasts available, allowing the reduction of overproduction and deadstock [21, 30, 31]. On-demand production has been used by many in the industry. For instance, in a project in collaboration with H&M, Clother devised a method whereby individuals could vote for 3D-printed product prototypes that allowed designers to analyse customer interest in order to produce precisely. On the other hand, the Post-Couture Collective in Belgium allows for the designs for 3D printing to be available online so they can be produced locally and assembled by the consumer [1].

3.3 3D Textile Technologies for Circular Economy

The adoption of 3D technologies in the fashion industry has permitted the emergence of a non-material economy in which garments exist beyond the material world and contribute to dematerialization by never being physically produced [32]. The use of 3D technologies for the virtual prototyping of products and made-to-measure clothing eliminates pre and post-production wastes and increases sustainability [29]. For instance, 3D visualization eliminated the need for logistics and resources to produce samples resulting in a reduced footprint [4]. One of the most significant aspects of this digitalization is the implementation of 3D tools in the computer-aided design (CAD) system during the design process. Contrary to the traditional method of modifying initial designs through seamstresses, simulations are produced for every required modification quickly and efficiently, while simultaneously generating detailed product specifications for manufacturing the 3D-designed garment. Industry leaders like Victoria's Secret and Nike have put 3D designs into effect. Browzwear provides software comparable to Lotta for 3D fashion design, whereas Clo3D offers 3D garment visualization which is a technology for designing 3D clothing directly on a 3D human model to simulate the effect of the garment on the human body without the need for repetitive fittings on a model. Using 3D body-scanning systems, it is common to construct digital body models and simulate new designs on these digital bodies. When biometric technologies are combined with body measurement data, it facilitates the custom garment market. It allows designers to validate their designs on computer-generated or body-scan models by incorporating technical data, fabric type, colour, drape, tensile, shearing, bending, and the effect of seams [33]. Similarly, V-Stitcher's style editing features 3D pattern editing capabilities that provide a straightforward correspondence of 3D alterations and 2D patterns [29]. The costs associated with consumer and supplier misunderstandings and manufacturing errors resulting in enormous waste have decreased as a result of 3D visualization's ability to accurately depict the products. Thus, the use of 3D software provides not only economic benefits through the conservation of resources but also environmental advantages. A great example of the use of this technology is by a Danish company, Son of a Tailor which only uses height, weight, age, and shoe size to create custom-fit garments which have led to zero inventory waste [13]. Similarly, companies like Atacac accept pre-orders while presenting the product using 3d technology [20].

3.4 Knit on Demand

The lack of accurate demand forecasting tools has left no choice but to manufacture depending on the demand that has been expressed. The Swedish School of Textiles worked along with a knitwear manufacturer known as Ivanhoe AB and a bespoke clothes shop known as SOM Concept. AB created a project known as Knit on Demand. The project produced a business model for mass customization which allows customers to design or personalize their own knitted clothing. It facilitates customization and localized production increasing emotional durability for the customer while reducing logistic emissions. In a supply chain like Knit on Demand, production does not start until a customer order is placed eliminating the possibility of a deadstock. Raw materials such as yarn, accessories, and threads are acquired and stashed away in the manufacturing plant for production once the order is received. With the utilization of computer-aided design (CAD) software, the project creates digital representations of knitted patterns in the form of virtual models. These designs can be altered and adapted to meet the requirements of individual users. Following the generation of the machine instructions, the digital design is used to direct automated knitting machines in the production of the required textile items. The process starts whenever a customer walks into the shop with the intention of buying a knitted item that can be personalized. The buyer is led through the design process and is given the opportunity to make modifications to four aspects of a garment: the model, the fit, the colour, and the details. The product is delivered to the customer within 1 to 3 weeks of the order [34, 35].

Table 8.2 Stakeholder data needs for circular economy

Stakeholders	Data needs for circularity
Designers	Detailed material content for designing with textiles of low environmental impact, high functional durability, and easy recyclability, or biodegradability
Textile producers	Ethical sources of raw material, demand forecasts to avoid overproduction, production data for process monitoring and improvement to reduce waste generation, and environmental impact of the end product
Textile product manufacturer	Ethical sources of textile fabric, demand forecasts to avoid overproduction, production data for process monitoring and improvement to reduce waste generation, and environmental impact of the end product
Retail and distributors	Ethical sources of textile products, demand forecasts to avoid deadstock, the environmental impact of inventory waste, and logistics
End of life processors	Locations of collection of post-consumer waste, reverse logistics data, Spectral or material/chemical composition data for sorting and recycling, and the environmental impact of the end-of-life processing

4 The Role of IoT and Big Data in Circularity

Industry 4.0 is made up of machines with embedded intelligence and these intelligent devices generate big data. This immense data collected from various devices and operations cannot be processed with conventional methods due to the speed of collection, required analysis, and its heterogeneous nature. The data generated by IoT devices is of great value therefore, it needs to be analysed accurately with the finest analytical tools such as machine learning in order to facilitate circularity. This data can be used to optimize all operations of the value chain and data-driven decision-making. Big data can fuse with exterior environmental data which enables informed decisions that empower improved value retention and decreased environmental impact [11]. With the world's focus on sustainability, companies are obligated to set circular economy goals and disclose their progress. To achieve the goals, they must have access to data measuring the efficacy of the circular economy in their products and operations. Comprehensive data to understand the sustainability of a product is often deficient, as it may not have been collected, or analysed, or there are no data-sharing policies. As illustrated in Table 8.2, reliable product data serves as the basis for an environmentally friendly supply chain and by utilizing reliable and accurate data from all value chain stakeholders, it enables them to make circularity-promoting data-driven decisions [36].

5 Supply Chain Transparency for Circular Economy

The increased awareness of the customer and the implementation of the circular economy has created a demand for transparency. When the customer wants to know how his food was grown, he is also increasingly interested in how his clothes were made. Having this knowledge increases customers' trust and educates them about the environmental impact of a single product discouraging them to buy more. Keeping track of products and transparency of material is of utmost importance in the circular economy and this has been made easy with technologies like RFID, QR codes, and Blockchain. Alignment with the processes across the value chain facilitates the identification of the materials in the system to enhance the recycling processes' output quality. It also creates the possibility of supply and demand to meet in real-time, making unthinkable supply chains possible where intermediaries are completely eliminated [21].

Transparency not only makes material reuse, recovery, and recycling a simple process, but it also enables circular business models to thrive. Several companies such as Content Thread are researching the application of RFID tags to keep track of the product from the manufacturing stage containing all the information about the product such as materials used, chemical input, and the manufacturing process making it easy for recycling and sorting companies. Transparency makes it incredibly easy to authenticate products, increasing the emotional durability of the product. IoT devices such as NFC, and RFID are common technologies used to trace the supply chain [37].

5.1 Traceability Tags

Traceability tags such as QR codes, NFC tags, and RFID tags are great tools for transparency in the circular economy. Barcodes and QR labels are physical tags that can be printed, embroidered, or woven. Once scanned, the coding opens a database with product, assembly, and material information. On the other hand, NFC exchanges information between physical objects and digital systems [38]. Similarly, radiofrequency identification (RFID) can wirelessly measure and monitor enormous amounts of reliable data [5]. The major component of an RFID system is a reader and a small radio frequency transponder (RF tag). In the read stage, the reader reads the data from the sensor and tags, which include valuable information such as the geographical location and geographical factors sensed during tracking. When an RFID tag comes within range of an RFID reader, the reader sends a radio signal to the tag that activates it. In the communication phase, the traced information is transmitted to the appropriate authority using the embedded communication protocol [37]. The likes of Fendi and Michael Kors use RFID tags for product authentication containing data from sourcing to point of sale [29]. Similarly, printed or organic electronic tags are inexpensive and can reduce the cost of logistics while collecting a large amount of data. POE can be produced with additive manufacturing techniques with minimal substances of concern and are biodegradable. These traceability tags can contain information on every aspect of the product [15].

5.2 Digital Twins

A digital twin refers to a cyber-physical equivalent of a physical product [39]. Digital twins manage product life cycle information and share it among different actors of the supply chain. The information it contains includes but is not limited to provenance, current location, material, and chemical composition. knowing the origin not only adds monetary value but provides enough information for repair and recycling which can be helpful throughout life cycle management. Initially, it can be used for digital prototyping and then focus on optimal product use while at the end provide with end-of-life options [11]. Digital twins are ultimately utilized in the design and development phase to improve resource use efficiency while integrating data from all stakeholders for supply chain transparency. Due to its real-time data monitoring, it can identify waste reduction, repair, reuse, and recycling opportunities. Digital twins along with blockchain technology can become tamper-proof solidifying supply chain transparency [39].

5.3 Blockchain Technology

Blockchain has come to light due to its use in cryptocurrency however it can be implemented in any supply chain. A supply chain consists of a collection of organizations that perform various operations and transactions. The information of these operations and transactions are stored in insecure, limited access systems. Data such as provenance, traceability, authenticity, and sustainability can be provided using blockchain [29]. Blockchain is based on distributed ledger configuration that offers data safety and validity. Due to its decentralized nature, there are numerous benefits associated with its use. It offers unparalleled security because once the data is recorded, it cannot be altered. Pseudonymization provides this additional layer of security to the blockchain. It records information in a chronological order thereby enhancing traceability. Data is saved and recorded in the form of blocks which are then encrypted using hash functions that turn the data into binary data. Blockchains have different access levels that can be public, consortium, and private. The public blockchains are open for everyone while the consortium blockchains have authorities with permissions. A private blockchain has restrictions and requires permission to operate [39]. Customers can access a public blockchain to assess a company's transparency while a consortium chain can be utilized for low-risk B2B transactions [37]. For the textile sector, consortium blockchain can allow all actors of the value chain to collaborate while ensuring data privacy and security [39]. Table 8.3 presents the data that may be kept by all players involved in the value chain, with the aim of enhancing transparency in their respective business operations.

IoT and blockchain combined provide real-time data on textile production and consumption, which facilitates the transition to a circular economy [37]. Aura blockchain is the most relatable example of blockchain technology in the field of

Table 8.3 Information to be stored in a blockchain adapted from [40]

Category of data	Data stored
Business and finance	Transactions, contracts, commodity prices, material costs, and additional financial data
Environmental	Material and chemical composition of the material, GHG emissions and environmental impact, Resources consumed, waste generated, certifications, and biodegradability
Societal	Ethical wages and hours for workers, Ethical instructions of care and disposal
Functional	Capability and utility, repair instructions, and quality control

textiles. Aura is a consortium blockchain of global luxury brands such as Cartier, and Louis Vuitton that guarantees transparency to its customers. It allows customers to authenticate their products from the beginning of the supply chain. These luxury companies realize the value that is added to their products through transparency [13]. Following the United Nations garment traceability project, a blockchain of luxury brands for the traceability of the cotton value chain has also emerged [29]. Similarly, blockchain paired with other technologies such as blockchain emission trading systems helps measure the environmental impact of a product and make suggestions to lower that impact [39].

6 Limited Use of Artificial Intelligence in the Circular Economy

In the textile industry, artificial intelligence is not novel however, utilizing AI for circularity is a novel concept. AI has been utilized for process enhancement throughout the industry. Using an AI "expert system" that employs "if–then" logic to make decisions on complex problems, the appropriate process and apparatus for environmentally friendly production can be selected resulting in a low negative environmental impact [39]. Using AI techniques such as artificial neural network (ANN) it can identify areas of dysfunction, defects, and opportunities for waste reduction [40]. Through similar facets of AI, the fashion industry has gained customer loyalty by improving customer experience by generating big data by monitoring customers' reactions to gauge their interest in goods, thereby enabling retailers to improve demand forecasting and reducing deadstock that ends up in landfills [39, 41].

7 Virtual and Augmented Reality for Dematerialization

Virtual reality and Augmented reality are powerful digital technologies that eradicate the need for physical products and promote dematerialization. It can be used to create digital showrooms instead of stores filled with inventory. Virtual reality can enable the customer to experience the product in a situation it will be used in, while standing in the showroom which will help the customer make an informed decision regarding the purchase.

AR technology has improved customer engagement in recent years and fashion labels use it for presentations and smart mirror try-ons [27]. For instance, the magic mirror concept presents the idea of 3d scanning the customer's surroundings and having them try on the clothing virtually. The magic mirror with an integrated ERP system can consist of all the information regarding the product and the customer can make the purchase directly from the mirror [41]. These magic mirrors in Japanese stores like Uniqlo allow customers to virtually try on clothes and take pictures to share on social media [29]. Similarly, going fully digital, DressX offers influencers virtual clothing just so they can share on social media [8]. This type of technology can aid in reducing returns and mitigating the negative effect of return shipping resulting in a sustainable society [20].

8 Benefits of Digitalization in the Circular Economy

The transition towards circularity in the textile sector has been helped tremendously by digitalization as it can close the realization gap between the theory and practice of the circular economy [4]. Digitalization has not only helped to increase productivity and effectiveness of resource utilization but also suggested new ways in which resources can be utilized to their full potential and retain full value [42]. Due to the rise of digitalization, supply chain transparency has become an essential component of the circular economy, which has led to the development of more effective supply chains. Digitalization-driven dematerialization of the supply chain has produced numerous benefits in terms of increasing resource efficiency, condensing, eliminating, and shortening the number of business activities, and reorganizing the operational model to be more collaborative [32]. It has enabled make-to-order production systems such as Additive Manufacturing and knit-on-demand that reduce material consumption and emissions related to component transportation. With big data, it is possible to address the complexities of decision-making while Implementing AI increases productivity by improving optimization, real-time data analysis, and design, which all contribute to fostering circularity [43].

9 Challenges of Digitalization in the Circular Economy

Unquestionably the greatest challenge faced by digitalization for a circular economy is in the form of economic constraint. Most organizations operate the circularity framework with economic considerations rather than environmental considerations as a primary focus. The majority of the current businesses are centred toward quick return on investment and uncertain gains from practicing a circular economy make it an unsuitable approach. For instance, instead of collecting, sorting, and recycling previously owned garments before remanufacturing them, it is more cost-effective to produce more from virgin material [44]. In a circular economy, the operational risk is transferred to the organization, and when the user is no longer in ownership of the products, the company may incur additional maintenance costs to enhance its sustainability. Consequently, the CE paradigm is still infrequently employed by businesses [45].

Scalability is also one of the greatest barriers to the complete integration of emerging technologies [27]. The incompatibility of these technologies with legacy systems is another significant obstacle to overcome which results in an increased cost of implementation. Training for staff is another area of concern which can be difficult due to the diverse range of equipment and technologies in the textile sector [39]. For example, 3D technicians are very difficult to find as they are in high demand due to the 4th industrial revolution [29].

Organizations are focused on individual growth and not collaborative growth towards circularity gatekeeping important data [6]. As data is the primary enabler for the digitalization of the circular economy, digital information carriers present significant recycling barriers at various stages of the value chain, despite their effortless ability to transport data. Additionally, the extent to which end-users are prepared to be monitored by RFID tags is debatable. It is also important to note that a substantial number of products currently in circulation cannot be tracked or analysed because they lack the necessary technology. The real issue is how to transition these products to circularity [40].

According to the decision-making trial and evaluation laboratory (DEMATEL) analysis, some of the problems with putting a circular economy into place are low consumer awareness, low amount of businesses that have successfully put a circular economy into place, lack of transparency between stakeholders, and lack of law pushing towards circularity. In China, for example, the 2002 law of cleaner production promotion and the 2009 law of circular economy promotion law has introduced circular economy into public policy. In addition, the European Commission has implemented a number of circular economy policies and introduced the circular economy package. The lack of urgency toward a circular economy is fatal to a sustainable world [46].

10 Future Work

Digitalization for circularity in the textile sector is a very new concept and requires extensive research. Some possible research topics are listed below:

- Industry 4.0 utilizes thousands of IoT devices generating enormous amounts of data. Thorough research is required to analyse which data is most influential for the implementation of the circular economy and how data sharing can be encouraged to prevent data monopoly [6].
- Traceability tags that can withstand washing cycles are required in the circular economy therefore, research on how to make traceability tags more durable is crucial.
- IoT devices such as 4D textiles can facilitate textile sectors' progression towards a circular economy but require comprehensive research [47].
- Sorting textiles for recycling requires research to a great extent which may include creating a system to identify drenched and soiled pieces of fabric as well as using colour scanners along with other technologies to help sort multi-coloured textiles. The sorting of blended textiles and the use of AI for this purpose also require research [48].
- The use of AI for design, trend, and demand prediction are great subjects of research [49].

11 Conclusions

The impact of the textile industry on the environment cannot be overlooked. The circular economy and Industry 4.0 have paved the way for the development of an environmentally responsible textile industry through the implementation of new ideas and procedures made possible by technological advancements. In the world of reuse, repair, and recycling, the application of new technologies has opened a previously unexplored frontier. Although it is impossible to ignore the environmental benefits of a circular economy, many industries choose to circumvent the system because doing so enables them to generate enormous profits. Before Industry 4.0 and a circular economy can be implemented, everyone with an interest in the project must be on board. From factories to the sale of finished goods to customers. Awareness, education, and the adoption of environmentally friendly practices are the only things that can save us from our own destruction. The results indicate that the textile industry must transition to a circular economy through digital transformation in order to reduce its environmental impact, optimize resource consumption, and achieve long-term sustainability.

References

1. A New Textiles Economy: Redesigning Fashion's Future. https://ellenmacarthurfoundation.org/a-new-textiles-economy. Accessed 18 Jul 2023
2. UNFCCC (2015) Adoption of the Paris agreement–Paris agreement text english
3. A circular economy vision for fashion and textiles industry. https://ellenmacarthurfoundation.org/our-vision-of-a-circular-economy-for-fashion. Accessed 18 Jul 2023
4. Wiegand T, Wynn M (2023) Sustainability, the circular economy and digitalisation in the german textile and clothing industry. https://doi.org/10.20944/PREPRINTS202305.0046.V1
5. Ghoreishi M, Happonen A (2022) The case of fabric and textile industry: The emerging role of digitalization, internet-of-things and industry 4.0 for circularity. Lect Notes Netw Syst 216:189–200. https://doi.org/10.1007/978-981-16-1781-2_18
6. Luoma P, Penttinen E, Tapio P, Toppinen A (2022) Future images of data in circular economy for textiles. Technol Forecast Soc Change 182:121859. https://doi.org/10.1016/J.TECHFORE.2022.121859
7. Angelova RA (2020) The circular economy: A new paradigm for the textile and clothing industries. In: E3S web of conferences. EDP Sciences
8. Rethinking business models for a thriving fashion industry. https://ellenmacarthurfoundation.org/fashion-business-models/overview. Accessed 18 Jul 2023
9. Terzioglu N, Brass C, Lockton D (2016) 3D printing for repair: a paradigm shift in fixing our relationships with things
10. Bloomfield M, Borstrock S (2018) Modeclix. The additively manufactured adaptable textile. Mater Today Commun 16:212–216. https://doi.org/10.1016/j.mtcomm.2018.04.002
11. Liu L, Ramakrishna S (2020) An introduction to circular economy. An introduction to circular economy. 1–631. https://doi.org/10.1007/978-981-15-8510-4
12. Joss S (2005) One planet living: a tale of three UK urban sustainability initiatives sustainable cities: the governance of urban innovation view project smart eco-cities for a green economy (SMART-ECO): a comparative study of Europe and China View project Sustainable Innovation 2014 2
13. Cruz F, Rosado Da Cruz AM, Colombi C, D'itria E (2023) Fashion digital transformation: innovating business models toward circular economy and sustainability. Sustainability 15:4942. https://doi.org/10.3390/SU15064942
14. Hansen EG (2019) Smart products as enabler for circular business models: the case of B2B textile washing services sustainability-oriented and circular innovation: products, product-service systems and business models-research stream view project responsible leadership systems-research project view project smart products as enabler for circular business models: the case of B2B textile washing services
15. Fagergren M, Junebrink M (2022) Applications of printed and organic electronics: How printed and organic electronics can facilitate circular business models in the fashion industry through traceability. Linköping University, Environmental Technology and Management
16. Riba JR, Cantero R, Canals T, Puig R (2020) Circular economy of post-consumer textile waste: classification through infrared spectroscopy. J Clean Prod 272. https://doi.org/10.1016/j.jclepro.2020.123011
17. Xing W, Xin B, Deng N, Chen Y, Zhang Z (2019) A novel digital analysis method for measuring and identifying of wool and cashmere fibers. Measurement (Lond) 132:11–21. https://doi.org/10.1016/j.measurement.2018.09.032
18. Peets P, Leito I, Pelt J, Vahur S (2017) Identification and classification of textile fibres using ATR-FT-IR spectroscopy with chemometric methods. Spectrochim Acta A Mol Biomol Spectrosc 173:175–181. https://doi.org/10.1016/j.saa.2016.09.007
19. Dahlbo H, Aalto K, Eskelinen H, Salmenperä H (2017) Increasing textile circulation—consequences and requirements. Sustain Prod Consum 9:44–57. https://doi.org/10.1016/J.SPC.2016.06.005

20. Textiles–from waste to resource by 2030 | European circular economy stakeholder platform. https://circulareconomy.europa.eu/platform/en/knowledge/textiles-waste-resource-2030-0. Accessed 18 Jul 2023
21. Matthes A, Beyer K, Cebulla H, Arnold MG, Schumann A (2020) Sustainable textile and fashion value chains: drivers, concepts, theories and solutions sustainable textile and fashion value chains: drivers, concepts, theories and solutions. 1–453. https://doi.org/10.1007/978-3-030-22018-1
22. Dissanayake DGK, Weerasinghe D (2022) Towards circular economy in fashion: review of strategies, barriers and enablers. Circ Econ Sustain 2:25–45. https://doi.org/10.1007/S43615-021-00090-5
23. Step by step towards a circular concept–WEAR2GO. https://wear2.com/en/corporate-workwear/step-by-step-towards-a-circular-concept/. Accessed 18 Jul 2023
24. Welcome to the World of WEAR2® Ecostitching Technology–WEAR2GO. https://wear2.com/corporate-workwear-nl/welcome-to-the-world-of-wear2-ecostitching-technology/. Accessed 18 Jul 2023
25. Kravchenko M, Pigosso DCA, McAloone TC (2020) Circular economy enabled by additive manufacturing: Potential opportunities and key sustainability aspects. In: Proceedings of the NordDesign 2020 Conference, NordDesign 2020. https://doi.org/10.35199/NORDDESIGN2020.4
26. Keefe EM, Thomas JA, Buller GA, Banks CE (2022) Textile additive manufacturing: an overview. Cogent Eng 9
27. Jin BE Palgrave Studies in Practice: Global fashion brand management series editors
28. Romani A, Rognoli V, Levi M (2021) Design, materials, and extrusion-based additive manufacturing in circular economy contexts: from waste to new products. Sustainability 13:7269. https://doi.org/10.3390/SU13137269
29. MacCarthy B (Bart L), Ivanov D the digital supply chain. 446
30. Boinepelli H (2015) Applications of big data. pp 161–179
31. Powell D, Riezebos J, Strandhagen JO (2012) Lean production and ERP systems in small- and medium-sized enterprises: ERP support for pull production. 51:395–409. https://doi.org/10.1080/00207543.2011.645954
32. Casciani D, Chkanikova O, Pal R (2022) Exploring the nature of digital transformation in the fashion industry: opportunities for supply chains, business models, and sustainability-oriented innovations. Sustain Sci Pract Policy 18:773–795. https://doi.org/10.1080/15487733.2022.2125640
33. Pal R, Jayarathne A (2022) Digitalization in the textiles and clothing sector. The Digital Supply Chain 255–271. https://doi.org/10.1016/B978-0-323-91614-1.00015-0
34. Peterson J, Larsson J, Carlsson J, Andersson P (2007) Knit on demand-simulation of an agile production and shop model for fashion products. In: ITMC 2007 International Conference
35. Larsson J, Peterson J, Mattila H (2012) The knit on demand supply chain. Autex Res J 12:67–75. https://doi.org/10.2478/V10304-012-0013-9/MACHINEREADABLECITATION/RIS
36. Niinimäki K, Cura K, Heikkilä P, Järvinen S, Mäkelä S, Orko I, Tuovila H-M (2023) How data can enhance circular economy of textiles: from knowledge and system understanding to actions
37. Akram SV, Malik PK, Singh R, Gehlot A, Juyal A, Ghafoor KZ, Shrestha S (2022) Implementation of digitalized technologies for fashion industry 4.0: opportunities and challenges. Sci Program https://doi.org/10.1155/2022/7523246
38. Cura K, Jain S, Niinimäki K (2022) Transparency and traceability in the textile value chain. Aalto University, School of Arts, Design and Architecture
39. Alves L, Ferreira Cruz E, Lopes SI, Faria PM, Rosado da Cruz AM (2022) Towards circular economy in the textiles and clothing value chain through blockchain technology and IoT: a review. Waste Manage Res 40:3. https://doi.org/10.1177/0734242X211052858
40. Laila C, Hanna N, Schliephake J (2021) Digital institutions to support data-driven circularity innovation-the improvement of textile and apparel recycling processes through blockchain technology. Thesis for Two-year Master, 30 ECTS Fashion marketing and management

41. von der Assen L (2023) Digitalization as a provider of sustainability?—the role and acceptance of digital technologies in fashion stores. Sustainability (Switzerland) 15. https://doi.org/10.3390/SU15054621/S1
42. Hedberg A, Šipka S (2021) Toward a circular economy: the role of digitalization. One Earth 4:783–785. https://doi.org/10.1016/J.ONEEAR.2021.05.020
43. Chauhan C, Parida V, Dhir A (2022) Linking circular economy and digitalisation technologies: a systematic literature review of past achievements and future promises. Technol Forecast Soc Change 177:121508. https://doi.org/10.1016/J.TECHFORE.2022.121508
44. Gowsiga M, Thayaparan M (2020) Incorporation of circular economy concept to the apparel industry: literature review. 117–125. https://doi.org/10.31705/FARU.2020.13
45. Liu Z, Liu J, Osmani M (2021) Integration of digital economy and circular economy: current status and future directions. Sustainability (Switzerland) 13
46. Chen WK, Nalluri V, Hung MC, Chang HC, Lin CT (2021) Apply DEMATEL to analyzing key barriers to implementing the circular economy: an application for the textile sector. Applied Sciences 11:3335. https://doi.org/10.3390/APP11083335
47. Turner C, Oyekan J, Camacho AM, Turner C, Oyekan J (2023) Personalised production in the age of circular additive manufacturing. Appl Sci 13:4912. https://doi.org/10.3390/APP13084912
48. Sikka MP, Sarkar A, Garg S (2022) Artificial intelligence (AI) in textile industry operational modernization. Res J Text Apparel
49. Giri C, Jain S, Zeng X, Bruniaux P (2019) A detailed review of artificial intelligence applied in the fashion and apparel industry. IEEE Access 7:95376–95396. https://doi.org/10.1109/ACCESS.2019.2928979

Circular Business Model

Ghazia Batool and Shumail Mazahir

Abstract This chapter explores the implementation of circular business models in the textile sector and the sustainability challenges in its implementation faced by the industry. It provides a wide-ranging outline of the key principles, strategies, and challenges related to adopting a circular business model in the textile sector. The chapter inspects the key components of circular business models, including product life extension strategy, product as a service model (Leasing or Renting), downcycling repurposing, and recycling. Further, it explores the strategies and initiatives employed by various textile companies to embrace circularity, such as closed-loop systems, product life extension, and material recycling. Furthermore, the chapter digs into the role of stakeholders, policymakers, and technological innovations in facilitating the transition to circularity.

1 Introduction

The tremendous waste generation, resource-intensive manufacturing operations, and scarcity of raw materials have propelled efforts for a circular economy. The circular economy aligns economic growth with environmental sustainability by decoupling resource consumption from economic activity. Over the years, circular economy initiatives have gained traction in business. Arguably, the pioneering business activity encompassing circular economy began with Caterpillar engines in 1970s. The caterpillar realized that the residual values of their large engines could be salvaged through product remanufacturing. Xerox advanced further by introducing a leasing model where the device would undergo multiple iterations in the hands of end-users. Both those applications involve highly durable business products. Soon the circularity principle penetrated consumer products. Today, there is a thriving secondary market

G. Batool
National Center for Composite Materials, National Textile University, Faisalabad 37610, Pakistan

S. Mazahir (✉)
SKEMA Business School, Univesite cote d'azure, Lille, France
e-mail: shumail.mazahir@skema.edu

S. R. Batool et al. (eds.), *Circularity in Textiles*, Textile Science and Clothing Technology, https://doi.org/10.1007/978-3-031-49479-6_9

for cars, computers, electronic devices, and furniture, and several firms, including Toyota, Apple, and Samsung, are exploiting its potential by offering reconditioned products. The scope of those initiatives is wider than just durable products with reasonably long lifecycles. Blockbuster showed commendable success in offering short-term rentals for videos. Their business model encouraged a wider circulation of CDs without the need for overproduction. Therefore, promoting a circular economy hinges upon identifying the correct business model for the product. European Environment Agency and The Organisation for Economic Co-operation and Development (OECD) defined the circular business model as.

A circular business model is defined as the implication of circular economy at the business level in which material loops are slowed down and which helps to achieve the sustainable development goals on micro, meso, and macro levels and focus on the creation and delivery of value with longevity economic, social and environmental impact [1–3].

2 Types of Circular Business Models

In the context of circular business models, four types of business models have gained attention.

2.1 *Product Life Extension Strategy*

This strategy involves product reoffering to another customer once a user has finished its use cycle. Under this strategy, a product passes at least two hands, and the firm charges for refurbishing the product before a product changes ownership [4]. Examples of this strategy include the reconditioned electronic product market. Because an efficient consumer-to-consumer selling system exists for used devices, a firm must design an efficient product retrieval and reconditioning mechanism lest it either not receive sufficient products for refurbishment or be unable to compete for the prices at consumer-to-consumer selling platforms. The better a firm's brand image or, the higher the variability in the used product's performance, the more a firm can leverage its reputation and warranty. Some firms proactively engage with consumers by locking in resale prices and trade-in programs.

2.2 *Product as a Service Model (Leasing or Renting)*

This kind of business model has been long in practice with business-to-business interactions but has recently gained popularity in business-to-consumer interactions. In this model, the firm retains the product ownership but passes the product to the consumers for a limited time, often holding the maintenance responsibility for the product [5]. There are two particular advantages to this strategy. Consider the case of a printer in a corporate office, which would want to avoid adding maintenance personnel for the device. Moreover, the office might want the flexibility to change and upgrade. A firm offering a leasing service can extract a premium for these tailor-made requirements of the corporate office by offering duration or print pages based on the contract. The product will likely generate more revenues in its operational life than the products in the selling model. In the context of consumer-to-consumer extraction, short-term leases can expand the market base. For example, a consumer may not be able to afford a high-end car but would happily pay a premium rent for a month of experience. Consumers often exhibit temporal distortion with respect to product valuations [6], where the first few days of product experience are valued much higher. The firms exploit this opportunity by offering a limited-term experience from a single product to multiple consumers. Consequently, unit profit could be higher than the pure selling model.

2.3 Down Cycling Repurposing

For a range of products, usage deteriorates the product performance and causes wear and tear in the components. In this case, a complete refurbishment might not be economically feasible [7]. Consider the example of a washing machine where the identification and repair of drum leakage might require exorbitant costs diminishing the potential for full-scale reuse. Downcycling and repurposing can be potential strategies in this case. An electric motor retrieved from a used product might not be good enough for another product, but it can find its usage in another product with substantially low power requirements. For example, [8] found that repurposing an old smartphone as a parking meter could save up to 55% of carbon emissions. However, significant challenges restrict the exploitation of this strategy. Primarily, the firm's product offerings might not include a product where the repurposed product can be incorporated. In this setting, finding an alternative for repurposed equipment/components is challenging. Even when a firm's product offerings include such a product, there are compatibility issues with respect to technical requirements. The reason is that separate product development teams manage most products that do not interact with each other. Establishing coordination among teams could resolve those inefficiencies.

2.4 Recycling

Recycling is the oldest circular economy strategy, with the long-term objective of waste diversion. The firms recycle products to neutralize their potentially hazardous environmental effects. However, with the rising prices of raw materials, firms are increasingly interested in value recovery by harvesting materials from the products [9]. Because value recovery on a material level is less preferable to a full-scale reuse, this strategy only kicks into effect when the rest of the options are exhausted. For example, product obsolescence reduces the attractiveness of a product in the market, and a high level of product obsolescence can favor recycling compared to reuse.

3 How do Consumers Value Used Products/Materials?

The success of the circular economy initiatives is contingent upon consumers' perception of used products/materials. Here we discuss some of the perceptions that influence the circular economy initiatives. The first and foremost is the perception of social responsibility and consciousness. At least some consumers perceive using recyclable/ recycled content as a signal for social responsibility. NIKE repurposes plastic bottles and uses them in making footwear for socially conscious consumers. However, not all consumers are socially responsible. A large segment derives utility from functional attributes, aesthetics, and novelty. The refurbished versions of a product made from recycled materials are considered inferior to the new version. This is because consumers perceive irreversible functional deterioration from using the products and ignore the fact that, unlike new counterparts, every refurbished product undergoes thorough testing. Finally, there can be negative sentiments such as disgust for the refurbished products. Consider the examples of food processors or furniture. People may feel disgusted knowing that the product was used by someone unknown and may never buy it [10]. These sentiments heavily bear on the consumer minds, and a firm must understand them to navigate a circular business model successfully.

According to The Economist Intelligence Unit, consumer's search for sustainable goods increased by 71% in past years. This trend of engagement with sustainable businesses is not just in first-world countries but also in developing countries globally.

Capgemini Research Institute reported in research in 2021 that around 72% of consumers want to be a part of circular practices such as buying the most durable products, reducing consumption, and enhancing the lifecycle of the product by maintaining and repairing the products. However, 50% of the consumers pointed out those companies are not putting their efforts into recycling, reusing, and reducing waste materials. Companies should take bold actions to adopt circular business models and teach consumers about these practices to create more awareness.

4 Circular Business Model in the Textile Sector

The textile sector is one of the most vital sectors of all because everyone around is involved in this in some way. But the sustainability in this sector is misunderstood in many ways so it is necessary to give training and knowledge to all the stakeholders who are concerned with textiles in the supply chain [11]. The global fashion

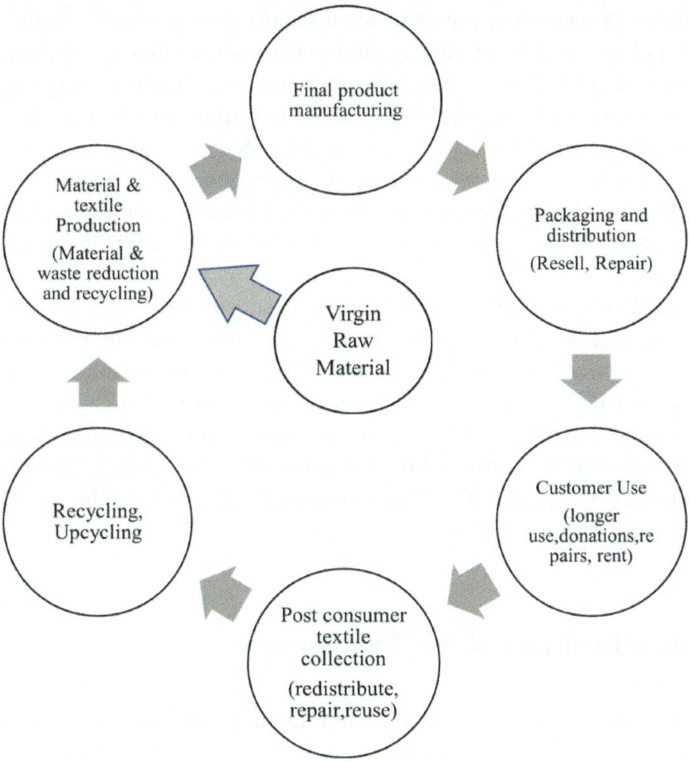

Fig. 1. Circular business model

and textile industry is known for its creativity and innovation among all sectors, it has a rapidly growing demand coupled with waste generation leading to huge environmental and social consequences. The linear "take-make" business model causes excessive usage of finite resources and pollution extraction leads to climate change (Fig. 1). Therefore, there is a dire need for time to implement the circular business models in this sector to cope with this scenario.

5 Goverment Legislation and Regulatory Challenges in Circularity

The UK and European legislations have a great impact on the end of life of materials especially integrating circular economy both in terms of industrial manufacturing and in terms of novel recycling methods. The end of life of textile products is a crucial stage at which the environmental impact of textiles can be substantially countered for example effective collection of discarded textile products can reduce the amount of virgin fibers utilization required to manufacture new products. At this step, the coordination of consumers, producers, and government bodies is required synergistically. Here it also needs to be emphasized that eco-designing is very important as the products pre-designed for circularity can be best fitted into the circular economy, after eco-designing, collection schemes of discarded textiles, fiber extraction and reuse of extracted materials need attention for implementing circular economy. It should also be noted that governments alone couldn't implement circular strategies subjected to huge amounts of waste generated so the role of producers and consumers is vital here [11].

There could be many policy options such as subsidies and tax breaks, reducing greenwashing, making the companies restricted to implementing the circular practices, more convenient circular options, and taking measures to make the consumers emotionally attached to the circular brands. European Commission's sustainable products initiative has already taken actions to make circular product choices convenient and attractive to the consumer by introducing labeling and digital product passports, which involve the particular product requirements about product durability, and recycling.

6 Circular Business Model Framework

Bocken (2018) developed the circular experimental model framework and discusses that as a company strives to experiment with a circular business model, sustainability improves. Application of circular business model should be in a sequence and the company can move up and down through that process and each step of the process needs learning to start the CBM and make it successful. When experimenting with the application of circular business models, it is necessary to involve internal and external stakeholders at each level. It is also identified in this research that the involvement of senior stakeholders is essential in the innovation and startup phase while is more acceptable if external stakeholders can be involved even after the process has been started so that a sustainable business model can work well. For example, H&M took the initiative to involve the end consumers in the recollection system of garments in which they allow the consumers to return the used clothes for recycling or repurposing. Mud Jeans is another example of involving the consumers in their closed-material loop systems by allowing them to lease their already recycled jeans only at €10/month which they rarely wear. Mud Jeans also attracts the customers by motivating them to send back their old jeans and get a free first-month lease with the option to continue the subscription for monthly new jeans or cancel after the initial month.

There must be sustainability measurement checks to measure the validity of the CBM implication experiment and measure if it is working in the right direction. For example, when people switch their clothing fashion trends each season, clothing waste will be prevented in the case of Mud Jeans which is a brand that has adopted eco-friendly practices and is working on sustainability. This is related to the term fast fashion which means that after the product is produced and shipped to the consumers, it has a high turnover as the trendy and in-fashion products are constantly purchased by the customers which can lead to clothing wastage. So, if Mud Jeans focuses on sustainability, then its fast fashion and trendy clothing items may have a contradiction with their claim of sustainable practices [12].

Maria Antikainen and Katri Valkokari (2016) talked about the circular business model framework by Valkokari who in 2014 said that the ecosystem of the whole business sector is changing and moving towards the circular economy that needs a multilevel analysis to bring systematic innovations. This must collaborate all the elements from the macro, meso, and micro level. A macro-level includes the global factors and the meso includes value co-creation and ecosystems while the micro includes the company itself, customers, and consumers [13].

Antikainen has given many suggestions for implementing the business model based on a circular economy to cope with the expected issues. For implementing the circular business model, the framework should be tested by implementing it in other cases in different companies and industries. Repeatedly examine the same situation by rearranging and redesigning the process involved in the circular business model to deal successfully with the required situation. As the framework includes the evaluation of circularity and sustainability it is necessary to use innovative methods, and novel methods to overcome the continuous utterance [13]. Seasalt clothing company publishes its annual sustainability report on its website to evaluate its progress and future planning. In their current report, they aim to have zero waste in the landfill by the end of 2023 by repurposing and recycling the products after recollecting the used products. The recycled products that are not able to be resold can be recycled into new products for the different industries. Seasalt is also aiming to reduce the samples for display by moving towards the 3D technology of displaying the products to reduce the usage of resources. Many companies like Renewcell, Infinite Fibers, and Nanollose accept secondhand garments and then separate the fibers, turn the cellulose into liquid, and also include fermentation. Then they turn the liquid into materials just like the cellulose regeneration process. It helps in avoiding the negative side of mechanical recycling [14].

6.1 Strategies for Designing Durable and Recyclable Textiles

Luca Coscieme suggested that due to change consumer behavior there should be a few policies that are needed and if businesses are adopting sustainable designs and production processes there should be incentives for them. There must be standards for the durability and implementation of taxes with short lifetimes on fast-fashion products, tax reductions on reparation, and slow-fashion companies should give financial incentives.

Circular business models need to be carefully designed to integrate strategies that incorporate the entire lifespan of a product and focus on innovative ways of remanufacturing garments, prosumerism, customization, and other emerging sustainable solutions in the fashion industry. These policies should stress the importance of education, skills development, and nurturing consumer behaviors so that they can be brought into line with different circular models.

Salmi and Kaipia also suggested some strategies to transition towards the circular business model. Three main points of their suggestions are: (1) Capability of opportunity sensing, that is in terms of adopting new regulations and technologies as well as declaring the incentives for the right customers who choose to use those durable products. (2) The company should have the capability to redesign its product and move towards a sustainable structure from the linear model. (3) The companies should have the capability to get involved in the supply chain process.

6.2 Educating Consumers About Proper Care and Longevity

Coscieme et al. [15] also called for implementation of some policies to change consumer behaviors and suggested giving more incentives for sustainable models and production methods. For example, policies like long-lasting products i.e., durability, long-life principles, and brands, taxes on short-life textile or fashion products, low taxes on repairs, and monetary incentives to the "slow fashion" businesses. Without all these policies it's difficult to change the stakeholder's behavior and innovative circular models in the textile and fashion industry which in turn is not favorable for achieving the circular economy. These policies should be implemented throughout the lifecycle of the production process. Implementing these policies in a specific life stage of a fashion lifecycle will not be impactful if it is not applied to the other stages [15]. Zara, a globally recognized clothing brand involving circular business practices, not only focuses on eco-efficient stores through recycling their non-textile stuff like furniture, hangers, and tags but also focuses on in-house training to the employees to create awareness about sustainability and circularity.

Akı et al., [14] discussed in a chapter about denim recycling in the book "Waste in Textile and Leather Sectors". The lifecycle of the denim starts with the chemicals and fibers and is processed to convert into fabric which in turn uses the raw material, energy, and water and causes the emission of air, water, and waste. Further in garment process goes through a series of steps like cutting, sewing, washing, and designing accessories like buttons, etc., and finally, the finished product is sent to the warehouses and stores where consumers can buy these products as per their use. They use it according to their need, wash or dry clean throughout its phase of usage. When it completes its life span at the consumer's place this garment has various end-of-life phases. These phases are discussed above many times as a part of a circular business model which includes recycling, reuse, refurbishment, and disposal. These all stages can be seen in Fig. 2.

Authors of this chapter accessed the lifecycle of the product through the Lifecycle assessment (LCA) through which the environmental impact of the product from raw material to end-life stage can be accessed. That they termed as the cradle to grave.

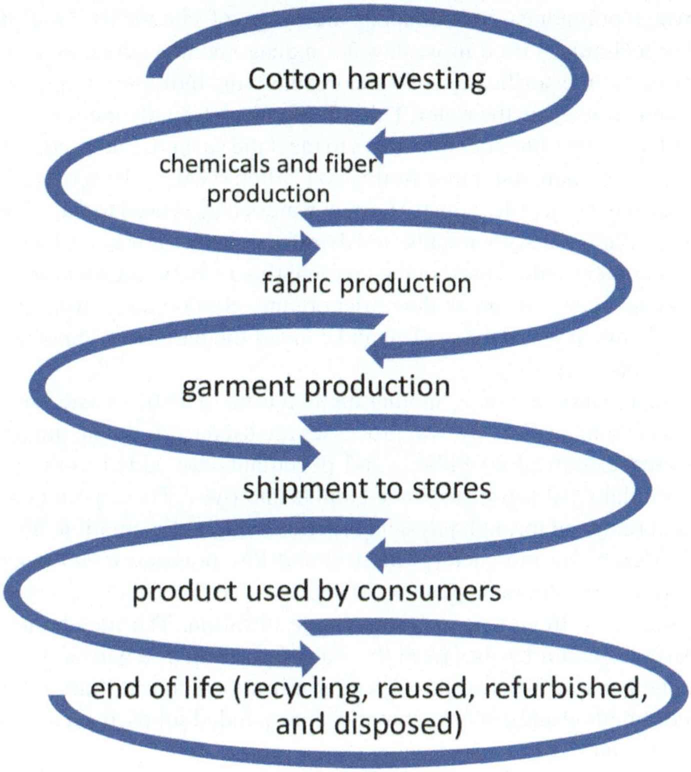

Fig. 2. Denim recycling phases

Through LCA the authors analysed the environmental impacts of the denim in a systematic way to locate the hotspots.

7 Water Management in the Textile Sector

Water management in textiles is planning, controlling, and managing the optimum use of water in textile manufacturing. Water management is the biggest challenge in sustainability. As the population is increasing, usage and wastage of water make it an obvious threat that the world will face a shortfall of water resources in the future. The textile industry uses excessive water for bleaching, coloring, dyeing, and washing processes and in turn discharges pollutant water with a huge number of impurities.

Ergas et al. [16] conducted research and described a few methods for wastewater management. One of the traditional strategies known as Zero discharge is a process of using the discharged water back into the manufacturing process which has many benefits minimizing the usage of water, reducing waste water management costs, reducing water pollutants, and reducing the usage of chemicals. One of the vast and valuable techniques used in wastewater management is ozoneation. It involves applying ozone for the purification of water by removing biological impurities, colors, and unpleasant odors from the water. This environment-friendly method is used in the textile industry along with other industries to meet the sustainability requirements for wastewater management. Granular Activated Carbon (GAC) adsorption is one of the techniques used in the textile industry to treat water GAC is used to trap all impurities like organic contaminants, chemicals, and dyes from the wastewater. Electrochemical oxidation is another method used in the textile industry for wastewater management. It uses electrical energy to break down the organic contaminants from the wastewater. Electrochemical reactions oxidize and convert the harmful composites into less harmful elements [16].

Physio-chemical treatment is another method used in textile wastewater management. Several chemical and physical processes are used to eliminate impurities from the wastewater. Chemical coagulants and flocculants are added to the wastewater to help accumulate and settle suspended solids and dyes. These pollutants are then separated and removed through physical processes like sedimentation, filtration, and adsorption. Membrane bioreactors (MBRs) and low-pressure membranes are also used in textile wastewater management. MBRs is a dual process of waste removal that combines biological treatment with membrane filtration. The membranes separate the solid particles and microbes from the water by acting as a barrier. Low-pressure membranes help in fine filtration through microfiltration or ultrafiltration membranes. It is a low-cost method and effectively removes suspended solids, bacteria, and viruses from the wastewater [17].

Nusrat Jahan in 2022 suggested a sequential approach for sustainable water management. First of all, companies should measure the utilization of water at every stage of the production process so that the areas of improvement can be identified. At the next stage identify the opportunities for water recovery by considering the available Zero Liquid Discharge techniques. Then current performance should be accessed through a standard line and set the KPIs for future performance measurements. Finally, continuous performance improvement should be set as the main focus [18]. Mud Jeans implements a water management strategy as their denim items contain 30% hemp which is an innovative and sustainable material because it does not need any pesticides and requires only small amounts of water. The fiber often requires only rainwater to grow. Every pair of MUD Jeans is recycled into new MUD Jeans, leaving no waste and using 92% less water than average jeans. They claim that they saved 311 million liters of water in 2021. To produce one pair of MUD Jeans the company uses 393 liters of water, compared to an industry standard of 7,000 liters. So, Mud Jeans save 6,607 liters of water. This is made possible by using recycled cotton, and through the water recycling plants and innovative washing techniques at their partner factories.

8 Challenges for Implementing the CBM in the Textile Sector

As the textile and clothing industry has huge environmental costs and the highest resource usage, it's challenging to transform from a linear business model to a circular business model, especially at the company's level. By born circular companies are not bound to face difficulty in changing their older linear business models instead they try to develop a new type of customer base by giving them incentives so they can develop their market. Those businesses who are involved in linear business and plan to move towards a circular business model, face plenty of challenges to change from their product orientation and portfolio and their position in the supply chain [19]. These businesses can adopt different strategies to make it easy for them to move towards complete circular business from a linear business model. For example, Outerknown is a sustainable clothing brand working worldwide. Outerknown is parenting with their circular program project named Vermont, which allows customers to buy pre-loved clothes that will be recycled or repurposed into new products. They aim to convert 100% into circular business by 2030.

Aki in 2020 has found many challenges regarding recycling. (1) Cotton, as many countries in the world have restricted or declined the import of second-hand textiles and recycled cotton. (2) Another challenge is the composition of the denim as it contains more than half elastane fiber instead of 100% cotton as it was in the original denim in history. Also, there are handmade fibers and regenerated cellulose fibers in the composition of the jeans which makes it complex and leads to hurdles in recycling the jeans. (3) One more challenge is about the consumer's mindset. If we analyse the measurements there are only 13% of the textile is recycled and most of the recycling is in the form of clothes that are used for wiping and insulation materials. After discarding more than half of the garments are not recycled as they are disposed of in the landfills with other household trash. It is also revealed in the chapter that an average consumer buys 60% more clothes per annum over the past 15 years and keeps them only half of the time. so consumer awareness is much more important in the whole circular model [14].

Abdelmeguid in 2022 discussed different challenges in implementing the circular business model. He talked about the fast-fashion trend which motivates the consumer to choose less durable and quickly disposed of products with short-life clothes. He also pointed out the challenge that there are less cost-effective and less efficient technologies to implement the circular models. The govt. regulations, legislation, and corporate social responsibility also affect the working pattern of the circular systems. Fashion companies prefer to outsource their operations to countries where there are fewer regulations regarding sustainability and circularity fewer ethical practices and where they can work with freedom. Sometimes customers have no interest in circular products and governments do not support these practices. The stakeholders have no links to coordinate with each other to support the circularity. Companies have to meet the requirements of all stakeholders which also creates hurdles for the implementation of CBM. So, there is a lack of collaboration among all stakeholders.

Keeping in mind the financial challenges, the companies focus on short-term financial benefits instead of focusing on long-term benefits. These investments are costly and risky so, the companies do not find any financial support for this purpose. As the CBM is a new concept for most companies, they have no availability of past data and it is difficult to estimate the data solely based on assumptions [20].

For example, Oasis Textiles, an Indian Textile company, faces the challenge of lacking advanced technology as the recycled fibers are of blended fibers and it is difficult to identify the content of the fibers so when clothes are recycled, they are turned into different products likes rags or insulator stuffs.

Dragomir and Dumitru [21] say that in the textile industry there not no specific model of circularity. Fashion retailers have different outlays and designs of the circular economy. H&M and Inditex are examples of different types of value chain configurations in downstream segments like retail and alliance. He discussed many other companies that focused on the upstream segments.

H&M is the biggest global textile brand that focuses on the circular business and conveys it to its stakeholders through its website and annual reports but [21] referred to a documentary [22] investigative reporter in which it showed that international brands H&M and Zara do not mention the disclosure of the recycled textiles percentage going into their 'sustainable' collections. The reporters criticized that their value chain cannot be included in the circular model because it is rare that used textiles are transformed into new garments. Hence, international brands must consider the concerns of stakeholders and start establishing their circular models by focusing on the transparency and traceability of fibers and materials.

Businesses do not define any quantitative target and benchmark to attain circularity at its best. Companies disclose their practices to the stakeholders and their reliability cannot be questioned by itself but there are many escape clauses in these disclosures which means that there is an ambiguity in their disclosures through which they try to evade many obligations and laws [21].

9 Conclusion

The transformation from a linear business model to a circular business is challenging and hard to achieve. By adopting circular practices, textile businesses can reduce resource consumption, waste generation, and environmental impact through the transformation of their operations and design processes. The regulations and legislations required pose unique hurdles for the implementation that must be overcome to acquire the full potential of circularity in the industry.

However, it is important to recognize the collective effort of all stakeholders to take part in the successful implementation of the circular business model. By implementing circular business models, textile businesses can not only minimize their environmental impact but also create a more vigorous, resilient, and profitable industry. Governments, industry players, consumers, researchers, and all other stakeholders should work together to deal with the challenges and take advantage of the opportunities gained by circular business practices.

References

1. Business models for the circular economy opportunities and challenges from a policy perspective RE-CIRCLE resource efficiency & circular economy project
2. Geissdoerfer M, Savaget P, Bocken NMP, Hultink EJ (2017) The circular economy—a new sustainability paradigm? J Clean Prod 143:757–768. https://doi.org/10.1016/J.JCLEPRO.2016.12.048
3. Transitioning to a circular business model with design. https://www.mckinsey.com/capabilities/mckinsey-design/how-we-help-clients/design-blog/transitioning-to-a-circular-business-model-with-design. Accessed 9 Oct 2023
4. Bakker C, Wang F, Huisman J, Den Hollander M (2014) Products that go round: exploring product life extension through design. J Clean Prod 69:10–16. https://doi.org/10.1016/J.JCLEPRO.2014.01.028
5. Johnson E, Plepys A (2021) Product-service systems and sustainability: analysing the environmental impacts of rental clothing. Sustainability 13:2118.https://doi.org/10.3390/SU13042118
6. Dittmar H, Bond R (2010) I want it and I want it now: Using a temporal discounting paradigm to examine predictors of consumer impulsivity. British J Psychol 101:751–776. https://doi.org/10.1348/000712609X484658
7. Assefa G, Ambler C (2017) To demolish or not to demolish: life cycle consideration of repurposing buildings. Sustain Cities Soc 28:146–153. https://doi.org/10.1016/J.SCS.2016.09.011
8. Zink T, Maker F, Geyer R et al (2014) Comparative life cycle assessment of smartphone reuse: repurposing vs. refurbishment. Int J Life Cycle Assess 19:1099–1109. https://doi.org/10.1007/S11367-014-0720-7/TABLES/3
9. Mudali UK, Patil M, Saravanabhavan R, Saraswat VK (2021) Review on E-waste recycling: Part I—a prospective urban mining opportunity and challenges. Trans Indian Nat Acad Eng 6:3 6:547–568. https://doi.org/10.1007/S41403-021-00216-Z
10. Abbey JD, Meloy MG, Guide VDR, Atalay S (2015) Remanufactured products in closed-loop supply chains for consumer goods. Prod Oper Manag 24:488–503. https://doi.org/10.1111/POMS.12238
11. Luján-Ornelas C, Güereca LP, Franco-García ML, Heldeweg M (2020) A Life cycle thinking approach to analyse sustainability in the textile industry: a literature review. Sustainability 12:10193. https://doi.org/10.3390/SU122310193
12. Bocken NMP, Schuit CSC, Kraaijenhagen C (2018) Experimenting with a circular business model: lessons from eight cases. Environ Innov Soc Transit 28:79–95. https://doi.org/10.1016/J.EIST.2018.02.001
13. Antikainen M, Valkokari K (2016) A framework for sustainable circular business model innovation. Technol Innov Manag Rev 6:5
14. Akı SU, Candan C, Nergis B, et al (2020) Understanding denim recycling: a quantitative study with lifecycle assessment methodology. Waste in textile and leather sectors.https://doi.org/10.5772/INTECHOPEN.92793
15. Coscieme L, Manshoven S, Gillabel J, et al (2022) A framework of circular business models for fashion and textiles: the role of business-model, technical, and social innovation 18:451–462. https://doi.org/10.1080/15487733.2022.2083792

16. Ergas SJ, Therriault BM, Reckhow DA (2006) Evaluation of water reuse technologies for the textile industry. J Environ Eng 132:315–323. https://doi.org/10.1061/(ASCE)0733-9372(2006)132:3(315)
17. Ahmed M, Mavukkandy MO, Giwa A et al (2022) Recent developments in hazardous pollutants removal from wastewater and water reuse within a circular economy. npj Clean Water 5:1–25. https://doi.org/10.1038/s41545-022-00154-5
18. Jahan N, Tahmid M, Shoronika AZ, et al (2022) A Comprehensive review on the sustainable treatment of textile wastewater: zero liquid discharge and resource recovery perspectives. Sustainability 14:15398.https://doi.org/10.3390/SU142215398
19. Salmi A, Kaipia R (2022) Implementing circular business models in the textile and clothing industry. J Clean Prod 378:134492. https://doi.org/10.1016/J.JCLEPRO.2022.134492
20. Abdelmeguid A, Afy-Shararah M, Salonitis K (2022) Investigating the challenges of applying the principles of the circular economy in the fashion industry: a systematic review. Sustain Prod Consum 32:505–518. https://doi.org/10.1016/J.SPC.2022.05.009
21. Dragomir VD, Dumitru M (2022) Practical solutions for circular business models in the fashion industry. Cleaner Logist Supply Chain 4:100040. https://doi.org/10.1016/J.CLSCN.2022.100040
22. The truth behind fast fashion—Are fashion retailers honest with their customers? | DW Documentary—YouTube. https://www.youtube.com/watch?v=23vUvQN-R1Y. Accessed 23 May 2023

Generation, Assessment, and Mitigation of Microplastics

Asif Hafeez, Aqib Saleem, and Khubab Shaker

Abstract This chapter provides a comprehensive overview of the issues related to microplastics including their generation, assessment, and mitigation techniques. Microplastics are available everywhere in the environment and pose a great challenge due to their potential environmental and health impacts. The sources and types of microplastics, as well as the methods for sampling and analysis of microplastics, is discussed first. Then the environmental and health impacts of microplastics are then described in detail, including potential impacts on different ecosystems and organisms. The chapter also discusses mitigation strategies for microplastics, including reducing microplastic generation, reducing microplastic release into the environment, and cleanup and remediation efforts. The chapter is concluded with a brief discussion of potential policy and technological interventions to address microplastic pollution.

1 Introduction

Microplastics are small solid particles with the diameter less than 5 mm however larger than 1 μm [1–3]. These particles produce due to breakdown of macro-plastics like bags, plastic bottles, food packages plastic, shedding of microfibers when textile clothes washed. Microplastics are also produce due to addition them into cosmetic and cleaning agents [4]. As the size of microplastics are too small, so they can enter the environment easily and become part of ecosystem. These particles found in lakes, rivers, ocean, soil as well as in air. Animals may ingest microplastics, which could be harmful to both their health and the wider ecosystem. People may be exposed to microplastics through their food and drink, according to studies, even though it is still unknown how they may affect their health [5].

The history of microplastics can be traced back to the development of plastic materials in the early twentieth century. Bakelite is one of the earliest polymers produced

A. Hafeez · A. Saleem · K. Shaker (✉)
Department of Materials, School of Engineering and Technology, National Textile University, Faisalabad, Pakistan
e-mail: shaker.khubab@gmail.com; khubab@ntu.edu.pk

S. R. Batool et al. (eds.), *Circularity in Textiles*, Textile Science and Clothing Technology, https://doi.org/10.1007/978-3-031-49479-6_10

from phenol and formaldehyde monomers in 1907. It was widely used for electrical insulation applications, and consumer items. Later other synthetic polymers were produced including polyolefins and polystyrene that have grown in popularity due to their low cost. With the increased consumption of these plastics and their environmental impact, the concerns grew about managing this plastic waste. The plastic trash in water was first reported in 1970 and was termed as "marine litter". It inspired the researchers to investigate the impact of plastic waste on marine ecosystems [6].

The term "microplastics" was first introduced in early 2000, to name the microscopic plastic particles found in marine ecosystems. Research on microplastics has since expanded to include freshwater and terrestrial habitats, as well as sources and pathways into the ecosystem. Microplastic detection and analysis technology have also evolved significantly, allowing for more detailed and precise evaluations of their quantity and dispersion. Research on microplastics can offer vital information and proof in favor of legislative measures limiting the manufacture, consumption, and disposal of plastics and support eco-friendly alternatives [7].

2 Classification and Generation of Microplastics

Primary and secondary microplastics are the main categories of microplastics. Plastic particles produced in small size for specific applications are termed as primary microplastics, e.g., personal care products (cosmetics), nurdles (small plastic pellets), and microfibers shed from clothing. Whereas the secondary microplastics are particles produced by the degradation of larger plastic products, such as water bottles, and plastic bags, into smaller plastic particles when exposed to environmental factors. It is the result of weathering and photodegradation of mismanaged plastic waste such as discarded plastic bottles, or from non-deliberate losses of fishing nets, etc. [8]. The life cycle of a plastic product, and microplastics generated at different steps are shown in Fig. 1.

2.1 Primary Microplastics

Primary microplastics refer to tiny plastic particles that are deliberately produced in small sizes and released directly into the environment as plastic particles. The main characteristic of primary microplastics is their small size (typically less than 5 mm), durability, and potential for harm to organisms in the environment. It has been estimated that the annual global release of microplastics into the environment is about 3 million tons [9], while others report it to be 3.2 million tons, of which 1.5 million tons is released into the oceans [10]. This corresponds to an average 400 g of microplastics released into the environment per person in a year. The three common primary microplastics include microbeads, microplastic pellets, and microfibers, as shown in Fig. 2.

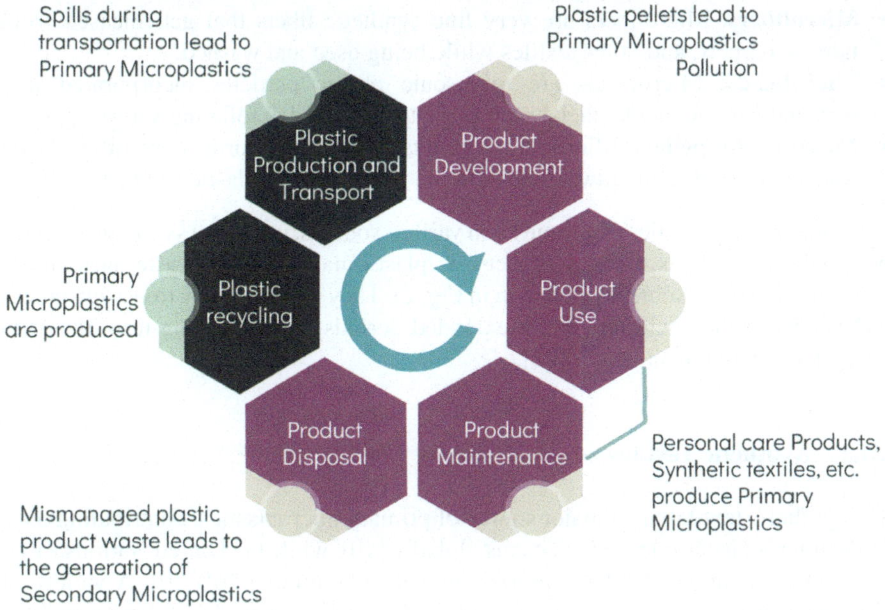

Fig. 1 Life cycle of a plastic product, as a source of microplastics

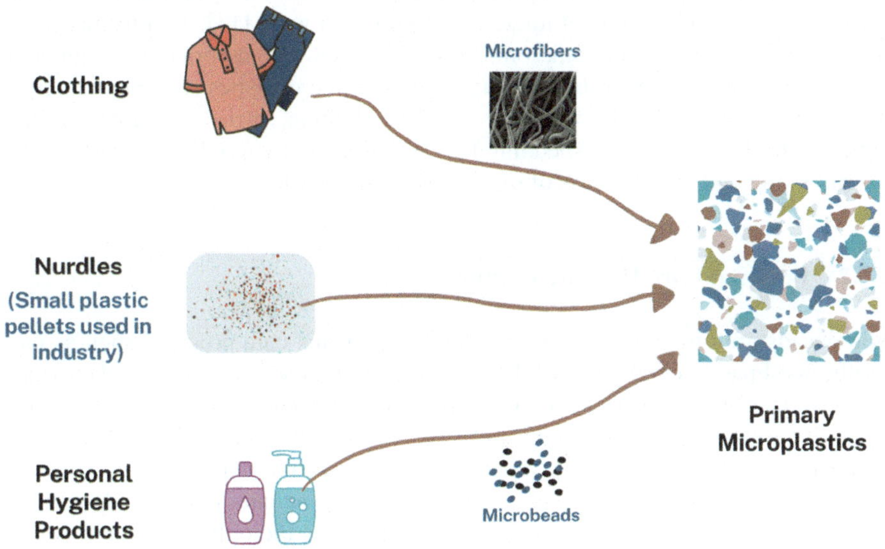

Fig. 2 Generation of primary microplastics from various sources

- **Microfibers**: Microfibers are very fine synthetic fibers that get released from fabrics, carpets, and other textiles while being used and washed.
- **Microbeads**: Microbeads are minuscule plastic particles incorporated into personal care items like facial scrubs, toothpaste, and exfoliating shower gels.
- **Microplastic pellets**: Microplastic pellets, also called nurdles are little plastic spheres utilized as the raw material in the production of plastic products [11].

Primary microplastics can come from various sources, including synthetic textiles, personal care products, industrial processes, plastic manufacturing waste, agricultural products, and other sources (as shown in Fig. 2). They can also carry toxic substances and persist in the environment for extended periods, posing a long-term threat to ecosystems and human health [12].

2.1.1 Synthetic Textiles

The synthetic textiles are a major source of primary microplastics, contributing 35% of the microplastics released to oceans globally [10], while the United Nations Environment Programme (UNEP) approximates it to be around 16% [9]. A variety of clothing items, including activewear and outdoor gear, are produced from synthetic fibers like polyester, acrylic, elastane, etc. During laundering of these clothing items, tiny plastic fibers are shed that enter the environment via wastewater. A single garment can shed thousands of microfibers with each wash [13]. Textiles produced from natural fibers also shed microfibers, but these have no reported environmental hazards. In addition to microfibers, other shapes of microplastics also originate from the various materials and accessories used in clothing, such as coatings, prints, buttons, etc. [14]. The fiber shedding/loss is estimated by number of wash cycles for a clothing and reported loss of microplastics per wash.

2.1.2 Personal Care/Hygiene Products

Microbeads are tiny plastic particles added to personal care products such as face scrubs, toothpaste, and body wash for a variety of purposes, e.g., to provide texture, for delivery of active ingredients, for exfoliation, etc. These microbeads represent up to 10% of the product weight and several thousand microbeads per gram of product [15]. Gouin et al. [16], reported that 93% of the microbeads in personal care products are polyethylene (PE), while remaining includes polypropylene, PET, and PA. Personal care products can be categorized as the only intentional source of microplastics and results in the direct introduction of the microplastics into the wastewater streams. United States has passed the Microbead-Free Waters Act and prohibited the use of plastic microbeads in personal care products.

2.1.3 Plastic Pellets

Plastic pellets are small granules (diameter: 2–5 mm) used as a raw material for the plastic industry. During the production of plastic products, these plastic pellets may be unintentionally released into the environment, resulting in the formation of primary microplastics. In addition, the manufacture and utilization of synthetic rubber, paints, and coatings can also lead to the generation of microplastics [14]. Plastic pellets may be introduced directly to the ocean as a result of some accident during transportation. In addition, they can be released during the production and disposal of plastic products [17].

The growing plastic production has led to a considerable increase in microplastic pollution. These microplastic pellets comprise of various polymers such as polyolefins, polystyrene, etc., have different surface structures, contain different additives, and offer different affinity to pollutants. The plastic pellets serve as a transport medium as well as a potential source of toxic chemicals in the marine environment [18].

2.1.4 Other Sources

In addition to these three sources, there are other sources for generation of microplastics including road markings, tire abrasion, marine coatings, city dust, etc. Different types of road marking (e.g., paint, thermosetting, thermoplastic, etc.) are applied during the development of road infrastructure. Weathering of these road markings or abrasion by vehicles may result in the release of microplastics into the environment. Meanwhile, the tires of vehicles are also eroded due to abrasion. The outer part of tire generates microparticles consisting of Styrene Butadiene Rubber (60%), along with natural rubber and other additives [19]. Coatings are applied to marine vessels for protection and result in microplastics generation during building, repair/maintenance, or use. City dust includes losses from the abrasion of objects (footwear soles, cooking utensils, etc.), abrasion of infrastructure (artificial turfs, building coatings, etc.), blasting of abrasives, etc.

Microplastics can be generated from certain types of artificial turf used in sports fields and landscaping, as the plastic fibers they are made from can break down into small particles as they wear down over time [20]. Microplastics can also be transported through the air and deposited in the environment. Sources of airborne microplastics include vehicle exhaust, plastic waste burning, and industrial emissions. These particles can travel long distances and can be deposited in remote areas, including the ocean. Agricultural practices involve the use of plastic mulch films for crop coverage and weed management. However, these films can degrade into tiny plastic particles that can infiltrate the environment through soil or water. In addition, the application of synthetic fertilizers and pesticides to crops can result in the generation of microplastics [21].

2.2 Secondary Microplastics

Secondary microplastics refer to plastic particles that form by the disintegration of larger plastic products, like bags, packaging films, and bottles, the abrasion of road markings and car tires, etc. as shown in Fig. 3. These particles can vary in size, from small fragments to more significant bits, but all are smaller than 5 mm.

Secondary microplastics can infiltrate the environment through different channels, such as littering, incorrect disposal of plastic waste, and domestic or industrial wastewater discharge. Once they get released, they can accumulate in ecosystems, posing a risk to wildlife and the environment. These microplastics can be mistaken for food by the marine creatures, leading to physical harm and affecting their reproduction and growth. Furthermore, the presence of secondary microplastics in the soil and freshwater systems endangers the terrestrial and freshwater organisms as well. These microplastics can also be carried with the wind and inhaled by humans and animals, causing serious health concerns.

These are made up of the same plastic materials as the larger items from which they originated, such as polyethylene, polypropylene, polystyrene, etc. and come in various shapes, such as irregular fragments, fibers, or beads. Secondary microplastics share some common characteristics with primary microplastics, such as their small size and potential to harm. Furthermore, presence of secondary microplastics in the environment might exacerbate the accumulation of plastic waste and pollution [22]. The secondary microplastics are mostly originated from mismanaged waste during the disposal of products containing plastics. The breakdown of larger plastic items releases tiny plastic pieces, and these smaller pieces can become secondary microplastics.

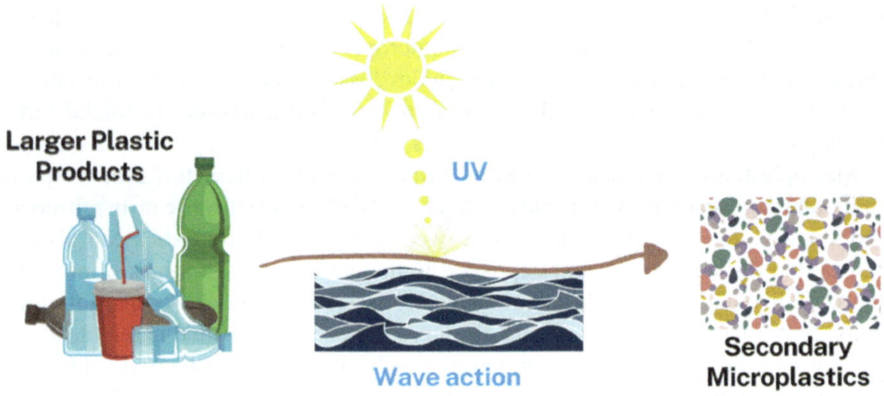

Fig. 3 Secondary microplastics from various sources

3 Pathways of Microplastics into the Environment

It has been estimated that the release of primary microplastics into the ocean is 3.01 million tons per year globally [9]. These Figures correspond to the release of 212 g of an empty plastic product or equivalent, thrown into the ocean per person in a week. The majority of primary microplastics losses (98.3%) are caused by land-based activities, while only 1.7% is generated from sea activities (weathering of marine coatings). The laundering of synthetic clothing and abrasion of tires with road are the largest contributors to land-based activities for generation of plastic particles. The majority of microplastic release to the oceans is caused by the use or maintenance of daily use items, as shown in Table 1.

These microplastics are transported in the environment through air or aquatic currents, depending on the size and density of particles. Microplastics can be transferred from the pedosphere to the hydrosphere by surface runoff, rainfall and ocean circulation, as shown in Fig. 4. Microplastics can be carried not only from land to water, but also from water to land as a result of ocean circulation. Furthermore, lighter, and smaller microplastics can be carried by the wind and thus delivered to isolated locations such as glacier zones and high mountains.

Microplastics can be transported to surface waters by heavy rainfall and surface runoff from agricultural regions and urban areas (the hydrosphere). Furthermore, stormwater runoff transports microplastics from the natural wear of tires on the road to nearby surface waters. Additionally, airborne microplastics made up of light fibers from clothing, landfills, and garbage incineration can be transported vast distances and deposited by atmospheric fallout. The details of sources for primary and secondary microplastics are given in Table 2.

Table 1 Global release of microplastics into the environment annually [9]

Source of microplastics	Amount (Million tons)	Weightage (%)
Microbeads from personal care products and cosmetics	0.01	0.3%
Abrasion of tires (rubber)	1.41	46.8%
Marine coating weathering	0.05	1.7%
Laundering of clothing (textiles)	0.26	8.6%
Markings on the road	0.59	19.6%
City dust	0.65	21.9%
During plastic production (Virgin plastic pellets)	0.03	1.0%
Total microplastics released into the environment	**3.01**	–

Fig. 4 Microplastics: generation and transport pathways [23]

4 Risk Assessment of Microplastics

As discussed earlier, microplastics are small plastic particles that measure less than 5 mm in size. These microplastics produced from diverse applications/processes give rise to environmental pollution. These microplastics can accumulate in the environment and have harmful consequences on the ecosystem. They can infiltrate water bodies through stormwater and wastewater runoff, and be consumed by aquatic organisms, thereby endangering their well-being and environmental roles. Primary microplastics further exacerbate the plastic pollution crisis, and their elimination from the environment after release is an arduous task. Some countries have prohibited the usage of microbeads in personal care items, and enterprises are working towards creating alternatives to plastic products and packaging.

Table 2 Primary and secondary microplastics and their details

Category	Source	Details
Primary microplastics	Primary Source	Plastic pellets, exfoliant beads in facial scrubs and cleansers, polymers used in air-blasting technologies, and sparkles in nail paint and make-up products
	Secondary Source: Water and wastewater treatment plants discharge	Smaller microplastics may escape the primary unit of the wastewater treatment plant and infiltrate the subsidiary units
	Secondary Source: Wear and tear from daily use plastics	Washing clothes and textiles, fishing activities, wear and tear on automotive tires, and degradation of household objects and plastic items
	Secondary Source: Airborne dust	It includes plastic dust produced by activities such as plastic manufacture, plastic waste incineration, traffic emissions, road, and street weathering, etc. Plastic goods used in the home, such as food packaging, plastic utensils, and plastic furnishings, contribute to indoor airborne microplastics
Secondary microplastics	Decomposition of Macro-plastics	Secondary microplastics are generated by the disintegration and weathering of macroplastics. For example, degradation of plastic trash such as disposable plastic cutlery, food containers, etc. that end up on coastal shorelines

4.1 *Environmental Impacts of Microplastics*

Microplastics can contaminate water sources which can have negative impacts on environmental health. The accumulation of microplastics in the environment, particularly in marine and freshwater ecosystems, can harm wildlife. They can be misunderstood as food by creatures, leading to physical harm. Discharge of poisonous chemicals from these microplastics can also negatively affect wildlife. The washing of synthetic textiles is considered to be a major source of water pollution in the environment. Similarly, the degradation and fragmentation of plastic products result in an increased amount of smaller microplastic particles that are bioavailable to a variety of species, including small organisms as well (Fig. 5) [24].

The microplastics also enter the atmosphere and contribute to air pollution. These airborne microplastics can be generated by a number of sources including wear and tear of plastic products, burning of plastics and during manufacturing in the plastics industry. Plastic products like packaging, tires, synthetic clothing, etc. release microplastics into the air, contributing to air pollution. Plastic waste burning and plastic processing are also a source of releasing microplastics into the environment. These microplastics are transported by air currents over a great distance and deposited in isolated areas. Therefore, it is essential to have a thorough understanding

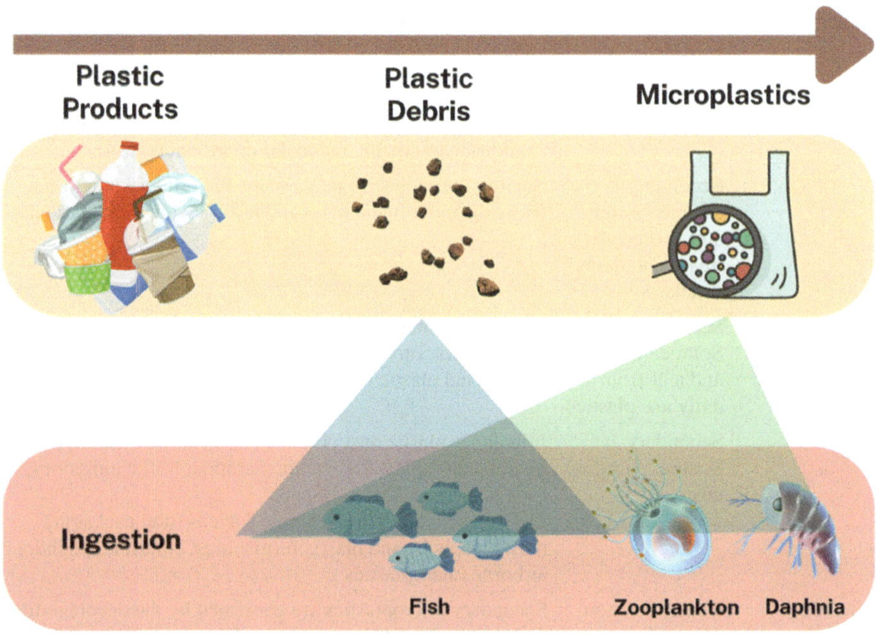

Fig. 5 Interactions of microplastic with aquatic ecosystem

of microplastic pollution and its impact on ecosystems to develop effective mitigation techniques [25].

4.1.1 Marine and Freshwater Ecosystem

Microplastics affect marine and freshwater organisms in two ways. Firstly, organisms such as fish, turtles, etc. might mistake microplastics for food and ingest it, resulting in physical damage to digestive system, bioaccumulation, malnutrition, suffocation and even death. The accumulation of microplastics also leads to increased exposure to potential toxins and pollutants attached to the microplastics. Secondly, small animals like plankton and larval fish can become entangled in microplastic fibers, preventing them from feeding or swimming.

4.1.2 Terrestrial Ecosystem

Although microplastics are more associated with marine and freshwater habitats, they also affect the terrestrial ecosystem. The deposition of microplastics in soil lowers the availability of nutrients for plants, inhibiting growth and production and potentially resulting in economic losses and food poverty. It will have a lasting impact on the terrestrial ecosystem, agriculture, and wildlife. Microplastics cause physical harm to the digestive systems of terrestrial animals such as birds and mammals, when ingested. The toxins released by the microplastics can disrupt hormonal balance and cause serious health issues. Similarly, inhalation of airborne microplastics may potentially lead to respiratory issues. The negative effects of microplastics on wildlife population includes changes in behavior, decreased fitness, and increased mortality rate.

4.1.3 Food Chain

When microplastics enter the environment, they can be mistaken as food by small species such as plankton, and ingested. These small species serve as prey for larger predators and this process continues up the food chain. The concentration of ingested microplastics also increases as we move up in the food chain. It is known as biomagnification and may lead to higher levels of microplastics accumulation in top predators, such as large fish and other marine animals. Microplastics attract and absorb other toxins like heavy metals and pesticides from the environment, and accumulate in the tissues of organisms, potentially leading to negative health impacts.

Microplastics can occupy space in the digestive systems of the organism, reducing its ability to digest nutrients. It may result in retarded growth and reproductive success in the affected organism. Behavioral changes in these organisms include their ability to hunt, evade predators, and carry out other essential activities. Overall, the effects of microplastics on the food chain are complicated and far-reaching, with potentially devastating impact on entire ecosystems.

4.2 Microplastics and Human Health

Microplastics, due to their slow degradation rate, have the potential to remain in the environment for centuries, implying that even if plastic production is stopped immediately, the microplastics currently existing in the ecosystem will continue to pose a difficulty. Microplastics are common in our drinking water and air, and therefore can have several potential pathways of exposure to humans, including ingestion, inhalation, and dermal contact. Microplastics can accumulate in the bodies of marine and terrestrial animals, and human beings are exposed to microplastics through this contaminated food.

Although the impact of microplastics on human health has not been thoroughly investigated, there is growing concern that they may be hazardous [26]. Some studies report that these accumulated microplastics can potentially cause harm including inflammation, hormone disruption, growth of harmful bacteria, infections, and even cancer development [27]. Microplastics can contain and absorb toxic chemicals, such as persistent organic pollutants (POPs), that can be released into the body upon ingestion.

4.2.1 Ingestion

Human beings can ingest microplastics through contaminated food and water sources. Typical examples of contaminated food are the seafood and bottled water. Seafood (e.g., fish, shellfish, etc.) may contain high levels of microplastics, bottled water and other beverages are also a source of microplastics. Microplastics having size less than 10 μm were recognized as the most abundant particles in PET bottled waters [28]. These ingested microplastics are accumulated in the body and release toxic chemicals, causing damage to internal organs. Therefore, minimizing the amount of microplastics in food and water sources is critical for human health protection.

4.2.2 Inhalation

Microplastics are tiny airborne plastic particles, and inhalation of this air during breathing results in a direct exposure to these particles. This is particularly concerning in regions prone to plastic contamination, such as industrial zones, landfills, and indoor environments where plastics are commonly used. Microplastics, when inhaled, can harm the respiratory system. The accumulation of these particles in the lungs can have long-term health consequences.

4.2.3 Dermal Contact

The tiny plastic particles come into contact with the human skin through various sources, e.g., personal care products, contaminated materials, or surfaces, etc. Dermal contact with microplastics can cause skin irritation, inflammation, and even penetration into the skin and enter the bloodstream. Some potential impacts include physical damage, chemical exposure, allergic reactions, microbial growth, etc. Microplastics could attract and absorb harmful chemicals from the environment. Encountering these microplastics results in exposure to these chemicals that can disrupt the hormonal balance or trigger allergic reactions.

4.2.4 Potential Health Effects

Microplastics cause an immunological reaction in the body, which leads to inflammation. Over time, this inflammation destroys tissues and organs, potentially raising the risk of a variety of health issues such as cardiovascular disease, type 2 diabetes, and cancer, as well as intensifying pre-existing disorders such as arthritis and asthma. According to study, microplastics can also cross the blood–brain barrier and accumulate in the brain, potentially causing neurological injury.

Exposure to microplastics negatively affects the immune system, causing its dysfunction, leading to an increased risk of infections. Furthermore, it can lead to changes in immune system cells, such as a decrease in white blood cell (WBC) count and changes in the production of cytokines (signaling molecules). These changes ultimately weaken the immune system and compromise the body's ability to fight off diseases.

Microplastics, being capable of adsorbing and releasing toxic chemicals (e.g., endocrine-disrupting chemicals, etc.) can interfere with the normal functioning of hormones in the body. For example, exposure to endocrine-disrupting chemicals during pregnancy can affect fetal development, and leads to decreased fertility, obesity, and diabetes in adults.

Microplastics entering the lungs via breath cause inflammation, scarring, and even blockages. Over time, it leads to respiratory problems such as asthma, chronic obstructive pulmonary disease (COPD), and even lung cancer. Animal studies have reported that inhalation of microplastics causes inflammation and fibrosis, indicating that human respiratory system is also susceptible to similar effects.

Microplastics on ingestion can have physical and chemical effects on the gastrointestinal tract causing irritation, inflammation, and damage to the lining of the intestine. It can result in a range of gastrointestinal problems, including changes in gut microbiota, decreased nutrient absorption, and increased risk of infection. Furthermore, certain toxic chemicals can leach from microplastics into the body and cause further harm.

5 Mitigation Strategies for Microplastic Pollution

The most common approaches practiced for mitigation of microplastic pollution can be summarized into three categories including policy approaches, technological approaches, and changing consumer behavior. As secondary microplastics are produced by the degradation of large plastic products, implementing an appropriate waste management practice is crucial. Encouraging the production of biodegradable and compostable plastics can also help decrease the amount of plastic waste and ultimately microplastics [29]. The addressal of microplastic pollution requires a multifaceted approach including technological innovations, policy interventions, and consumer behavior changes. A combination of these approaches can effectively mitigate the negative impacts of plastic pollution on the environment.

5.1 Policy Approaches

Microplastic pollution can be minimized by a number of policy approaches, including ban on single use plastics, extended producer responsibility, deposit return schemes, plastic taxes, regulations on microbeads, public awareness campaign, etc.

5.1.1 Ban on Single-Use Plastics

The generation of secondary microplastics can be mitigated by restricting the production and sale of single-use plastics (e.g., straws, plastic bags, utensils, etc.). It will help to reduce the amount of mismanaged plastic waste that ends up in the environment. Furthermore, it can motivate businesses to develop and use more eco-friendly materials for such products. However, the effectiveness of these policies is subject to efficient enforcement, public support, and awareness.

5.1.2 Extended Producer Responsibility (EPR)

It is an environmental policy approach that extends the producers responsibility to the post-consumer stage of the product, including the collection and recycling of plastic waste. It will shift the responsibility (partially or fully) towards the producer and reduce the load on municipalities. Furthermore, the producers may be incentivized for considering the environmental considerations, encouraging to produce less plastic and use more sustainable materials. It is a widely practiced policy globally and shows promising results in mitigating pollution.

5.1.3 Deposit Return Schemes

Deposit return schemes are established to reduce plastic waste and tackle climate change across many countries. This scheme focusses on changing the throwaway culture and helping protect the environment for generations to come. Customers in this scheme pay a small deposit when purchasing single-use plastic bottles or containers and receive a refund when they return the empty bottle or container for recycling. The system incentivizes people to recycle plastic waste and decrease littering while simultaneously serving as a source for high-quality material recycling.

5.1.4 Plastic Taxes

Plastic product manufacture and use can be taxed to reduce demand and encourage the adoption of sustainable alternatives. These levies raise the cost of plastic products while also raising funds for waste management and research into the development of sustainable materials and recycling systems. However, the government should evaluate the potential effects of growing non-plastic manufacturing and consumption.

5.1.5 Regulations on Microbeads

Microbeads are microplastic particles that are widely found in personal care products such as face wash and toothpaste. These microbeads are emptied straight into the water supply and end up in oceans and other bodies of water, contributing to microplastic pollution. It is avoidable by prohibiting the use of microbeads in personal care products. Such laws will necessitate product reformulation by firms and should be implemented after consultation with stakeholders. However, by lowering microplastic pollution, these rules serve to protect the environment and human health in the long run.

5.1.6 Raise Public Awareness

Educating the public about the negative impacts of microplastic pollution and the significance of recycling can also aid in the reduction of microplastic pollution. A combination of these policy options is expected to effectively reduce microplastic contamination. The measures that target the core causes of plastic pollution, such as overproduction and consumption of plastic products, will be the most effective.

5.2 Technological Interventions

Another important component in reducing microplastic pollution is technological intervention. These technological interventions include R&D financing, wastewater treatment, the creation of biodegradable polymers, plastic product alternatives, clean-up/recycling technology, and so on.

5.2.1 Funding for R&D

Research and development projects should be funded to create innovative technologies and eco-friendly materials that are less harmful to the environment. Investment in sustainable solutions will accelerate the shift to a more circular economy, with less waste, better resource conservation, and less environmental impact. Advanced recycling technologies and the usage of renewable energy sources can also help to promote sustainable production and minimize manufacturing's carbon footprint.

5.2.2 Wastewater Treatment

Wastewater treatment is one of the most important technological interventions to reduce microplastic pollution because it prevents microplastics from entering the environment via wastewater. Microplastics can be effectively removed from wastewater treatment facilities that use modern filtration methods such as membrane bioreactors and activated carbon filters. Particles as fine as 0.1 μm could be filtered out by such systems (far less than the size of microplastics). Wastewater treatment can safeguard aquatic environments and the creatures that rely on them by preventing microplastic pollution [30].

5.2.3 Biodegradable Plastics

Using biodegradable plastics is a viable approach for minimizing the accumulation of plastic waste in the environment. When exposed to the environment, these plastics are engineered to decompose into natural components (sunlight, heat, and moisture). These plastics are made from renewable resources, such as plant-based polymers, and are more environmentally friendly and sustainable than petroleum-based traditional plastics. Biodegradable polymers are popular in the packaging, agriculture, and medical industries due to their environmental friendliness.

Biodegradable food packaging can help to minimize plastic waste while also encouraging sustainable consumption. Biodegradable films can also be used to improve soil quality in agriculture. Biodegradable implants, like non-biodegradable implants, can lower the risk of infection and facilitate faster recovery. However, these biodegradable plastics also require specific environmental conditions to degrade properly, can still contribute to pollution if not handled properly. Therefore, proper disposal and management of biodegradable plastics is crucial to ensure the effectiveness [31].

5.2.4 Alternatives for Plastic Products

Promoting and developing alternatives to plastic products can lead to a reduction in the production and use of plastic products. One example is the use of reusable containers and bags made from sustainable materials like bamboo, glass, and metal. These materials are more environmentally friendly and can be used multiple times [32].

5.2.5 Clean-Up Technologies

Developing clean-up technologies can prove to be an effective way of removing plastic pollution from the oceans. These ocean clean-up systems include floating barriers and nets, beach cleaning machines, etc. However, these techniques are expensive and address only the symptoms rather than the root cause.

5.3 Changes in Consumer Behavior

In addition to the policy and technological approaches, consumer behavior changes also play a key part in mitigating microplastic pollution. Some noteworthy examples of consumer behavior changes are penned below.

5.3.1 Reduce Plastic Use

The most effective way for consumers to mitigate microplastic pollution is by reducing the use of plastic products, specifically single-use plastic items. Opting for reusable alternatives, such as stainless-steel and bamboo utensils (water bottles, straws, etc.) can significantly reduce the plastic consumption. It minimizes the amount of plastic waste entering the environment [33].

5.3.2 Proper Disposal

A proper disposal of plastic waste, instead of littering, is an effective way of reducing plastic pollution. The improper disposal of plastic waste will result in litter, and end up in the environment, where its breakdown over time will lead to generation of microplastics. Proper disposal will result in either composting to create nutrient-rich soil for gardening or recycling of plastic waste into a new product. Disposing of plastic waste in trash bins helps prevent it from ending up in waterways or being ingested by wildlife [34]. Composting not only helps reduce the amount of plastic packaging generated from food waste but also creates nutrient-rich soil that can be used for gardening and farming [35].

5.3.3 Choose Environment Friendly Products

Consumers can help to reduce microplastic pollution by selecting environment-friendly products. Natural textiles like cotton, flax, hemp, etc. are more eco-friendly as compared to their synthetic counterparts like polyester and nylon, which release microfibers during washing. Consumers can also look for products labelled as "microplastic-free" or "zero waste", that are designed to minimize the environmental impact and packaged in eco-friendly materials. Selection of ecofriendly products will send a message to the industry that environmentally responsible practices are important [36].

5.3.4 Support Sustainable Brands

Another way to support environmentally friendly products is to seek out brands that prioritize sustainability. It would involve some research and choosing products from companies that use eco-friendly materials, production processes, and packaging. Supporting these brands can help incentivize more companies to adopt sustainable practices and reduce their reliance on plastics. Consumers should stay informed and spread awareness, to play their role in reducing microplastic pollution and protecting the environment [37].

5.3.5 Buy in Bulk

Buying in bulk is another way consumers can help reduce the amount of plastic waste generated from their purchases. Purchasing larger quantities of a product reduces the amount of plastic packaging used per unit item. This approach can also save money in the long run by lowering the frequency of trips to the store, which reduces carbon emissions from transportation. Consumers can have a huge impact on decreasing plastic waste and saving the environment by making conscious shopping decisions [38].

6 Conclusions

Microplastics with a diameter of less than 5 mm are a growing concern due to their possible influence on the environment and human health. This chapter covered the classification, sources, and creation of microplastics, including primary and secondary microplastics. It has been estimated that 3.01 million tons of primary microplastics are released into the ocean per year via different pathways discussed in the chapter. Risk assessment of microplastics was done in terms of impact on environment and human health. These microplastics have a significant impact on marine and freshwater creatures, terrestrial organisms, and the food chain. Humans are also vulnerable to harm if microplastics are swallowed, breathed, or come into contact with the skin.

All these challenges highlight the need for mitigation techniques to limit microplastic formation and release into the environment. Policy initiatives, technical interventions, and changes in consumer behavior were identified as possible mitigation strategies. These techniques addressed microplastic mitigation strategies, such as lowering plastic manufacture, seeking biodegradable alternatives, selecting environmentally friendly items, reducing microplastic release during use and disposal, and prospective cleanup and remediation initiatives.

References

1. Hartmann NB, Hüffer T, Thompson RC et al (2019) Are we speaking the same language? Recommendations for a definition and categorization framework for plastic debris. Environ Sci Technol 53:1039–1047. https://doi.org/10.1021/ACS.EST.8B05297/ASSET/IMAGES/MEDIUM/ES-2018-05297K_0006.GIF
2. Blackburn K, Green D (2021) The potential effects of microplastics on human health: what is known and what is unknown. Ambio 51:3 51:518–530. https://doi.org/10.1007/S13280-021-01589-9
3. Zhang K, Hamidian AH, Tubić A et al (2021) Understanding plastic degradation and microplastic formation in the environment: a review. Environ Pollut 274:116554. https://doi.org/10.1016/J.ENVPOL.2021.116554
4. Sobhani Z, Lei Y, Tang Y, et al (2020) Microplastics generated when opening plastic packaging. Scientific Reports 10:1–7. https://doi.org/10.1038/s41598-020-61146-4
5. Zolotova N, Kosyreva A, Dzhalilova D, et al (2022) Harmful effects of the microplastic pollution on animal health: a literature review. PeerJ 10https://doi.org/10.7717/PEERJ.13503
6. Zhang K, Hamidian AH, Tubić A et al (2021) Understanding plastic degradation and microplastic formation in the environment: a review. Environ Pollut, 274
7. Mitrano DM, Wohlleben W (2020) Microplastic regulation should be more precise to incentivize both innovation and environmental safety. Nat Commun 11:1–12. https://doi.org/10.1038/s41467-020-19069-1
8. An L, Liu Q, Deng Y, et al (2020) Sources of microplastic in the environment. In: Handbook of environmental chemistry. Springer Science and Business Media Deutschland GmbH, pp 143–159
9. Ryberg MW, Laurent A, Hauschild M et al (2018) Mapping of global plastics value chain and plastics losses to the environment
10. Boucher J, Friot D, Primary microplastics in the oceans: a global evaluation of sources
11. Andrady AL (2017) The plastic in microplastics: a review. Mar Pollut Bull 119:12–22. https://doi.org/10.1016/j.marpolbul.2017.01.082
12. Zhang Y, Wang H, Xu J et al (2021) Occurrence and characteristics of Microplastics in a wastewater treatment plant. Bull Environ Contam Toxicol 107:677–683. https://doi.org/10.1007/s00128-021-03142-6
13. Acharya S, Rumi SS, Hu Y, Abidi N (2021) Microfibers from synthetic textiles as a major source of microplastics in the environment: a review. Text Res J 91:2136–2156. https://doi.org/10.1177/00405175211991244
14. Microplastics from textiles: towards a circular economy for textiles in Europe
15. Lassen C, Hansen SF, Magnusson K et al (2015) Microplastics : occurrence, effects and sources of releases to the environment in Denmark. Danish Environmental Protection Agency
16. Gouin T, Avalos J, Brunning I et al (2015) Use of micro-plastic beads in cosmetic products in europe and their estimated emissions to the north sea environment. SOFW J 141:40–46
17. Acosta-Coley I, Olivero-Verbel J (2015) Microplastic resin pellets on an urban tropical beach in Colombia. Environ Monit Assess 187.https://doi.org/10.1007/s10661-015-4602-7
18. Mato Y, Isobe T, Takada H et al (2001) Plastic resin pellets as a transport medium for toxic chemicals in the marine environment. Environ Sci Technol 35:318–324. https://doi.org/10.1021/es0010498
19. Peter Sundt, Per-Erik Schulze, Frode Syversen (2014) Sources of microplastic-pollution to the marine environment
20. Regnell F (2019) Dispersal of microplastic from a modern artificial turf pitch with preventive measures-Case study Bergaviks IP, Kalmar Status: Report
21. Lwanga EH, Beriot N, Corradini F et al (2022) Review of microplastic sources, transport pathways and correlations with other soil stressors: a journey from agricultural sites into the environment. Chem Biological Technol Agric 9:1–20. https://doi.org/10.1186/s40538-021-00278-9

22. von der Esch E, Lanzinger M, Kohles AJ et al (2020) Simple generation of suspensible secondary microplastic reference particles via ultrasound treatment. Front Chem 8:1–15. https://doi.org/10.3389/fchem.2020.00169
23. Frost H, Bond T, Sizmur T, Felipe-Sotelo M (2022) A review of microplastic fibres: generation, transport, and vectors for metal(loid)s in terrestrial environments. Environ Sci Process Impacts 24:504–524
24. Koelmans AA, Redondo-Hasselerharm PE, Nor NHM et al (2022) Risk assessment of microplastic particles. Nat Rev Mater 7:138–152
25. Mammo FK, Amoah ID, Gani KM et al (2020) Microplastics in the environment: Interactions with microbes and chemical contaminants. Sci Total Environ 743:140518. https://doi.org/10.1016/j.scitotenv.2020.140518
26. microplastics-are-in-our-bodies-how-much-do-they-harm-us @ www.nationalgeographic.com
27. Mohamed Nor NH, Kooi M, Diepens NJ, Koelmans AA (2021) Lifetime accumulation of Microplastic in children and adults. Environ Sci Technol 55:5084–5096. https://doi.org/10.1021/acs.est.0c07384
28. Oliveri Conti G, Ferrante M, Banni M, et al (2020) Micro- and nano-plastics in edible fruit and vegetables. The first diet risks assessment for the general population. Environ Res 187. https://doi.org/10.1016/j.envres.2020.109677
29. Syberg K, Khan FR, Selck H et al (2015) Microplastics: addressing ecological risk through lessons learned. Environ Toxicol Chem 34:945–953. https://doi.org/10.1002/etc.2914
30. Sol D, Laca A, Laca A, Díaz M (2021) Microplastics in wastewater and drinking water treatment plants: Occurrence and removal of microfibres. Appl Sci (Switzerland) 11
31. Maier N (2018) Biodegradable plastics: approaches and experiences from 16 members of the EPA network, 1–40
32. Schmaltz E, Melvin EC, Diana Z, et al (2020) Plastic pollution solutions: emerging technologies to prevent and collect marine plastic pollution. Environ Int, 144.https://doi.org/10.1016/j.envint.2020.106067
33. Marazzi L, Loiselle S, Anderson LG et al (2020) Consumer-based actions to reduce plastic pollution in rivers: a multi-criteria decision analysis approach. PLoS ONE 15:1–15. https://doi.org/10.1371/journal.pone.0236410
34. Jallaludin NSK, Sukarno NS, Md Nasir SNB, et al (2021) A systematic review on consumer behavior toward plastic consumption in Asian countries. Adv Bus Res Int J 7:150. https://doi.org/10.24191/abrij.v7i1.14298
35. Hou H, Lu W, Liu B, et al (2023) Exploring the role of fossil fuels and renewable energy in determining environmental sustainability: evidence from OECD countries. Sustainability (Switzerland) 15.https://doi.org/10.3390/su15032048
36. Simões FD (2016) Consumer behavior and sustainable development in China: the role of behavioral sciences in environmental policymaking. Sustainability (Switzerland) 8.https://doi.org/10.3390/su8090897
37. de Sousa FDB (2023) Consumer awareness of plastic: an overview of different research areas. Circular Econ Sustain.https://doi.org/10.1007/s43615-023-00263-4
38. Northen SL, Nieminen LK, Cunsolo S et al (2023) From shops to bins: a case study of consumer attitudes and behaviours towards plastics in a UK coastal city. Sustain Sci. https://doi.org/10.1007/s11625-022-01261-5

Regulation in Recycling and Circularity: Future Prospective

Ghazia Batool and Yasir Nawab

Abstract With a particular focus on the legal frameworks in various countries of Europe and the USA, this chapter offers a thorough analysis of circularity and recycling in the textile sector. It looks at how these regulations, including the New York Fashion Sustainability and Social Accountability Act, EU's Carbon Border Adjustment Mechanism (CBAM), EU Strategy for Sustainable and Circular Textiles, The UN climate agreements and Extended Producer Responsibility (EPR) schemes, etc., are transforming the textile industry by encouraging sustainable practices and waste minimization. This chapter clarifies their possible effects on textile producers, consumers, and the environment by examining the essential components of these legislative frameworks. A future where the textile industry embraces circularity is also envisioned, emphasizing sustainable practices and recycling from textile to textile. The textile industry's stakeholders, politicians, and researchers may seek a deep insight from this chapter about the promising future of circularity and recycling in the textile industry as well as the changing regulatory landscape.

1 Introduction

From a circular perspective, the current linear flow of materials needs to change, where products and processes are designed to create closed-loop systems, keeping materials circulating like nutrients instead of being wasted. Textile reuse and recycling are considered better for the environment than incineration or landfilling, but there are challenges in the textile sector that limit recycling. Circular Economy (CE) is a method that transforms supply chain operations into circular models, reusing, recycling, and remanufacturing materials to reduce waste and negative environmental impacts. Implementing CE strategies can significantly reduce manufacturing waste and barriers to sustainable business models. Transforming the textile industry from

G. Batool · Y. Nawab (✉)
National Center for Composite Materials, National Textile University, Faisalabad 37610, Pakistan
e-mail: ynawab@ntu.edu.pk

S. R. Batool et al. (eds.), *Circularity in Textiles*, Textile Science and Clothing Technology, https://doi.org/10.1007/978-3-031-49479-6_11

linear to circular requires commitment, collaboration, and innovation from all stakeholders. While prior studies identified barriers, few suggested ways to transform the textile system [1].

This chapter aims to fill that gap by providing intervention strategies and regulations for a circular textile industry. This chapter also examines different forms of governing circularity, from state coordination to societal autonomy, and emphasizes the importance of policy mixes in governance.

Textile companies are aware of the need for both state intervention and self-steering to achieve circularity. However, the expectations of companies regarding interaction and collaboration with the public sector are not clear. Companies seem more familiar with traditional, top-down instruments, and some circular measures are seen as less urgent. The fragmented governance expectations of companies highlight the challenges in transitioning to a circular economy in the textile sector.

2 Regulations in Promoting Sustainable Textile Production

Combining Circular Economy (CE) implementation with sustainability measures is more effective than using either approach alone. However, introducing CE in the textile sector can be challenging due to various obstacles, including the sector's reliance on customer orders and transportation costs for raw materials. To overcome these challenges, the government needs to create favorable laws and regulations that support the textile industry's transition towards circular practices [1]. China's circular law is the world's first circular economy law that claims to be different from the linear take-make-waste model. Before China, Germany and Japan had started the laws related to recycling and waste management, but they were different from the circular economy laws. So China is the world's first country to bring the laws and regulations for circularity [2].

According to a study in 2017 "Make Fashion Circular" by the Ellen MacArthur Foundation, the fast-fashion sector has grown significantly which resulted in the global annual disposal of 48 million tonnes of clothes. These figures are rising and are anticipated to rise dramatically, mainly as a result of the adoption of the Waste Framework Directive. By requiring separate collection of textile waste from other garbage by 2025, this directive emphasizes the urgent need to address the fashion industry's environmental effect and encourage environmentally friendly methods for apparel production and disposal [3].

Regulations play a crucial role in promoting sustainable and circular textile production. These are like robust tools to provide guidance, set standards, and incentivize businesses to implement socially and environmentally responsible methods. Some of the keyways regulations that can promote sustainability recycling and circularity in textile production will be discussed further in this chapter.

Effective regulations and policies for the circular economy (CE) need to target both the supply and demand sides. On the supply side, investment in environmentally protective activities, known as "green finance," is crucial. Transforming traditional business models to circular business models is essential for a successful transition to a CE. Policymakers should focus on the challenges of attracting investment and ensuring the financial sustainability of these initiatives to foster further investment in the CE.

For the demand side, policymakers should make circular materials more appealing to consumers, including older individuals, those with lower education, and high-income groups. Incentives through fiscal policy can encourage circular practices. Future policymakers could analyze the impact of these consumer incentives on adopting sustainable behaviors in line with CE principles [4].

3 Key Legislation and Policy Frameworks

3.1 Environmental Regulations and Sustainability Standards

The development of a circular economy (CE) is mostly driven by environmental awareness and regulation. Policymakers should prioritize addressing environmental challenges and encouraging behavior changes that promote sustainability. To encourage the change towards a CE, extremely prominent acts and campaigns in the media can increase interest among consumers and exposure. The importance of waste reduction, encouraging reutilization, and minimizing unnecessary use of materials in packing must be made clear to consumers, who must know about its significance. Both private consumers and businesses should be encouraged to adopt practices that reduce waste generation and promote circular practices. By emphasizing the importance of environmental consciousness and taking proactive steps, we can accelerate the transition toward a more sustainable and circular economy.

To move economies beyond the conventional linear paradigm of the process of extraction, manufacturing, consumption, and waste, environmental regulation is essential. It should be created with strong economic and social objectives and focus on particular demographics, such as senior citizens, those with less education, and the wealthy. To promote sustainability, new products should be designed for durability, and planned obsolescence should be prohibited. Product tags or QR codes indicating the degree of sustainability could help consumers identify goods with recycled inputs. Higher taxes on products containing only primary raw materials can encourage environmental sensitivity. On the other hand, products with high sustainability levels could benefit from reduced value-added taxes. Strict regulations and targeted taxation can influence both demand and supply in the market [4].

3.2 The UN Climate Agreements

International agreements on climate change: UN Framework Convention on Climate Change (UNFCCC) in 1992 first time addressed climate change with 179 countries including the USA. They conducted conferences to stabilize the atmospheric content of greenhouse gases. Further, it led to the Kyoto Protocol in 2005 and the Paris Agreement in 2015 which required developed countries to reduce emissions but did not bind action from developing countries and aimed to control the global temperature and net zero emissions respectively [5].

3.2.1 Kyoto Protocol

Kyoto Protocol was adopted on 11 December 1997 but came into force on 16 February 2005. It is an international treaty that aims to reduce the emission of greenhouse gases that cause global warming. it had two commitment periods for the targets of GHG emissions stability. The first period was from 2008–2012 and the second period was from 2013–2020.

3.2.2 Paris Agreement

Paris Climate Agreement was adopted on 12 December 2015 and came into force on 4 November 2016 which aims to further reduce greenhouse gas (GHG) emissions and reduce the temperature by 2 °C above pre-industrial level in this century.

3.2.3 European Green Deal

The EU Commission formally launched the European Green Deal in December 2019. The EU has introduced the European Green Deal, a comprehensive plan to direct the transition to a greener future with the ultimate goal of reaching carbon neutrality by 2050. It emphasizes how crucial it is to take a coordinated, cross-sectoral strategy where different policy areas collaborate to address climate-related concerns. This effort recognizes the interconnection of the fields of climate, environment, energy, transport, industry, agriculture, and sustainable finance [6].

3.2.4 European Climate Law

The EU is required by law to become carbon neutral by 2050, according to the European climate law. It mandates that greenhouse gas emissions be cut by at least 55% from 1990 levels by 2030 in the EU and its member states.

4 EU Strategy for Sustainable and Circular Textiles

With the increase in population, the demand for textiles also increases which led to the increase in wastage and environmental impact. According to the European Commission, the average European throws 11 kg of textile waste annually. Those wasted textiles go to landfills and in turn have a negative impact on the resources, energy, water, and environment. EU Strategy for Sustainable and Circular Textiles presented various approaches to meet these challenges. This strategy defines solutions for textile consumption, innovative business models, and its environmental impact to align with sustainability. Some of the measurements of the strategy include the new design requirements under **the Eco-design for Sustainable Products Regulation**. These rules make it necessary for textile companies to use a certain amount of recycled material in their productions. They also emphasize that the product is lasting and easily repairable. Additionally, the product at the end should be recyclable.

Digital product passports and clear information on textiles are necessary to release the required information regarding sustainability and other environmental aspects. Under these strategies, there is a tight control on greenwashing (when companies make false or misleading claims to be environment friendly) which helps in protecting the consumers from being deceived. This strategy is also helpful in upcoming Green Claims Initiatives, which will help companies that claim they are environmentally friendly are genuine. One major issue it tackles is fast fashion, where low-cost and low-quality clothes are made speedily, often in poor working conditions outside the EU. To address this, the Strategy focuses on both the demand and supply sides of the problem [7].

Producer responsibility

On the textile producers' side, there will be new rules that they will take more responsibility for designing the product to make clothes more durable. The Commission will involve different stakeholders in circular business models in which they find better ways to make clothes using fewer resources and encourage practices like reusing and repairing clothes instead of throwing them away. It also suggests that Member States offer support and tax benefits to businesses that promote reusing and repairing clothes.

Consumer responsibility

On the consumers' side, this textile strategy will focus on encouraging them to change their buying habits towards valuing quality, durability, and longer use of clothes, as well as promoting repair and reuse. The European Circular Economy Stakeholder Platform will bring together designers, producers, retailers, advertisers, and citizens to work together in redefining fashion in the EU, making it more sustainable and responsible [7].

4.1 Fashion Industry Charter for Climate Change

The 2018 IPCC report emphasized the urgent need to address climate change and the fashion industry's obligation to support the Paris Agreement's objectives to reach net-zero emissions by 2050. UN Climate Change is driving efforts to bring together participants in the fashion industry, from companies to those who generate the raw materials, in a coordinated climate action plan. To successfully tackle climate change, this program aims to discover new initiatives and broaden existing ones throughout the fashion value chain.

4.2 EU's Carbon Border Adjustment Mechanism (CBAM)

The EU's Carbon Border Adjustment Mechanism (CBAM) is an important tool that aims to charge a fair price for the carbon emissions generated during the production of carbon-intensive goods imported into the EU. Its purpose is to incentivize cleaner and more sustainable industrial production in countries outside the EU. The CBAM will be introduced gradually, alongside the reduction of free allowances given to industries within the EU under the EU Emissions Trading System (ETS). This transition supports the EU's efforts to reduce greenhouse gas emissions and promote the decarbonization of its industries. The Carbon Border Adjustment Mechanism (CBAM) will become effective in a transitional phase starting from October 1, 2023. Initially, it will be applied to specific goods and precursor materials that have carbon-intensive production processes and are at a high risk of carbon leakage. These include cement, iron and steel, aluminum, fertilizers, electricity, and hydrogen. As the CBAM is gradually implemented, it will cover more than 50% of emissions in sectors already covered by the EU Emissions Trading System (ETS). During this transitional period, the objective is to pilot and learn from the implementation, involving importers, producers, and authorities, and gather valuable information on embedded emissions. This data will be used to refine the methodology for the definitive phase of the CBAM [8].

4.3 The United Nations Alliance for Sustainable Fashion

The United Nations Alliance for Sustainable Fashion is a collaborative initiative of UN agencies and partner organizations to advance the Sustainable Development Goals (SDGs) in the fashion industry. It promotes coordinated efforts among UN bodies working in fashion and supports projects and policies to ensure that the fashion value chain aligns with the SDGs' targets. The Alliance's work encompasses clothing, footwear, and accessories made from textiles and related materials, covering everything from production to distribution, consumption, and disposal. Sustainability is at the core of their efforts, addressing both social aspects like better working conditions and environmental issues such as waste reduction, water pollution, and greenhouse gas emissions [9].

In 2019, the United Nations introduced The United Nations Alliance for Sustainable Fashion to work together towards sustainable development goals in the fashion industry. Several cities were already supporting this initiative, like Stockholm, which canceled Fashion Week to find eco-friendly alternatives, and Paris, aiming to become a sustainable fashion capital by 2024. Many countries have also launched various national initiatives to promote sustainable fashion worldwide. While the list is not exhaustive, it highlights the range of efforts in this growing area. The initiatives and regulations are mapped according to the different stages of production in a circular economy, showing how effectively each one is contributing to sustainability [10].

To advance the Sustainable Development Goals (SDGs) in the fashion industry, UN agencies, and affiliated organizations have joined forces to form the United Nations Alliance for Sustainable Fashion. This project focuses on several fashion-related issues, including raw materials, manufacturing, distribution, consumption, and disposal. It pursues sustainability by tackling both social and environmental challenges, such as reducing waste and greenhouse gas emissions, as well as social ones like equitable working conditions. The Alliance's ultimate goal is to make the fashion industry a force for attaining the SDGs, reducing its negative effects, and encouraging positive change [11].

4.4 Sustainable Clothing Action Plan 2020 Commitment

By bringing together UK fashion retailers, charity shops, and textile recyclers, the Sustainable Clothing Action Plan (SCAP) made significant progress from 2012 to 2020. The (Waste & Resources Action Program) WRAP-led, industry-driven project produced significant environmental and financial gains. They aim to achieve a 15% reduction in carbon and water footprints, a 15% reduction in waste to refill, and a 3.5% reduction in overall waste arising from the whole lifecycle of the product. The carbon footprint, water use, and waste connected with clothes were significantly reduced as a result of WRAP's facilitation of collaboration, guidance of improvement actions, setting of goals, tracking of progress, and sharing of experience. SCAP's nine-year journey cemented its position as a pioneer in sustainable fashion by giving businesses of all sizes the knowledge and resources they need to transform the textile sector [12].

4.5 Textiles 2030

Leading sustainability professionals from the UK fashion and textiles industries have joined the Waste & Resources Action Program (WRAP) in the ground-breaking initiative Textiles 2030. It is intended to push this sector of the economy towards a circular one in which things are recycled and reused. It's not a rigid mandate; rather, it's a voluntary agreement supported by funds from the government and participating organizations. The signatories pledge to cooperate in the fight against climate change, water conservation, and the advancement of circular textiles. They can influence UK policies about textiles. The Sustainable Clothing Action Plan (SCAP 2020) is built upon this plan, which intends to engage the majority of UK fashion and textile companies in collaborative climate-friendly projects [13].

5 New York Fashion Sustainability and Social Accountability Act

The New York Fashion Sustainability and Social Accountability Act is a proposed state law that, if it is approved, would require any fashion retailer or manufacturer doing business in the state of New York to disclose their environmental and social supply chain due diligence policies and create a community benefit fund with the money taken in from fines and penalties. The measure has not yet been passed and is still pending in the consumer protection committee of the New York State Senate. By establishing stringent supply chain due diligence and reporting requirements, the proposed Act aims to hold fashion producers and retailers responsible for their social and environmental implications. It requires thorough due diligence, which obliges companies to find, stop, and correct negative effects on the environment and human rights. The proposal also includes a fashion cleanup fund that will be financed by fines imposed on non-compliant vendors. Brands are required to set greenhouse gas reduction goals and declare material negative effects. The obligations in the fashion supply chain may be dramatically altered by this legislation, which is part of a larger trend of increased industry regulation. Regardless of whether it is passed, businesses should examine their supply chain diligence programs to be in line with the increased scrutiny [14].

6 Legal Obligations for Manufacturers, Retailers, and Consumers

As of 2019 in France, fashion sellers and retailers aae prohibited from throwing away or burning unsold clothes. Instead, they are obligated to collect and donate the unsold clothing, preventing it from ending up in regular landfills. This law places responsibility on clothing companies to manage their production quantities and implement proper logistics to handle surplus and unsold items.

New York State established a new rule extending manufacturer warranties in February 2021. The textile industry is likewise subject to this law, which requires businesses that generate over 10% of their waste as textiles to recycle the leftovers.

While the fines for non-compliance are relatively low, the real impact comes from the operational requirements that encourage businesses to adopt recycling practices and embrace circular economy principles. These measures are expected to contribute to extending the lifespan of garments. Moreover, the regulation not only increases manufacturers' responsibility but also encourages markets to think innovatively and consider sustainable business models that minimize waste [10].

Currently, major clothing brands are working with collectors like I: CO to manage textile waste efficiently. To enable textile-to-textile recycling, this collaboration generates an abundance of old textiles. Together, large and small businesses must control their waste streams and generate recyclable textiles and a supply of textiles from recycled materials. This cooperation is essential to build logistics, guarantee materials' availability across the board, and generate sufficient demand for recycling technology [15].

7 Extended Producer Responsibility (EPR)

Extended Producer Responsibility (EPR) is viewed as a politically driven facilitator for circular practices. It involves holding companies financially accountable for the environmental impacts of their products beyond their use phase. The purpose of EPR schemes is to create incentives for companies to reduce waste generation and promote more circular activities. This concept has been incorporated into the Circular Economy (CE) action plans, both in 2015 and more recently in 2020. In this study, EPR is considered an enabler not only when companies bear the responsibility for their pollution but also when they actively manage take-back processes as part of circular projects. Other politically driven framework conditions, like blending quotas, were not included in the theoretical framework, as companies are more likely to highlight their own success stories rather than attribute compliance to circular economy initiatives [16].

The Extended Producer Responsibility (EPR) principle, according to the OECD (Organization for Economic Co-operation and Development), requires importers and manufacturers to bear a considerable amount of responsibility for the environmental impacts of their products throughout the product's life cycle. This involves taking into account how the choice of resources, the manufacturing process, and the usage and disposal of the products will affect the environment. Manufacturers should deliberately develop their products to minimize any negative environmental effects while also accepting responsibility for any effects that cannot be completely eradicated via design on a legal and socioeconomic level. In essence, EPR encourages producers to be accountable for the environmental consequences of their products and to work towards minimizing their overall environmental footprint [17].

Extended Producer Responsibility (EPR) has the potential to help the textile sector transition to a zero-waste, circular model. Countries including the United States and New Zealand have endorsed textile EPR in policy recommendation reports like the

California Product Stewardship Council (CPSC) and Textile Product Stewardship Project respectively.

Recover™ is a top manufacturer of high-quality, low-impact recycled cotton fiber and cotton fiber blends on a global scale. Closing the loop on fashion is accomplished by Recover™, which converts textile waste into durable recycled fibers. Recover™ acknowledges the benefits of the EU Textiles Strategy, such as encouraging the circular design of textiles and possibly utilizing EPR tax revenue to increase post-consumer garment recycling. To prepare for policy implementation, the industry must complete several challenging activities. This entails creating and scaling new technologies, tools, and processes, upending the status quo in terms of business models, and influencing customer behavior [3].

The establishment of obligatory and standardized (EPR) programs for textiles across all EU Member States is being proposed by the European Commission. Producers will be responsible for paying the costs associated with managing textile waste under this effort, which was motivated by successful models for products like packaging and electronics. By implementing the idea of "eco-modulation," these expenses will be modified under the textiles' environmental impact. By 2025, these regulations will also make it easier to collect separate textiles, and they will invest producer contributions in infrastructure for collection, reuse, and recycling. Prioritizing textile reuse and recycling while fostering social entrepreneurial opportunities is the goal. By defining trash more clearly and assuring ecologically friendly waste shipments, the plan promotes research into circularity technologies and solves the issue of unlawful textile waste exports [18].

8 New Dutch Rules for Waste Management, Recycling, and Reuse of Textile Products

New rules for textile products have been implemented in the Netherlands, with implications for producers, importers, and suppliers. Consumer clothing, work clothing, and home textiles are all covered by these regulations. To maintain compliance, foreign producers must appoint a representative in the Netherlands. Important responsibilities include disclosing the kind and quantity of textiles to be sold, setting up year-round, no-cost textile collecting programs, and reaching recycling and reuse targets, with a focus on textile fiber recycling. Although they must abide by competition regulations, producers might join organizations to fulfill these commitments collectively. Penalties for non-compliance include both administrative and criminal consequences. These restrictions affect many facets of the textile sector, from design to waste management, and are consistent with EU sustainability principles and the Circular Economy Action Plan. [19]

9 Measurement of Circular Products

9.1 *Material Circularity Indicator (MCI)*

The Material Circularity Indicator (MCI) is a framework used to measure how well a product or company follows circular practices, which means using resources efficiently and reusing materials. This approach is gaining attention from both industry experts and researchers. It includes circular economy requirements, making it more comprehensive than other indicators. One unique feature is that it considers how long a product can be used or how many times its materials can be reused, which is not well considered in other indicators. This makes the MCI a promising and ambitious tool for assessing circularity and guiding government policies [20].

9.2 *Product Circularity Indicator (PCI)*

Soo et al., (2021) provided one more comprehensive way to measure how circular a product is, called the Product Circularity Indicator (PCI). This new method improves upon the previous Material Circularity Indicator (MCI) by considering more factors. It now includes material losses during the production process and highlights the benefits of reusing product components. Additionally, the PCI focuses on the amount of recycled content based on different materials used, rather than just looking at the overall product. This allows for a better understanding of how to increase the use of recycled materials, especially in complex products like vehicles made from multiple materials. This new approach offers a more detailed and accurate way to assess and improve circularity in products. These indicators are chosen because they provide a comprehensive assessment based on circular economy requirements, even though there is no standard approach for evaluating circularity. The PCI method extends the MCI approach by considering manufacturing losses and the exchange of recycled materials and reused components with external systems. [20].

10 Potential Impact of Future Legislation on the Textile Sector

The global production and consumption of textiles are at an all-time high, leading to increased waste and environmental pollution. To address this, the textile industry needs to adopt sustainable and circular practices. This involves eliminating hazardous substances, using resources more efficiently, extending the lifespan of products, and improving recycling at the end of life. To achieve this, stronger governance and policies are needed to incentivize innovative business models and reduce consumption. Collaboration and financial support are essential to scale up sustainable solutions, while changes in consumer habits can be encouraged through education, labeling, and awareness campaigns to promote product longevity and reduce climate impacts. Adopting these strategies will create a textile industry that benefits both businesses and the environment, fostering a more sustainable and circular value chain [21].

Despite efforts to extend the life of garments, the majority of clothes (82%) still end up being incinerated or thrown into landfills, showing that life-extending practices alone are not enough to tackle the issue of unwanted textiles. To address this, enhancing recycling practices becomes crucial to divert textile waste from landfills and use it as a resource. Recycling can be done mechanically by cutting and shredding materials or chemically by breaking down synthetic materials for repolymerization. However, textile-to-textile recycling is still in the early stages and faces economic challenges. To encourage stakeholders in the fashion industry to recycle textiles, there needs to be economic incentives. Implementing textile recycling challenges, the linear production model and requires changes throughout the value chain, including new design practices, collection systems, reverse logistics, and supportive legal frameworks. Collaboration between clothing companies, government, researchers, collectors, and recycling companies is essential to establish an effective textile recycling system. Applying sustainability initiatives and a circular approach to large supply chains can be challenging because all stakeholders involved need to align with circular principles. The size and complexity of supply chains make this task difficult. There is a gap between designers and production, making it hard to design products for recyclability. Designers often lack influence over product development, even though a significant portion of a product's environmental and economic impact is determined during design. Additionally, there is currently no well-established system for recycling textiles, and there is a lack of a sufficient market for recycled materials. This creates barriers in accessing recycled materials and ensuring a stable supply. To overcome these challenges, regulations can play a crucial role in providing a framework for change, and incentives should be identified to create demand for textile recycling [15].

11 Changes Needed to Realize Textile Recycling

Sandvik and Stubbs suggested two types of changes are required to realize recycling and circularity in textiles. (1) Technological changes (2) Systematics changes.

11.1 Technological Changes/Digital Tools

New technologies and digitalization offer exciting possibilities for the fashion industry's sustainable transformation. Innovations like 3D printing and knitting technology can reduce clothing waste by producing garments based on real-time demand and incorporating designs for easy disassembling and recyclability. Digital receipts can enhance transparency by documenting the materials and chemicals used in products, making recycling more efficient. Automating the sorting process of textiles and clothes is essential for textile-to-textile recycling to become a viable business opportunity. Advanced sorting technologies can identify different fiber types quickly and inexpensively, aiding in recycling efforts. Additionally, incorporating digital tags or ID codes can enable easy identification of garment materials for efficient recycling and circular supply chains. The essential functions to consider when adopting new technologies for textile recycling are traceability, transparency, standardization, automation, and seamless connections between stakeholders and processes. These advancements have the potential to revolutionize the fashion industry's approach to sustainability and circularity [15].

11.2 Systematics Changes

Collaboration is crucial for transitioning towards a circular fashion industry. To ensure a continuous flow of materials, the relationship with consumers needs to be redefined. Involving consumers in the sorting process through collection campaigns could be a way to achieve this, requiring changes in the current legal framework. Educating consumers about the environmental impacts of their clothes is essential to increase their awareness and encourage them to return used garments. Although strategic alliances can benefit everyone involved, collaborating with competitors for significant industry-wide shifts can be challenging. The transition towards circularity may begin with large companies due to their resources and influence over producers and regulations, inspiring other corporations to follow suit. While large corporations play a vital role in driving change, full systemic change requires participation from SMEs and luxury brands. It's like turning an entire field green; a few green spots representing large companies won't be enough to achieve sustainability throughout the whole industry [15].

12 Stakeholders and Industry Collaboration in Recycling

According to Neves & Marques, (2022), the population's age distribution plays a role in the transition to a circular economy. Older individuals may be less interested in adopting circular practices, while younger people actively support them. High numbers of older people compared to the working-age population can hinder circular economy efforts, whereas high numbers of young people can encourage them. Environmental awareness, measured by CO2 emissions per person, is also essential. The more informed a population is about environmental issues, the more likely they are to embrace circular practices. Moreover, environmental regulations play a vital role in promoting circularization, emphasizing the importance of government rules and policies in transitioning from a linear to a circular economy.

Nerves & Marques further suggested that individuals with higher levels of education are more likely to recycle and adopt environment-friendly behaviors. They are more inclined to purchase products made from recycled materials and actively engage in recycling practices. People with tertiary education are particularly aware of environmental issues, which drives their proactive approach to sustainability. Policymakers should concentrate on educating those with less formal education to ensure that information is available to everyone and to constantly encourage recycling throughout the population. Further promoting sustainable behavior is the provision of financial incentives or monetary compensation for purchasing goods that are made of primarily recycled materials. By implementing these strategies, barriers related to social acceptability can be overcome, leading to a more sustainable society.

Due to a high old-age reliance ratio, older people are reluctant to adopt circular practices that are helpful for the Circular Economy (CE). Policymakers should create strong policies and procedures that actively engage those in this age bracket in circularity to get around this obstacle. Implementing third-degree price discrimination, in which different prices are assessed to various consumer groups for the same product, is one strategy. Creating discount coupons for seniors who buy products made from recycled materials or rewarding persons who reach rigorous recycling targets can also serve as strong incentives. On the other side, a high young-age dependence ratio promotes circularization, and young people are more aware of the value of reutilization. In the body of extant knowledge, there is still disagreement regarding the connection between a person's age and their propensity to recycle. To effectively engage both age groups in supporting circular practices, policymakers need to customize their approach.

Implementing a special income tax regime for high-income people could be a good strategy to promote the use of items with a high percentage of materials that have been recycled. This policy would make these eco-friendly products more appealing to consumers in higher income brackets, driving demand for circular economy (CE) products. To fulfill the rising customer demand for such items, it would also encourage companies to develop and incorporate recycled materials into their production processes. Plans for post-COVID-19 recovery may include actions to ease this transition. Policymakers can adopt rules and policies that specifically target high incomes to encourage the use of recycled materials in luxury items, such as high-end cars. These initiatives will support a sustainable economy in addition to helping the environment.

Reintroducing products into the economy and using recycled resources in place of virgin ones depend on recycling rates. Effective regulations must be put in place by policymakers to increase consumer interest in items made from recycled materials. Encouraging waste separation for recycling is essential, and the collection and valorization of waste are critical for breaking the link between waste generation and economic growth. Local authorities can play a significant role in promoting circularity by increasing the availability of recycling stations and making waste separation easier for consumers. Decentralized actions and awareness campaigns can sensitize populations to adopt behaviors and attitudes that support the circular economy [4].

13 Conclusion

Regulations in recycling and circularity hold significant promise for shaping a more sustainable and responsible future. Throughout the chapters, various aspects of regulations and their potential impact on advancing recycling and circular practices in the textile industry are explored. The transition towards a circular economy requires a well-coordinated effort from various stakeholders, including governments, businesses, consumers, and society. Environmental awareness and regulation are key drivers in promoting circularity. By setting ecological standards, restrictions on harmful chemicals, and promoting resource efficiency, regulations can steer industries towards more sustainable practices. Extended Producer Responsibility (EPR) policies further hold manufacturers accountable for their product's life cycle, promoting product design with recycling and reusability in mind. Despite the challenges and complexities, regulations can be a powerful instrument for steering industries and societies toward a circular and sustainable future. However, clear communication, education, and awareness-building are essential to overcome resistance and foster cooperation from various stakeholders.

The journey towards a circular economy is a complicated process that requires innovative policies, combined efforts, and continuous research. By exploiting the potential of regulations in recycling and circularity, we can pave the way for a more resilient, efficient, and environmentally responsible world. This requires a collective commitment from governments, businesses, consumers, and researchers to embrace sustainable practices and work together towards a circular and regenerative future.

References

1. Chen WK, Nalluri V, Hung HC, et al (2021) Apply DEMATEL to analyzing key barriers to implementing the circular economy: an application for the textile sector. Appl Sci 11:3335.https://doi.org/10.3390/APP11083335
2. Ogunmakinde OE (2019) A review of circular economy development models in China, Germany and Japan. Recycling 4:27.https://doi.org/10.3390/RECYCLING4030027
3. Fashion Industry, We Must Get Prepared for New Regulations—Sourcing Journal. https://sourcingjournal.com/sustainability/sustainability-news/recover-texiles-recycling-fashion-industry-regulations-waste-337870/. Accessed 17 Sep 2023
4. Neves SA, Marques AC (2022) Drivers and barriers in the transition from a linear economy to a circular economy. J Clean Prod 341:130865. https://doi.org/10.1016/J.JCLEPRO.2022.130865
5. Global Climate Agreements: Successes and Failures | Council on Foreign Relations. https://www.cfr.org/backgrounder/paris-global-climate-change-agreements#chapter-title-0-7. Accessed 17 Sep 2023
6. A European Green Deal. https://commission.europa.eu/strategy-and-policy/priorities-2019-2024/european-green-deal_en. Accessed 18 Sep 2023
7. EU Strategy for Sustainable and Circular Textiles. https://ec.europa.eu/commission/presscorner/detail/en/QANDA_22_2015. Accessed 17 Jul 2023
8. Carbon Border Adjustment Mechanism. https://taxation-customs.ec.europa.eu/carbon-border-adjustment-mechanism_en. Accessed 20 Jul 2023

9. Home—The UN Alliance for Sustainable Fashion. https://unfashionalliance.org/. Accessed 20 Jul 2023
10. Mizrachi MP, Tal A (2022) Regulation for promoting sustainable, fair and circular fashion. Sustainability (Switzerland) 14.https://doi.org/10.3390/SU14010502
11. Home—The UN Alliance for Sustainable Fashion. https://unfashionalliance.org/. Accessed 18 Sep 2023
12. Sustainable Clothing Action Plan 2020 Commitment | WRAP. https://wrap.org.uk/taking-action/textiles/initiatives/scap-2020. Accessed 18 Sep 2023
13. Textiles 2030 | WRAP. https://wrap.org.uk/taking-action/textiles/initiatives/textiles-2030. Accessed 18 Sep 2023
14. New York's Fashion Sustainability and Social Accountability Act. https://www.natlawreview.com/article/new-york-s-fashion-sustainability-and-social-accountability-act-round-2-underway. Accessed 18 Sep 2023
15. Sandvik IM, Stubbs W (2019) Circular fashion supply chain through textile-to-textile recycling. J Fash Mark Manag 23:366–381. https://doi.org/10.1108/JFMM-04-2018-0058/FULL/XML
16. Stumpf L, Schöggl JP, Baumgartner RJ (2021) Climbing up the circularity ladder?—A mixed-methods analysis of circular economy in business practice. J Clean Prod 316:128158. https://doi.org/10.1016/J.JCLEPRO.2021.128158
17. Fact Sheet: Extended Producer Responsibility—OECD. https://www.oecd.org/env/waste/factsheetextendedproducerresponsibility.htm. Accessed 23 Jul 2023
18. Circular economy for textiles. https://ec.europa.eu/commission/presscorner/detail/en/ip_23_3635. Accessed 17 Sep 2023
19. New Dutch rules for waste management, recycling and reuse of textile products | Kennedy Van der Laan. https://kvdl.com/en/articles/nieuwe-regels-voor-afvalbeheer-recycling-en-hergebruik-van-textielproducten. Accessed 18 Sep 2023
20. Soo VK, Doolan M, Compston P et al (2021) The influence of end-of-life regulation on vehicle material circularity: a comparison of Europe, Japan, Australia and the US. Resour Conserv Recycl 168:105294. https://doi.org/10.1016/J.RESCONREC.2020.105294
21. Textiles—UNEP circularity platform. https://buildingcircularity.org/textiles/. Accessed 20 Jul 2023

The manufacturer's authorised representative in the EU is Springer Nature Customer Service Centre GmbH, Europaplatz 3, 69115 Heidelberg, Germany. If you have any concerns regarding our products, please contact ProductSafety@springernature.com

Printed and bound by CPI Group (UK) Ltd, Croydon, CR0 4YY
30/09/2026
02225322-0005